The Biology of Crustacea

VOLUME 7

Behavior and Ecology

Edited by

F. JOHN VERNBERG

*Belle W. Baruch Institute for Marine Biology
and Coastal Research
University of South Carolina
Columbia, South Carolina*

WINONA B. VERNBERG

*College of Health
University of South Carolina
Columbia, South Carolina*

ACADEMIC PRESS 1983

A Subsidiary of Harcourt Brace Jovanovich, Publishers

New York London
Paris San Diego San Francisco São Paulo Sydney Tokyo Toronto

ACADEMIC PRESS, INC.
111 Fifth Avenue, New York, New York 10003

United Kingdom Edition published by
ACADEMIC PRESS, INC. (LONDON) LTD.
24/28 Oval Road, London NW1   7DX

Library of Congress Cataloging in Publication Data
Main entry under title:

The Biology of Crustacea.

Includes bibliographies and indexes.
Contents: v. 7. Behavior and ecology
Dorothy E. Bliss, editor-in-chief / edited by
F. John Vernberg, and Winona B. Vernberg.
1. Crustacea. I. Bliss, Dorothy E. II. Vernberg, F. John.
III. Vernberg, Winona B. IV. Series.
QL435.B48 vol. 7   595.3s   82-20556
ISBN 0-12-106407-7   [595.3'045]

PRINTED IN THE UNITED STATES OF AMERICA

83 84 85 86   9 8 7 6 5 4 3 2 1

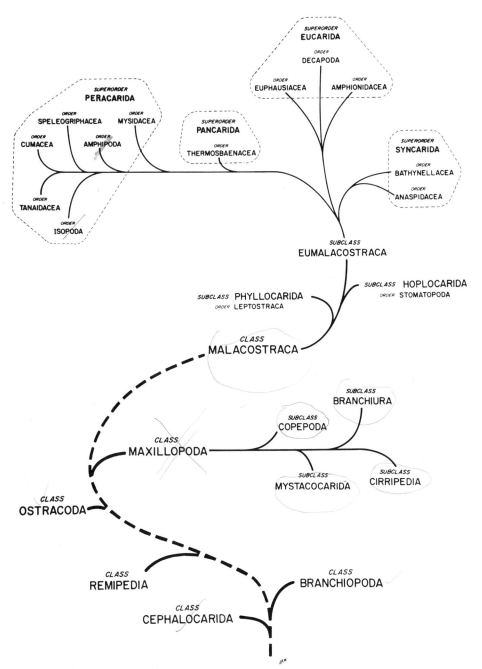

A visual representation of the Bowman and Abele classification of Crustacea (see Chapter 1, Volume 1). This is not intended to indicate phylogenetic relationships and should not be so interpreted. The dashed line at the base emphasizes uncertainties concerning the origin of the five classes and their relationships with one another.

Contents

1 Communication

MICHAEL SALMON AND GARY W. HYATT

2 Movement Patterns and Orientation

WILLIAM F. HERRNKIND

3 Biological Timing

PATRICIA J. DeCOURSEY

4 Symbiotic Relations

D. M. ROSS

5 Pelagic Larval Ecology and Development

A. N. SASTRY

6 Biotic Assemblages: Populations and Communities

BRUCE C. COULL AND SUSAN S. BELL

Contributors

Numbers in parentheses indicate the pages on which the authors' contributions begin.

Susan S. Bell (283), Department of Biology, University of South Florida, Tampa, Florida 33620

Bruce C. Coull (283), Belle W. Baruch Institute for Marine Biology and Coastal Research, and Department of Biology, University of South Carolina, Columbia, South Carolina 29208

Patricia J. DeCoursey (107), Belle W. Baruch Institute for Marine Biology and Coastal Research, and Department of Biology, University of South Carolina, Columbia, South Carolina 29208

William F. Herrnkind (41), Department of Biological Science, Florida State University, Tallahassee, Florida 32306

Gary W. Hyatt (1), 568 Mount Hope Avenue, Cincinnati, Ohio 45202

D. M. Ross (163), Department of Zoology, University of Alberta, Edmonton, Alberta T6G 2E9, Canada

Michael Salmon (1), Department of Ecology, Ethology, and Evolution, University of Illinois, Champaign, Illinois 61820

A. N. Sastry (213), Graduate School of Oceanography, University of Rhode Island, Kingston, Rhode Island 02881

General Preface

In 1960 and 1961, a two-volume work, "The Physiology of Crustacea," edited by Talbot H. Waterman, was published by Academic Press. Thirty-two biologists contributed to it. The appearance of these volumes constituted a milestone in the history of crustacean biology. It marked the first time that editor, contributors, and publisher had collaborated to bring forth in English a treatise on crustacean physiology. Today, research workers still regard this work as an important resource in comparative physiology.

By the latter part of the 1970s, a need clearly existed for an up-to-date work on the whole range of crustacean studies. Major advances had occurred in crustacean systematics, phylogeny, biogeography, embryology, and genetics. Recent research in these fields and in those of ecology, behavior, pathobiology, comparative morphology, growth, and sex determination of crustaceans required critical evaluation and integration with earlier research. The same was true in areas of crustacean fisheries and culture.

Once more, a cooperative effort was initiated to meet the current need. This time its fulfillment required eight editors and almost 100 contributors. This new treatise, "The Biology of Crustacea," is for scientists doing basic or applied research on various aspects of crustacean biology. Containing vast background information and perspective, this treatise will be a valuable source for zoologists, paleontologists, ecologists, physiologists, endocrinologists, morphologists, pathologists, and fisheries biologists, and an essential reference work for institutional libraries.

In the preface to Volume 1, editor Lawrence G. Abele has commented on the excitement that currently pervades many areas of crustacean biology. One such area is that of systematics. The ferment in this field made it difficult for Bowman and Abele to prepare an arrangement of families of Recent Crustacea. Their compilation (Chapter 1, Volume 1) is, as they have stated, "a compromise and should be until more evidence is in." Their arrangement is likely to satisfy some crustacean biologists, undoubtedly not all. Indeed, Schram (Chapter 4, Volume 1) has offered a somewhat different arrangement. As generally used in this treatise, the classification of Crustacea follows that outlined by Bowman and Abele.

Selection and usage of terms have been something of a problem. Ideally, in a treatise, the same terms should be used throughout. Yet biologists do not agree on certain terms. For example, the term *ostracode* is favored by systematists and paleontologists, *ostracod* by many experimentalists. A different situation exists with regard to the term *midgut gland,* which is more acceptable to many crustacean biologists than are the terms *hepatopancreas* and *digestive gland.* Accordingly authors were encouraged to use *midgut gland.* In general, however, the choice of terms and spelling was left to the editors and authors of each volume.

In nomenclature, consistency is necessary if confusion as to the identity of an animal is to be avoided. In this treatise, we have sought to use only valid scientific names. Wherever possible, synonyms of valid names appear in the taxonomic indexes. Thomas E. Bowman and Lawrence G. Abele were referees for all taxonomic citations.

Every manuscript was reviewed by at least one person before being accepted for publication. All authors were encouraged to submit new or revised material up to a short time prior to typesetting. Thus, very few months elapse between receipt of final changes and appearance of a volume in print. By these measures, we ensure that the treatise is accurate, readable, and up-to-date.

Dorothy E. Bliss

General Acknowledgments

In the preparation of this treatise, my indebtedness extends to many persons and has grown with each succeeding volume. First and foremost is the great debt owed to the authors. Due to their efforts to produce superior manuscripts, unique and exciting contributions lie within the covers of these volumes.

Deserving of special commendation are authors who also served as editors of individual volumes. These persons have conscientiously performed the demanding tasks associated with inviting and editing manuscripts and ensuring that the manuscripts were thoroughly reviewed. In addition, Dr. Linda H. Mantel has on innumerable occasions extended to me her advice and professional assistance well beyond the call of duty as volume editor. In large part because of the expertise and willing services of these persons, this treatise has become a reality. Also deserving of thanks and praise are scientists who gave freely of their time and professional experience to review manuscripts. In the separate volumes, many of these persons are mentioned by name.

Thanks are due to all members of the staff of Academic Press involved in the preparation of this treatise. Their professionalism and encouragement have been indispensable.

Finally, no acknowledgments by me would be complete without mention of the help provided by employees of the American Museum of Natural History, especially those in the Department of Invertebrates and in the Museum's incomparable library.

Dorothy E. Bliss

Preface to Volume 7

Since the earlier treatise, "The Physiology of Crustacea," edited by Waterman in 1960–1961, a tremendous upsurge in investigation of many phases of crustacean biology has occurred. The need to concentrate many of these studies into a series of volumes dealing with crustacean biology was evident: this one deals with their ecology and behavior.

Because no single volume can summarize all facets of the complex subjects of ecology and behavior of crustaceans, we present here an update and overview of most of the dominant lines of research. The development and organization of this volume are closely intertwined with Volume 8, entitled Environmental Adaptations, but other volumes in this series also have relevance. One of the healthy developments in biology has been the breaking down of the artificial boundaries separating the various disciplines. Hence, the ecologically oriented biologist has learned that knowledge of internal machinery is necessary to explain the ability of a crustacean to exist in a complex environment.

In Chapter 1, Salmon and Hyatt deal with the rapidly advancing topic of how crustaceans communicate with members of the same species as well as on an interspecific basis. Not only do they give an overall view of communication in an ecological setting, but they review the important problem of communication in terms of quantitative measurement, functions and mechanisms, and evolution.

This leads into Chapter 2, which provides a synthesis and review of patterns of movement and orientation of crustaceans in nature. Coupled with communication is the necessity for an organism to perceive the various components of its environment and to be able to be in the "right" place at the "right" time. The cyclic nature and timing of many biological functions is dealt with by DeCoursey in Chapter 3. She reviews the basic concepts in the regulation of biological rhythms, surveys rhythms in Crustacea, and then analyzes the data from an ecological perspective.

A crustacean does not live in isolation, but interacts with other organisms. Ross (Chapter 4) updates and summarizes our knowledge of symbiotic relationships of crustaceans with other crustacean and noncrustacean hosts. He challenges future researchers to go beyond the description of symbiotic interactions and to analyze and understand these relationships.

Not only is it necessary to study the ecology of adult organisms, but we must know how larval stages interact with the various biotic and abiotic factors; Sastry (Chapter 5) comprehensively reviews our knowledge of this subject. He cites work on adaptation of egg and development to the environment. Dispersal and recruitment are also discussed.

The orientation of Chapter 5, by Coull and Bell, is toward assemblages of organisms into populations and communities. This subject is within the natural progression of this volume from the individual organism to the more complex interactions that are associated with biotic assemblages.

In summary, the ever-increasing body of knowledge on the ecology and behavior of crustaceans attests to the challenge this diverse group of animals has had for past and present investigators. We hope this volume will stimulate the new generation of students to continue to accept the "Crustacean Challenge."

We gratefully acknowledge the invaluable and willing contributions made by our numerous colleagues who critically reviewed all of the chapters. Ms. Anne B. Miller aided in the editing and Ms. Nanette C. Baker assisted in many ways in the preparation of this volume.

F. John Vernberg
Winona B. Vernberg

Classification of the Decapoda*

Order Decapoda Latreille, 1803
  Suborder Dendrobranchiata Bate, 1888
      Family Penaeidae Rafinesque, 1815, *Penaeus, Metapenaeus, Penaeopsis,*
                    Trachypenaeopsis
          Aristeidae Wood-Mason, 1891, *Gennadus, Aristeus*
          Solenoceridae Wood-Mason and Alcock, 1891, *Solenocera,*
              Hymenopenaeus
          Sicyoniidae Ortmann, 1898, *Sicyonia*
          Sergestidae Dana, 1852, *Sergestes, Lucifer, Acetes*
  Suborder Pleocyemata Burkenroad, 1963
    Infraorder Stenopodidea Claus, 1872
      Family Stenopodidae Claus, 1872, *Stenopus*
    Infraorder Caridea Dana, 1852
      Family Procarididae Chace and Manning, 1972, *Procaris*
          Oplophoridae Dana, 1852, *Oplophorus, Acanthephyra, Systellaspis*
          Atyidae De Haan, 1849, *Atya, Caridina*
          Nematocarcinidae Smith, 1884, *Nematocarcinus*
          Stylodactylidae Bate, 1888, *Stylodactylus*
          Pasiphaeidae Dana, 1852, *Leptochela, Parapasiphae*
          Bresiliidae Calman, 1896, *Bresilia*
          Eugonatonotidae Chace, 1936, *Eugonatonotus*
          Rhynchocinetidae Ortmann, 1890, *Rhynchocinetes*
          Campylonotidae Sollaud, 1913, *Bathypalaemonella*
          Palaemonidae Rafinesque, 1815, *Palaemon, Palaemonetes, Macrobrachium*
          Gnathophyllidae Dana, 1852, *Gnathophyllum*
          Psalidopodidae Wood-Mason and Alcock, 1892, *Psalidopus*
          Alpheidae Rafinesque, 1815, *Alpheus, Synalpheus, Athanus*
          Ogyrididae Hay and Shore, 1918, *Ogyrides*

*Prepared by Lawrence G. Abele

Hippolytidae Dana, 1852, *Hippolyte, Thor, Latreutes, Thoralus, Lysmata*
Processidae Ortmann, 1896, *Processa*
Pandalidae Haworth, 1825, *Pandalus, Parapandalus, Heterocarpus*
Thalassocarididae Bate, 1888, *Thalassocaris*
Physetocarididae Chace, 1940, *Physetocaris*
Crangonidae Haworth, 1825, *Crangon, Pontophilus*
Glyphocrangonidae Smith, 1884, *Glyphocrangon*
Infraorder Astacidea Latreille, 1803
　Family Nephropidae Dana, 1852, *Nephrops, Homarus*
　Thaumastochelidae Bate, 1888, *Thaumastocheles*
　Cambaridae Hobbs, 1942, *Cambarus, Orconectes, Procambarus, Cambarellus*
　Astacidae Latreille, 1803, *Astacus, Pacifastacus, Austropotamobius*
　Parastacidae Huxley, 1879, *Euastacus, Cherax, Astacopsis, Engaeus*
Infraorder Thalassinidea Latreille, 1831
　Family Thalassinidae Latreille, 1831, *Thalassina*
　Axiidae Huxley, 1879, *Axius, Calocaris*
　Laomediidae Borradaile, 1903, *Naushonia, Jaxea, Laomedia*
　Callianassidae Dana, 1852, *Callianassa*
　Callianideidae Kossmann, 1880, *Callianidea*
　Upogebiidae Borradaile, 1903, *Upogebia*
　Axianassidae Schmitt, 1924, *Axianassa*
Infraorder Palinura Latreille, 1903
　Family Glypheidae Zittel, 1885, *Neoglyphea*
　Polychelidae Wood-Mason, 1874, *Polycheles*
　Palinuridae Latreille, 1803, *Panulirus, Palinurus, Jasus*
　Scyllaridae Latreille, 1825, *Scyllarus, Scyllarides*
　Synaxidae Bate, 1881, *Palinurellus*
Infraorder Anomura H. Milne Edwards, 1832
　Family Pomatochelidae Miers, 1879, *Pomatocheles*
　Diogenidae Ortmann, 1892, *Paguristes, Dardanus, Diogenes, Clibanarius*
　Coenobitidae Dana, 1851, *Coenobita, Birgus*
　Lomisidae Bouvier, 1895, *Lomis*
　Paguridae Latreille, 1803, *Pagurus*
　Lithodidae Samouelle, 1819, *Lithodes, Paralithodes*
　Parapaguridae Smith, 1882, *Parapagurus*
　Galatheidae Samouelle, 1819, *Galathea, Munida, Pleuroncodes*
　Aeglidae Dana, 1852, *Aegla*
　Chirostylidae Ortmann, 1892, *Chirostylus*
　Porcellanidae Haworth, 1825, *Porcellana, Petrolisthes*
　Albuneidae Stimpson, 1858, *Albunea*
　Hippidae Latreille, 1825, *Hippa, Emerita*
Infraorder Brachyura Latreille, 1803
　Section Dromiacea De Haan, 1833
　　Family Homolodromiidae Alcock, 1899, *Homolodromia*
　　Dromiidae De Haan, 1833, *Dromia*
　　Dynomenidae Ortmann, 1892, *Dynomene*
　Section Archaeobrachyura Guinot, 1977
　　Family Cymonomidae Bouvier, 1897, *Cymonomus*
　　Tymolidae Alcock, 1896, *Tymolus*

Homolidae De Haan, 1839, *Homola*
Latreilliidae Stimpson, 1858, *Latreillia*
Raninidae De Haan, 1839, *Ranilia, Ranina*
Section Oxystomata H. Milne Edwards, 1834
  Family Dorippidae MacLeay, 1838, *Ethusina, Dorippe*
      Calappidae De Haan, 1833, *Calappa*
      Leucosiidae Samouelle, 1819, *Persephona, Randallia*
Section Oxyrhyncha Latreille, 1803
  Family Majidae Samouelle, 1819, *Maja, Hyas*
      Hymenosomatidae MacLeay, 1838, *Hymenosoma*
      Mimilambridae Williams, 1979, *Mimilambrus*
      Parthenopidae MacLeay, 1838, *Parthenope*
Section Cancridea Latreille, 1803
  Family Corystidae Samouelle, 1819, *Corystes*
      Atelecyclidae Ortmann, 1893, *Atelecyclus*
      Pirimelidae Alcock, 1899, *Pirimela*
      Thiidae Dana, 1852, *Thia*
      Cancridae Latreille, 1803, *Cancer*
Section Brachyrhyncha Borradaile, 1907
  Family Geryonidae Colosi, 1923, *Geryon*
      Portunidae Rafinesque, 1815, *Portunus, Carcinus, Callinectes, Scylla*
      Bythograeidae Williams, 1980, *Bythogrea*
      Xanthidae MacLeay, 1838, *Rhithropanopeus, Panopeus, Xantho, Eriphia,*
            Menippe
      Platyxanthidae Guinot, 1977, *Platyxanthus*
      Goneplacidae MacLeay, 1838, *Frevillea*
      Hexapodidae Miers, 1886, *Hexapodus*
      Belliidae, 1852, *Bellia*
      Grapsidae MacLeay, 1838, *Grapsus, Eriocheir, Pachygrapsus, Sesarma*
      Gecarcinidae MacLeay, 1838, *Gecarcinus, Cardisoma*
      Mictyridae Dana, 1851, *Mictyris*
      Pinnotheridae De Haan, 1833, *Pinnotheres, Pinnixa, Dissodactylus*
      Potamidae Ortmann, 1896, *Potamon*
      Deckeniidae Bott, 1970, *Deckenia*
      Isolapotamidae Bott, 1970, *Isolapotamon*
      Potamonautidae Bott, 1970, *Potamonautes*
      Sinopotamidae Bott, 1970, *Sinopotamon*
      Trichodactylidae H. Milne Edwards, 1853, *Trichodactylus, Valdivia*
      Pseudothelphusidae Ortmann, 1893, *Pseudothelphusa*
      Potamocarcinidae Ortmann, 1899, *Potamocarcinus*
      Gecarcinucidae Rathbun, 1904, *Gecarcinucus*
      Sundathelphusidae Bott, 1969, *Sundathelphusa*
      Parathelphusidae Alcock, 1910, *Parathelphusa*
      Ocypodidae Rafinesque, 1815, *Ocypode, Uca*
      Retroplumidae Gill, 1894, *Retropluma*
      Palicidae Rathbun, 1898, *Palicus*
      Hapalocarcinidae Calman, 1900, *Hapalocarcinus*

Contents of Previous Volumes

Volume 3: Neurobiology: Structure and Function
Edited by Harold L. Atwood and David C. Sandeman

Volume 4: Neural Integration and Behavior
Edited by David C. Sandeman and Harold L. Atwood

Volume 5: Internal Anatomy and Physiological Regulation
Edited by Linda H. Mantel

1

Communication

MICHAEL SALMON AND GARY W. HYATT

I. GENERAL CONSIDERATIONS

At first glance, the Crustacea appear to be poor candidates for the analysis of communication. The majority of species are solitary, nocturnally active, and secretive. Their best known aquatic environment, commonly the shallow water bordering continents or islands, is often turbid, acoustically noisy, and characterized by patterns of current flow which are altered unpredictably as storms and tidal erosion transform the structure of the bottom. Thus, one would not expect to find chemical, acoustic, or visual commu-

1

THE BIOLOGY OF CRUSTACEA, VOL. 7

nication systems as complex or highly developed as those of insects. On the other hand, the Crustacea exhibit striking diversity. Some species remain sessile throughout their entire lives. Others are active as swimmers, crawlers, climbers, or burrowers. Some remain aquatic throughout life, while others are semiterrestrial or live permanently on land. A few species are highly social, living in dense aggregations, or in male–female pairs until death. The majority interact with conspecifics only on rare occasions, either to fight for some resource or to mate.

This diversity of ecology, behavior, and structure makes crustaceans appealing as subjects for proximate and evolutionary studies of communication because certain species represent interesting products of particular arrays of selection pressures. Some of these, such as the hermit and fiddler crabs or the stomatopods and alpheid shrimps, have proven particularly amenable to study.

Our purpose in this chapter is not to provide an all-inclusive review; several of these reviews have already appeared in the recent literature (Schöne, 1961, 1968; Reese, 1964; Frings and Frings, 1967; Salmon and Atsaides, 1968; Crane, 1975; Hazlett, 1975a; Warner, 1977; Weygoldt, 1977). Rather, we propose to concentrate on central methodological techniques and evolutionary issues, to emphasize studies which have been particularly illuminating, and to suggest avenues of future research which hold special promise.

A. Approaches to the Study

There are almost as many definitions of communication as there are authorities on the subject. Living organisms gather information not only from other living organisms, but also from inanimate objects. If one defines communication broadly, it can include such exchanges as those between the sun or stars and an organism using these objects for orientation. At the other extreme, one can limit communication to intraspecific situations, thereby excluding not only interspecific interactions, but also symbiotic relationships between plants and animals. Qualitative distinctions serve to further narrow the phenomenon by suggesting that information must be exchanged reciprocally through a common, shared code, or that the consequences of such exchanges are adaptive to one or both participants (Marler, 1968; Wilson, 1975). The problem is that none of these restrictive definitions can adequately account for all forms of communication, or for all of the selection pressures which have shaped them.

Alternative approaches, recently suggested by Barnett (1977) and Dawkins and Krebs (1978) view communication as a paradigm with properties identified by further empirical investigation. The model (1) consists of a sender or actor, a signal, and a receiver or reactor, (2) requires some measurement of the information transferred, (3) designates the probability of a change in response state by the receiver, (4) demonstrates the survival value of the communication to the sender, and (5) shows how these properties presumably represent an evolutionarily stable strategy. The latter is accomplished by an analysis of how alternative behaviors would fare, coupled with a consideration of how other actors/reactors are behaving.

B. Channel Selection

Crustaceans are known to use visual, chemical, tactile, and acoustic channels in communication. Visual, chemical, and acoustic channels are potentially useful at long range as well as at short distances. Unless both sender and receiver are close together, turbid water, a tendency toward nocturnal activity, and predator pressure probably combine to preclude long range visual signaling in the strictly aquatic crustaceans. Virtually all the studies of visual display in aquatic crustaceans are described with reference to agonistic or sexual encounters between individuals who are no more than a few body lengths apart. Light production occurs in a few aquatic species and could be used to facilitate aggregation or to prevent dispersion. Experimental analyses of behavior associated with light production remain to be done (Weygoldt, 1977).

Acoustic signals are useful for distance communication (as well as at close range). The use of "calling" signals is widespread in insects. Yet, refined temporal patterning of, and spontaneous "calling" with, sonic signals are virtually unknown for the aquatic Crustacea. Rather, the sounds are loud and variable "rasps," "snaps," or "rattles" produced at close range during agonistic interactions or in response to predators (e.g., Meyer-Rochow and Penrose, 1976). High background noise under water and lack of significant resonating structures to act as acoustic amplifiers may account for the absence of refined sonic communication among aquatic forms.

Chemical signals are potentially useful at both long and short ranges, and have the advantage of efficiency both during the day and at night, and transmission around objects via currents or diffusion. Furthermore, chemical stimuli can be energetically inexpensive to produce, as in cases where metabolites acquire signal value as attractants, repellents, or synchronizers of reproductive activities, especially in sessile forms. Bossert and Wilson (1963) introduced the concept of "active space," a zone around the point of emission within which the concentration of a pheromone is at or above threshold for a behavioral response. The shape of this zone varies depending upon the diffusion properties of the active compound(s), the current velocity, the amount of pheromone released, and the sensitivity of the receptor system used in detection. The principles outlined by Bossert and Wilson for evolutionary "adjustments" in fading time, temporal patterning, and response changes as a function of concentration, are all applicable under water. In several aquatic species there is evidence, ranging from suggestive to conclusive, for the existence of sex pheromones as attractants at a distance or for sex discrimination and individual recognition at close range (see below).

Some semiterrestrial and terrestrial crustaceans, in contrast to the aquatic forms, have exploited both visual and acoustic channels for long-range signaling. The extent to which these communication systems have developed depends on a complex of factors: spacing between individuals, extent of diurnal activity, environmental noise ("openness" of the habitat, and freedom from wind and surf disturbance), and exposure to climatic extremes, to name a few. There are no known examples of long-range chemical communication by any terrestrial forms, though a systematic search has never been made (Dunham, 1978).

Once in proximity, all potentially available channels of communication may be used by both aquatic as well as terrestrial species, either alone or in combination.

C. Signals and Noise

As signals pass through a channel, they are invariably distorted by noise. Hailman (1977) distinguishes between transmission noise, which physically alters the signal en route, and detection noise, which overwhelms the signal so that it cannot be distinguished from background noise. The extent to which noise effects crustacean communication systems and the adaptations by senders to circumvent noise have rarely been studied. Yet, experiments and measurements along these lines are essential for efficient signaling.

Horch's (1971, 1975) and Horch and Salmon's (1969) work with ghost crabs (*Ocypode* spp.) has revealed adaptations to both categories of noise. Males employ relatively loud acoustic signals to attract females to their burrows. As the sounds radiate from the burrow, they attenuate due to spherical spreading and absorption. Horch found that higher frequencies of the call attenuated much more rapidly than lower frequencies, and that moist sand was a better conductor than dry sand. Background noise was confined primarily to the lower frequencies, i.e., 1000 Hz and below.

The acoustic signals produced by several species (*O. ceratophthalmus, O. cordimana,* and *O. laevis*) differ in species-typical tempo, which presumably allows females to discriminate between heterospecific groups of males calling simultaneously. However, sound spectral energy distribution is remarkably similar. Most of the energy is concentrated between 1 and 3 kHz or just above prevalent background frequencies. Thus, males increase their probability of detection by transmitting in frequency bands just above noise but still low enough to minimize frequency-dependent attentuation effects. The hearing curves for several species are tuned to the 1–3 kHz band and rapidly decrease in sensitivity below 800 Hz. This adaptation further improves detection by filtering out competing low-frequency noise. Finally, neighboring males interact acoustically so that their sounds alternate and thus avoid overlap. Crabs at Enewetok (Marshall Islands) alternate calls within single species choruses (*O. ceratophthalmus, O. cordimana*) as well as between males of both species when the choruses are mixed.

Adaptations which make signals clearer may also function, in part, as aids to counteract noise. For example, some fiddler crab species display in areas where vegetation is dense. Their wave displays are long in duration and often performed after the male climbs on high mounds near his burrow. The males will also pursue females while in display or incorporate acoustic signals into their bouts of waving (von Hagen, 1962; Salmon and Atsaides, 1968). In contrast, males displaying on open beaches rarely leave the burrow sites while waving, are typically lighter in color than the background when displaying, and produce acoustic signals only when females have been attracted close to the burrow entrance. The wave displays of these males also tend to be rapid and brief in duration.

The relationship between signals, noise, and transmission during chemical communication in crustaceans is poorly understood. Such work would be particularly applicable to communication between widely-spaced individuals where these problems are exacerbated. Presently, no crustacean sex pheromones have been characterized as to molecular structure. Thus, these substances may be general metabolic products, or they might be produced by special glands (Dunham, 1978). Christofferson (1970) showed that the molecular weight for the sex pheromone of *Portunus sanguinolentus* was close to 1000, or considerably larger than most known insect pheromones. Wilson (1970) predicted that either an increase in molecular weight, a lowering of receptor threshold, or some combination of the two would be needed under water to achieve comparable signal qualities of an analogous pheromone in air. Since receptor sensitivities in such systems already tend to be high, alteration in molecular weight is the most likely alternative, as Christofferson found. Will similar findings also characterize other crustacean sex pheromones? It would be particularly interesting to compare the molecular properties of these substances in species where communication occurs under water, as opposed to those where it occurs on land (as in some amphipods) (Williamson, 1951, 1953), or between pheromones which operate at a distance and those which are contact signals.

One way in which detection noise can be avoided is by evolving compounds which are distinct, i.e., which minimize the probability of confusion with other substances in the environment. Eales (1974) showed that the sex pheromones released by females of two portunid crabs, *Carcinus maenas* and *Macropipus holsatus,* elicited a "searching response" in conspecific but not in heterospecific males. These experiments showed that sex pheromones can have two important features: they are a species-typical signal for a conspecific crab, but an irrelevant component of background for heterospecific males which might be predaceous.

D. Description and Stimulus Analysis

In the initial stages of study, the motor patterns of signaling individuals (or release sites, active space, and chemical composition of pheromones) must be accurately described. The crustacean literature abounds with verbal descriptions, many of which are rarely precise enough or sufficiently quantitative to provide measures of central tendency or variability. An effort must also be made to sample populations so that the data pool is representative. The investigator can be aided in this regard by instrumentation, such as tape recorders and portable video systems, which can now be used even under the most extreme field conditions.

Hazlett (1972a) used film analysis to describe visual agonistic displays in a number of anomuran and brachyuran species. In all of these crabs, violation of personal space results in a movement of the legs and/or chelae, coupled with an elevation of the body. The initial phase of the description involves filming these movements, sometimes from several angles, and measuring angular appendage changes in final position with respect to the body (Fig. 1A,B). A comparison to nondisplay movements of the same appendages (during locomotion) revealed that the display movements were much less variable, as one would expect if the movement had signal value.

Visual displays used in courtship can also be described in quantitative terms. Hyatt (1977) developed film analysis techniques for describing complex, repetitive waving movements of fiddler crabs (Fig. 1C,D). Salmon et al. (1978) used this technique to describe differences between the displays of sibling species of fiddler crabs whose waves appeared nearly identical to the casual observer.

Quantitative techniques are particularly valuable in those species where courtship consists of changes in the probabilities with which acts shown in agonistic or investigatory behavior occur. Courtship in such cases is defined by a spectrum of act frequencies and linkages shown when males and females interact and eventually form pairs. These distributions are quite different from those shown by male–male or female–female pairs which fight until one animal dominates and drives the other away.

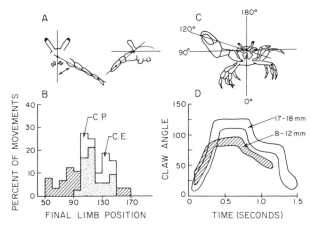

Fig. 1.   Techniques and applications of film analysis of crustacean displays. (A) Ambulatory raise (left) and cheliped movement (right) during hermit crab agonistic displays. (B) Distribution of angles during nondisplay (lined) and display (clear) movements. Overlap areas are stippled. C.P., cheliped presentation; C.E., cheliped extension (from Hazlett, 1972b). (C) Major chela angle during waving display of a male fiddler crab. (D) Distribution of angles for several waves shown by smaller (8–12 mm in carapace width) and larger (17–18 mm in carapace width) males of *Uca pugilator* (from Hyatt, 1977).

Once the display properties of behaviors are described, the investigation can turn to an analysis of the "effective stimuli," i.e., the components which carry information necessary for eliciting a given response. The position of select portions of the body and/or appendages can be manipulated in models and presented to another animal to ascertain their role in display. The most complete study of this kind is Hazlett's (1972b) work with the double ambulatory raise (DAR) agonistic display of a hermit crab, *Calcinus tibicen*. This display involves a lateral elevation of both first ambulatory legs to a position slightly below the horizontal, where they may be held for a few seconds to more than 1 min. A total of 21 models were presented to 176 test crabs, whose responses were classified into four categories: no reaction, retreat, duck in shell, and display (usually a cheliped movement). The models themselves fell into several categories, some of which are shown in Fig. 2. Hazlett found that all of the major features of the display (horizontal extent, a central portion though not necessarily shaped like a body, contrasting coloration of the leg tips) were important for an effective display. However, both leg width and body shape were apparently unused cues. The natural model was not the maximally effective one. Models with all white legs were classified as "supernormal" because they evoked greater responses than "normal" models possessing white only at the leg tips.

In other cases, analytical tests are needed to define the effective stimulus

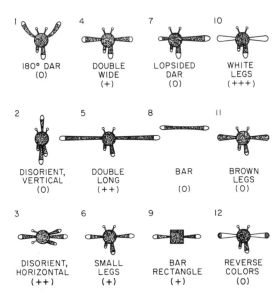

Fig. 2. Models used by Hazlett (1972b) to determine effective stimulus components in the double ambulatory leg raise (DAR) display of *Calcinus*. Models 1–3 test ambulatory and body orientation; 4–6, leg length and thickness; 7–9, body presence, position, and shape in relation to the legs; 10–12, ambulatory coloration. The DAR model (not shown) was as effective as a displaying conspecific in evoking agonistic responses. 0, neutral response; +, equivalent response to DAR model; ++ or +++, stronger response than DAR model.

channel. For example, playback experiments to fiddler (Salmon, 1965) and ghost (Horch and Salmon, 1971) crabs showed that the subjects were sensitive to the pulse repetition rate of conspecific signals. These stimuli could have been detected as substrate- or airborne energies. A resolution of this issue was important because the two media have very different properties with regard to sound velocity, attenuation, and background, which affect the potential for localization, discrimination, and other biologically relevant variables. In a series of measurements and experiments, Horch (1971) and Salmon et al. (1977) found that both ghost and fiddler crabs detect acoustic stimuli with Barth's organ, a receptor located on the meral segment of each walking leg. In ghost crabs the receptor was equally sensitive to both substrate- and airborne sound; in fiddler crabs the receptor responded primarily to substrate-borne energies.

A particularly elegant analysis of chemical communication in an aquatic amphipod (*Gammarus duebeni*) is provided by Dahl et al. (1970). Females initiated courtship by releasing a pheromone believed to operate as a sex attractant. The males possessed dimorphic antennae on which special receptors were located. When females were fed radioactive-labeled food, label was incorporated into the pheromone. Subsequent tests with males revealed that the attraction to females occurred concurrently with an accumulation of labeled pheromone on the antennae.

II. QUANTITATIVE MEASUREMENT OF COMMUNICATION

A search of the literature reveals over a dozen observational studies (see Tables I–III) of crustacean species wherein interindividual aggressive behavior was recorded as dyadic couplets and then summed into i by j contingency tables. These tables, which at a glance provide a quantitative picture of what one animal does in response to another, are then used in several ways. First, they are used to calculate transitional probabilities so that kinematic graphs can be drawn. These charts can provide a representation of the relative temporal and sequential ordering of the acts in the behavioral repertoire. Second, the tables are partitioned in suitable ways to test (with chisquare) the effects of preceding acts on the distribution of following acts, or

TABLE I

Survey of Aggressive Acts, Based on Published Ethograms of over 20 Crustacean Species

Act pattern	Number of species showing pattern
Body postures:	
Elevated	7
Crouched	8
Directed movements:	
Move away	11
Move toward	15
Enter shelter	3
Antennation	4
Ambulatory legs:	
Unilateral	3
Bilateral	4
Contact opponent	1
Chelipeds:	
Noncontact display	
Unilateral	9
Bilateral	12
Contact display	
Unilateral	7
Bilateral	8
No response, "freeze," etc.	6

the effect of a particular preceding act on a particular following act in a dyad. Third (see below), row, column, and cell totals are used to estimate measures of uncertainty reduction and information transmission, based on the Shannon-Wiener equation as modified in Losey (1978).

A. Statistical Techniques

Three null hypotheses are addressed by testing these tables with chi-square. (1) The overall distribution of receiver (following) acts is not affected by the overall distribution of sender (preceding) acts. (2) The distribution of receiver acts in a given row is not affected by the sender act heading that row. (3) A given dyad of acts (act by sender; act by receiver) does not occur more or less frequently than expected by chance.

Not one study (Tables II and III) failed to reject the first hypothesis: the behavior of the receiver was never random. The second and third hypotheses have been treated in different ways by different authors, and we have discussed some of the problems elsewhere (Hyatt and Salmon, 1979). In the final analysis, most studies have shown a selection of following acts and dyads which occur more or less frequently than chance expectation (Maxwell, 1961). Usually the statistical findings agree well with what one would intuitively expect by watching the animals.

B. Information Theory

The basic assumptions and formulations of information theory as applied to animal communication have been outlined and explained in some detail by several authors (Quastler, 1958; Dingle, 1972; Steinberg, 1977; Losey, 1978). We refer readers to these sources for the discussion that follows.

Fortunately, most information analyses of animal communication have been performed with crustaceans. Unfortunately, little uniformity exists in the design of observational experiments: mixed sexes, mixed sizes, and other factors (such as variable size of test arena and time of observation) preclude some across-species comparisons.

Despite the nonuniformity of design, there are some interesting qualitative and quantitative findings. Emergent patterns appear in the following variables: (1) the number (and sometimes type) of aggressive displays; (2) the average information (uncertainty reduction) about participants' behavior; (3) the nature and relative magnitude of uncertainty reduction related to participant status; and (4) the patterns of information transmission changes at different times during the interaction. A more extensive analysis of these findings than we present here is forthcoming (G. W. Hyatt, in preparation).

TABLE II

Summary of Information Analyses of Communication in the Crustacea

Organism/context	Size of repertoire	$H(X)$[a]	$X(Y)$	$t(X,Y)$ (%)	Source
Assorted hermit crabs (mixed-sex fights)	11–16	2.72–3.20	2.99–3.35	9.8–15.7	Hazlett and Bossert (1965, 1966) Hazlett and Estabrook (1974)
Alpheus heterochaelis (intermale fights during daytime)	8	2.29–2.51	2.26–2.96	8.85–20.95	Schein (1975a)
Menippe mercenaria (mixed-sex fights; H measures calculated from published data)	8	2.25–2.46	2.11–2.56	10.14–14.69	Sinclair (1977)
Alpheus rapax and *A. rapacida* (shrimp–goby symbiosis)	13	1.31	2.21–2.40	13–15	Preston (1978)
Uca pugilator (intermale fights)	12	2.24–2.42	2.33–2.45	21.01–35.20	Hyatt and Salmon (1979)
Uca pugnax (intermale fights)	13	3.01–3.06	3.15–3.27	19.45–33.74	Hyatt and Salmon (1979)
Uca pugilator (male–female courtship)	11 (♀), 12 (♂)	1.55–2.44	1.61–2.45	15.83–18.60	G. W. Hyatt (in preparation)

[a] All uncertainty measures are in bits.

1. DISPLAYS

The number of displays used by an assortment of decapod crustaceans during social encounters is shown as part of Table II. Most of these values are within range of behavioral displays performed by fishes (e.g., *Poecilia reticulata*, guppy: $n = 15$), birds (e.g., *Parus major*, great tit: $n = 17$), and mammals (e.g., *Cynomys ludovicianus*, prairie dog: $n = 18$) (Moynihan, 1970; cited in Wilson, 1975).

For a general picture of the kinds of display patterns visible across species, we evaluated over 20 examples (see references) of aggressive behavior ethograms shown by various crustaceans and constructed Table I. Based on this table, a typical aggressive encounter of the "average" decapod might follow the pattern: (1) approach the opponent, (2) perform a unilateral or bilateral noncontact cheliped display, (3) perform a unilateral or bilateral contact cheliped display, and (4) move away from the opponent. This would be done with the body in a crouched position and with little contact between ambulatory legs. The importance of "No Response" should not be underestimated since it was shown to contribute substantially to information transmission in crustaceans (Hyatt and Salmon, 1979) and insects (Steinberg and Conant, 1974).

At a different level, how to decide what to observe, and how to define, describe, and measure displays has been a separate topic of interest among behavioral biologists (Barlow, 1968, 1977; Altmann, 1974; Beer, 1977). Most discussion has revolved around observational sampling and variability measurements, both within and between animals. The subjectivity involved and the use of repetitive measures on the same animals (violating the assumption of independence) lead to interesting problems. Yet ethologists must, by definition, accept what animals present on their terms. Plugging them into factorial designs can disrupt the essence of what we study: the behavior of unrestrained animals under "natural" conditions.

2. UNCERTAINTY REDUCTION AND INFORMATION PRESENT

Table II also shows information measures calculated from contingency tables based on crustacean social interactions. Most involve aggressive behavior. Values for H(X) range from 2.25 bits in *Menippe mercenaria* to 3.20 bits for *Pagurus marshi* (hermit crab). $H(Y)$ ranges from 2.11 to 3.35 bits. Normalized transmission, $t(X, Y)$, ranges from 8 to 35%. These findings suggest that the "typical" crustacean studied to date can, on theoretical average, "comfortably" discriminate four to eight equiprobable signals with average transmission efficiencies of about 15–20%. All species show remarkably similar quantitative measures of communication.

3. STATUS/ROLE AND INFORMATION TRANSFER

During aggression, opponents can be construed as "haves" and "have nots," e.g., dominant versus subordinate or winner versus loser. They are thereby cast into different roles. Some statements have been made in this regard, and a general pattern (with exceptions) can be seen. For example, Hazlett (1982) states that for *Calcinus tibicen,* "The amount of information transferred was higher from losers to winners than from winners to losers. Eventual winners modified their behavior more predictably based upon the eventual loser's acts than vice versa."

Why should information transmission be higher from loser to winner? The fiddler crabs follow this pattern, as do the hermit crabs studied by Hazlett and mantis shrimp studied by Dingle. However, snapping shrimp (Schein, 1975a) and stone crabs (Sinclair, 1977) do not show greater transmission from loser to winner. In fact, the opposite is true. The reasons for these differences remain unknown.

There are only two examples of the quantitative analysis of crustacean communication where aggression was not the behavior studied. The first case involves an analysis of the communicative exchange in a goby–shrimp cleaning symbiosis (Preston, 1978). To our knowledge this is the only example of an interphyletic study. Over 170 encounters between *Alpheus rapax* (n = 106) or *A. rapacida* (n = 66) with *Psilogobius mainlandi* were recorded. Table II shows that uncertainty measures are within the ranges found for the aggressive encounters of other species. However, in this case there is no real disparity in normalized transmission. Both shrimp and goby are sharing and receiving about the same amount of information.

In the second example, the courtship behavior of *Uca pugilator* was examined by G. W. Hyatt (personal communication). The pattern of uncertainty measures again follows that of the aggressive behavior analyses, except there is no large difference in normalized transmission.

4. NONSTATIONARITY OF ACTS AND COMMUNICATION
   MEASURES

The nonstationarity of act probabilities from one time to another during a communicative interaction has been a major problem in assessing the value of information measures in animal behavior. Every investigator is cognizant of nonstationarity, but few have addressed it (but see Oden, 1977; Losey, 1978). Nonstationarity of crustacean display probabilities across aggressive encounters has been cursorily examined in three ways. Dingle (1969, 1972) matched *Gonodactylus bredini* and *G. spinulosus* in paired encounters for 60 min. He then arbitrarily divided the sequential data into the first 10 min, the second 10 min, the middle 20 min, and the final 20 min for analysis.

TABLE III

Information Analyses Partitioned to Demonstrate Effects of Nonstationarity of Act Probabilities on Uncertainty Measures[a]

A.	First 10 min	Second 10 min	Middle 20 min	Last 20 min
$H(X)$	2.99*	2.94*	2.87*	2.76*
$H(Y)$	2.73	3.00 (2.72*)	2.58	2.61
$H(X,Y)$	5.09*	4.91*	4.67*	4.67*
$T(X;Y)$	0.63	1.03 (.75*)	0.78	0.70 (0.69*)
$t(X,Y)$	23.22%	27.63%	30.08%	26.78%

B.	Act 2 (-3)	Act 4 (-5)	(n-1 to) nth Act
$H(Y)$	2.59	3.10	1.25
$T(X;Y)$	0.90	0.99	0.55
$t(X,Y)$	28%	32%	44%

C.	Act 2 (-3)	Act 4 (-5)	(n-1 to) nth Act
$H(Y)$	2.5	3.3	0.8
$T(X;Y)$	0.5	1.6	0.25
$t(X,Y)$	17%	55%	37%

D.	Before retreat		After retreat		E.	Before retreat		After retreat	
	W→R	R→W	W→R	R→W		W→R	R→W	W→R	R→W
$H(X)$	2.42	2.16	0.79	0.63		2.99	3.00	1.82	2.53
$H(Y)$	2.18	2.59	0.86	0.63		3.21	3.17	2.72	1.67
$H(X,Y)$	4.20	3.97	1.39	1.02		5.59	5.10	3.78	3.48
$T(X;Y)$	0.40	0.79	0.26	0.24		0.61	1.06	0.76	0.72
$t(X,Y)$	18.34%	30.40%	30.65%	38.02%		18.96%	33.64%	27.83%	43.10%

[a] All values are in bits, except $t(X,Y)$. Asterisks indicate our calculations. (A) *Gonodactylus bredini* (Dingle, 1969, Table III; see also Dingle, 1972); (B) *Paguristes grayi* (Hazlett and Estabrook, 1974, Table 2; *M. bicornutus* not included); (C) *Orconectes virilis* (Rubenstein and Hazlett, 1974; values estimated from Fig. 1); (D and E) *Uca pugilator* and *U. pugnax* (Hyatt and Salmon, 1979, Tables 3–6).

Table III shows (condensed from Dingle, 1972) that normalized transmission changes throughout the encounters, being low early in the fight, higher in the middle, and not quite so high late in the fight.

The second example is provided by Hazlett and Estabrook (1974). Sequential data for *Paguristes grayi* (hermit crab) and *Microphrys bicornutus* (spider crab) were manipulated so that information values were calculated for acts early and late in the fight. Rubenstein and Hazlett (1974) performed essentially the same analysis with *Orconectes virilis*. For all three species the normalized transmission rose throughout the fight (Table IIIB,C), to where knowing the next to last act almost completely eliminated uncertainty about how the fight would end.

The third example uses the sequential data of over 800 fights between fiddler crabs (*Uca pugilator* and *U. pugnax;* Hyatt and Salmon, 1978, 1979). We have reanalyzed our data for this chapter by dividing each fight (where appropriate) into two parts at the point where the Resident (burrow-owning) crab submitted to the Wanderer by retreating into his burrow. For each species a 10–30% increase in normalized transmission occurs, regardless of whether the measurement is made from wanderer (W) to resident (R) or from resident to wanderer (Table IIID,E). Once the resident submits, the uncertainty about who will win decreases.

Regardless of the method used to address nonstationarity, viz. an arbitrary time division (mantis shrimp), analysis of uncertainty reduction by acts (hermit and spider crab), or a "natural" behavioral division chosen by the animal (fiddler crab), the pattern is clear that as the contest goes on, the uncertainty of the fight outcome is reduced.

III. FUNCTIONS AND MECHANISMS

A. Agonistic Behavior

Spacing systems are major features of the social organization of animal groups. The Crustacea, especially the decapods, have proven particularly amenable for spacing studies because of their generally pugnacious nature. A vast qualitative and anecdotal literature exists which describes how animals fight and under what conditions contestants win or lose (Table I). A few more detailed and quantitative studies comparing closely related species (or genera) have produced educated speculation, if not hard data, regarding the structure and operation of agonistic systems. There is a surprising lack of information concerning what resources are contested under natural conditions, except for the hermit crabs, which fight over shells, stomatopods and alpheid shrimp, which fight over cavities or shelters, and the terrestrial crabs, which contest burrows or burrow sites.

1. INDIVIDUAL DISTANCE AND TERRITORY

The few studies of individual distance in crustaceans suggest that it shares a key characteristic with the same phenomenon in vertebrates: it consists of a complex array of multiple, overlapping probability zones of attack and retreat (Hazlett, 1975c) rather than a "single sphere," which varies from animal to animal and with ecological conditions. Hazlett found that for *Clibanarius tricolor,* personal space varied in the field from about 50 cm for crabs at low densities to 5 cm when the crabs were aggregated. He obtained results supporting these findings when density was experimentally varied under laboratory conditions. The general shape of the space was eliptical, with the animal located at one focus. The selective advantage of such a malleable system is clear. "Where the population density is high, an animal would spend all its time in agonistic interactions if their individual distance was too large, i.e., did not adjust downward. Where the density is low, it is probably advantageous to partition the acquisition of environmental resources (which are apparently more limiting since the population density is lower) via social spacing activities" (Hazlett, 1975c).

Jacobson (1977) extended Hazlett's findings to lobsters (*Homarus americanus*). In these tests, the subjects were housed under different space/density regimens to determine if their agonistic behavior would change in form or intensity. At low densities, the lobsters were primarily territorial, but at higher densities (and limited space) they existed in a dominance order. Molting during June and July resulted in a loss of status and decrease in territoriality, while larger males which did not molt became highly dominant (see also Tamm and Cobb, 1978).

Changes in spacing behavior as a function of environment (behavioral scaling) have also been measured along ontogenetic gradients. In spiny lobsters (*Jasus lalandii*), the smaller individuals are solitary, living in separate holes; big lobsters are also solitary, but intermediate-sized individuals form aggregations inside caves or under ledges (Fielder, 1965). These few studies suggest that surprising plasticity may characterize spacing behavior in the Crustacea.

The construction of earthen "hoods" (shelters) or "pyramids" is a manifestation of territorial behavior which occurs in several ocypodid crabs (Fig. 3A–D). Zucker (1974) demonstrated that the shelters of *U. terpsichores* males actually decrease territory size. Several correlational and experimental results are of interest in this (one of few) quantitative study of crustacean territoriality. (1) Shelter construction peaked during or just after the new or full moon, when the number of receptive females is also higher. (2) Shelters were significantly more numerous in areas of higher population density. (3) Males oriented their shelters randomly with regard to physical parameters. However, the number of other displaying males was always greatest behind the shelter. (4) Presentation of tethered males elicited more aggressive claw "shakes" from test males when stimulus crabs were presented in front of the shelter than from the same distance behind the shelter. It appears that owners of burrows with shelters defend a smaller, oval-shaped territory with most of the area in front of the shelter. The result is a decrease in the amount of time spent in agonistic interactions with other males and an increase in courtship time.

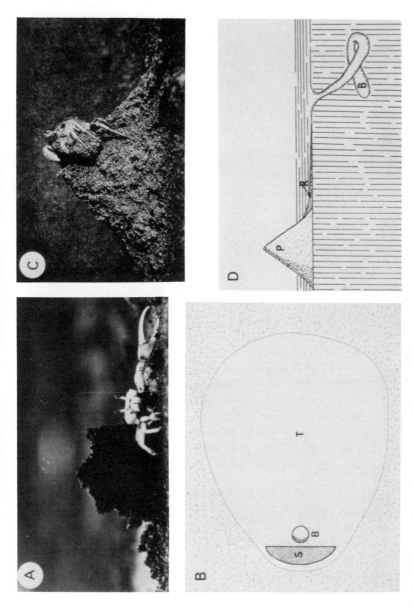

Fig. 3. Substratum formations constructed by ocypodid crabs. (A) Male *Uca terpsichores* displaying before his "shelter," which surrounds the burrow entrance. (B) Diagrammatic view from above of shelter (S), burrow (B), and defended territory (T). The territory of males without shelters is larger in area and contains a centrally located burrow (after Zucker, 1974). (C) Male *Ocypode saratan* building a sand pyramid. (D) Position of the pyramid (P) in relation to the spiral burrow (B) and route (R) between pyramid and burrow. The pyramids attract females and act as territory markers (from Linsenmair, 1967).

The sand pyramids of ghost crabs (*O. saratan* and *O. ceratophthalmus*) are constructed by sexually active males and attract females to the vicinity of the male's spiral burrow (where mating takes place). The pyramids also function to repel other males or serve as foci for fights between males (Lighter, 1974), which result in spacing. Linsenmair (1967) found that Egyptian populations of *O. saratan* built sand pyramids and were diurnally active; populations of this species in Ethiopia were strictly nocturnal, and though they constructed pyramids, these were nonfunctional except on moonlit nights.

2. DOMINANCE

Dominance, as well as the behaviors associated with dominance (such as noncontact threat and overt combat), are known to occur in several crustacean groups. Fights to determine status range from highly ritualized "display only" formats to extended, often vigorous fights resulting in partial dismemberment of one or both animals. Most experiments have been performed as "round robin" or multianimal encounters in buckets, basins, or tanks. The data have not been treated uniformly across studies.

The most extensive work on crustacean dominance has been done by Hazlett (Table IV) with hermit crabs. To quantify dominance behavior, Hazlett (1966a) developed a Dominance Hierarchy Statistic to test if the outcome of intraspecific agonistic interactions was random (the null hypothesis) or ordered as a straight-line dominance hierarchy. None of the three genera of hermits used for experiments (*Pagurus, Clibanarius,* and *Calcinus*) exhibited randomness in fight outcome. Sex of combatants was unimportant in determining fight outcome, but size of the crab and its shell were. Left unanswered was the mechanism by which order was maintained. At least two hypotheses are possible. (1) Every interaction was decided on the basis of opponent "aggressive state," that is, its readiness to engage in agonistic interactions, or (2) the outcome was determined after learning the specific attributes of another crab (individual recognition).

The question was recently resolved for one species of hermit crab (*P. longicarpus*) by Winston and Jacobson (1978). They established dominance hierarchies in several populations, consisting of four crabs each. The crucial tests involved comparisons between populations in which strange crabs of the same rank (subordinates or alpha crabs) were exchanged (controls) with groups in which a second alpha crab from a strange group was substituted for a familiar subordinate (experimental). Thus, all groups had a single stranger, but only in the latter was the stranger of a different rank. There was a significant increase in agonistic interactions only in the experimental condition. Therefore, the crabs were discriminating status on the basis of behavioral cues (aggressive state) rather than individual recognition. These results are to be expected in populations which aggregate in large numbers (too many for individual recognition) and in which group composition changes

TABLE IV

Summary of Factors Affecting the Outcome of Aggressive Contests (Victory or Defeat), Territorial Behavior, and Dominance Relationships (Position in Rank)

Factor	Animal and source	Comment
Animal size	*Pagurus longicarpus* (Allee and Douglis, 1945)	Larger crab wins. Paired, shell-less crabs fight continuously. Presence of shell important? "Hierarchy" limited to two ranks.
	Orconectes virilis (Bovbjerg, 1953)	"Larger" animal wins. No quantitative size information. Straight-line dominance orders (two-rank, with third and fourth ranks weakly detectable).
	Procambarus alleni (Bovbjerg, 1956)	Larger animal, male or female, wins. Male dominance a function of sexual maturity. Two-rank "hierarchy." Both visual and tactile cues function in dominance fights.
	Jasus lalandii (Fielder, 1965)	Larger animal dominant in contests over limited shelter, but at least two-rank order emerges in contests between equal-sized animals.
	Pagurus miamensis *Calcinus tibicen* (Hazlett, 1966a)	Strong evidence for hierarchal social order, but marginal evidence that size is a determinant (best evidence for *C. tibicen*).
	Clibanarius vittatus (Hazlett, 1968b)	Crab with larger cephalothorax length wins 89% of time; probability of large animal winning increases with increased size difference.
	Gonodactylus bredini (Dingle and Caldwell, 1969)	Larger shrimp wins. Very large animals ignore small animals.
	Potamon edulis (Vannini and Sardini, 1971)	Larger crab wins, regardless of sex, but correlation between relative size and probability of dominance not significant, except for young crabs. No evidence of individual recognition.
	Callinectes sapidus (Jachowski, 1974)	Smaller avoids larger.
	Petrolisthes spp. (Molenock, 1976)	Initiator wins 50% of time, regardless of size. Larger crabs win 60% of time. Large initiators win 70% of time.

(continued)

TABLE IV *Continued*

Factor	Animal and source	Comment
	Alpheus heterochaelis (Schein, 1975b, 1977)	Large animals "snap" more and are more dominant.
	Birgus latro (Helfman, 1977)	Smaller crab avoids larger (no quantitative data supplied).
	Menippe mercenaria (Sinclair, 1977)	Large animal more dominant ($p < 0.001$).
	Uca pugilator (Hyatt and Salmon, 1978)	Larger wins, but smaller burrow-owning crabs enjoy competitive advantage.
Sex	*Jasus lalandii* (Fielder, 1965)	No effect of sex on access to limited shelter.
	Pagurus miamensis *Clibanarius tricolor* *Calcinus tibicen* (Hazlett, 1966a,b)	Rank order correlations for intrasexual size rankings indicate sex unimportant in determining position in social order. Sex not important in determining outcome of shell fights.
	Clibanarius vittatus (Hazlett, 1968b)	No effect of sex on winning fights.
	Callinectes sapidus (Jachowski, 1974)	Males dominate females; mated males dominate unmated males.
	Menippe mercenaria (Sinclair, 1977)	Males more dominant ($p < 0.001$).
	Uca pugilator (G. W. Hyatt, unpublished)	Males displace females from burrows, independent of size relationship.
Shell shape	*Calcinus tibicen* (Hazlett, 1966c)	Discrimination of front or back of shell based on two-dimensional outline of shell. Model presentation experiments.
Shell size/weight	*Clibanarius vittatus* (Hazlett, 1970)	Empirical study. Wax and plastic model presentation. Larger size dominant.
Color phase	*Calcinus tibicen* (Hazlett, 1966b)	Turquoise victorious more often than brown phase.
Density	*Calcinus tibicen* *Clibanarius tricolor* (Hazlett, 1966a,b, 1975c)	Isolated crabs more aggressive, but probability of winning not changed. Dimensions of personal space vary with density of population.
	Pagurus acadianus (Grant and Ulmer, 1974)	Isolated crabs more dominant to crabs kept in higher density.

TABLE IV *Continued*

Factor	Animal and source	Comment
Hunger	*Calcinus tibicen* (Hazlett, 1966b)	Starved crabs initiate and win more fights.
Molt stage	*Calcinus tibicen* (Hazlett 1966b)	Recently molted crabs more subject to attack and loss of shell fight.
Previous experience	*Calcinus tibicen* (Hazlett, 1966b)	Past winners win more; past losers lose more.
Status	*Uca pugilator* and *U. pugnax* (Hyatt and Salmon, 1978)	Burrow-owning "resident" more likely to win, regardless of size.
Behavioral state	*Calcinus laevimanus* (Reese, 1962) *Pagurus longicarpus* (Winston and Jacobson, 1978)	Submissive animal inactive. Two-rank order. Dominance mediated via recognition of aggressive state of size-matched crabs. The more aggressive animal is dominant. No evidence of individual recognition.

frequently. A similar situation has also been described in birds which form temporary flocks (Murton et al., 1966).

Table IV summarizes the many factors that are known to affect aggressive behavior and dominance relationships of some crustaceans. While the table is not an exhaustive listing, animal size appears to be a consistent predictor of contest victory and/or superior hierarchical position. This finding supports the proposals suggested by Dawkins and Krebs (1978). "What sort of cues would be good to use in assessment of an opponent's fighting ability? Clearly the cues used should be closely linked to fighting ability and give reliable information. Assessment signals which are easily mimicked by weak individuals will, in the course of evolution, soon come to be ignored . . . while reliable cues will become established to the benefit of both sender and receiver. . . . Cues such as size, which are obviously linked to fighting ability, will tend to be resistent to bluff and should be used to settle contests."

B. Sexual Behavior

Crustaceans exhibit a bewildering variety of reproductive activities, reflecting differences in their spacing and abundance, preferred habitat, feeding patterns, activity, and other ecological variables. Some specialized forms, such as sessile species, parasites, or highly dispersed populations, are hermaphroditic (Cirripedia) and/or protandric (some Natantia, Anomura, Tanaidacea, and Isopoda). In other groups (Ostracoda, Notostraca), males may be absent from part of the species range but present in other areas. More typically, the sexes are separate and capable of movement over relatively long distances. We will not elaborate upon all these variations, but rather point out their importance as factors which can affect when spatial attraction between the sexes may occur (only in the mobile juvenile stages, as in the pinnotherid crabs), how often individuals will be able to select mates (once in some species, such as some *Emerita*), and under what kinds of constraints (predator pressure, habitat complexity, tidal events) in time and space attraction, copulation, and parental care must take place. A consideration of these variables can indicate whether sexual signals will involve only the "basics" (species and sex identification) or more elaborate choices (particular *classes* of mates possessing special attributes; information which can identify *individual* males or females).

1. ATTRACTION AND LOCALIZATION

In many crustaceans, the sexual stages are separated by moderate to large distances. Under these conditions, the broadcast of chemical, acoustic, and visual signals is believed to facilitate spatial localization. The following generalizations are applicable to these communication systems at the present time. First, sex pheromones are emitted by many aquatic species. They appear to operate at close range in the smaller, less mobile forms (amphipods, isopods, copepods), while in the larger decapods, such as the lobsters, crayfishes, and swimming (portunid) and some anomuran crabs (*Paralithodes*), stimulating effects occur at lower concentrations, and the individuals are usually separated by greater distances (see below). Second, among the terrestrial and semiterrestrial crustaceans, chemical signals operate at close range and again predominate in the smaller, less mobile forms; in larger species, localization depends on the elaboration of visual and/or acoustic signals. However, much basic information remains to be determined with regard to the effective distances over which the signals operate, the nature of the response elicited, the receptors and neuronal networks which respond, and the kinds of information extracted.

a. Chemical Attraction. In a particularly insightful review, Dunham (1978) has summarized the situation with regard to sex pheromones in general and their utility as attractants in particular. The most convincing evidence for their usage comes from those species where there is a temporal linkage between molting by females and mating (portunid crabs, caridean shrimps, and gammarid amphipods). In these species, sex pheromones have "releaser" properties, at least at the concentrations presented under laboratory conditions. Such responses are different from those associated with general arousal or food seeking and are often identical to those observed just before males seize females and mate.

The difficulty comes when one tries to relate these response patterns to how the pheromone may operate in nature. For instance, no pheromones have been characterized as to threshold concentrations and release rates. This makes it impossible to gauge the dimensions of the active space under which detection and response occur. Therefore, the behavioral responses elicited in the laboratory may bear little resemblance to those encountered in nature, especially since they are usually evoked in buckets, small aquaria with standing water, or in olfactometers which can restrict behavioral choice and confuse chemokinetic and chemotaxic responses.

Studies dealing with chemical localization in fishes (Kleerekoper, 1969; Bardach and Todd, 1970) and insects (Shorey, 1973) are particularly illuminating, as the problems encountered are similar. These studies show that localization is rarely accomplished by response to a chemical gradient. More commonly, chemical stimuli trigger movement upstream or upwind; that is, chemical stimuli act as releasers, but the orientation to the source is directed by currents. Thus, when a crustacean (usually the male) is presented with a sex pheromone in still water, the responses evoked may be distorted in two ways. First, the lack of current may evoke a kinesis, when normally an upstream movement would occur. Secondly, the response can be changed by stimulus concentration. If high stimulus levels are employed, the evoked behavior may be typical of situations occurring during the terminal portion of the search (just before a mate is located), leaving unknown the response during most of the orientation.

It has been presumed that females release the pheromones and males respond to them. Yet in one of the more sophisticated series of experiments, Teytaud (1971) showed that it was the male blue crab (*Callinectes sapidus*) which released stimulating compounds and the female which responded. His experiments were also important in that they demonstrated that pheromone, alone, was ineffective in eliciting courtship; chemical stimuli only evoked female courtship when coupled with the sight of the male.

Much of the ambiguity described above can be clarified by field observations or by experiments in large tanks mimicking as closely as possible natural conditions with regard to animal density, water currents, and other important variables. In addition, it is essential that paths taken by animals, from the point of initial stimulus contact to the source of the pheromone, are carefully measured (Dunham, 1978).

b. Acoustic Attraction. Spontaneous acoustic signaling occurs only in the ghost and fiddler crabs, though the degree to which acoustic "calling" is developed varies considerably. *Ocypode ceratophthalmus, O. laevis,* and

O. cordimana show some activity predominantly during nocturnal new moon periods, while *O. quadrata* calls briefly at dawn. In the fiddler crabs, some males of *Uca pugilator* and *U. panacea* produce sounds every night. Sonic calling is much less frequent in *U. pugnax* and apparently absent in *U. minax* (Salmon and Horch, 1973; Horch, 1975). Why these species differences occur is largely unknown.

The evidence that these sounds act as "calls" which attract females is indirect. First, calling sounds are produced exclusively by males, and calling occurs only during the breeding season. Second, diurnal sound production by fiddler crabs is invariably elicited from courting males by females, while intruding males are repelled with nonacoustic, aggressive displays. And third, males response to change in acoustic calling which occur when neighbors are stimulated by nearby females. These "chorusing" responses are consistent with the notion that males are acoustically competing with one another for access to females.

Physiological measurements have shown that both *Ocypode* and *Uca* can detect the calls at considerable distances (1 m in *Uca;* 10 m in *Ocypode*). How might such sounds be localized? Simultaneous "bi-ambulatory" comparisons of stimulus intensity seem unlikely, as attenuation gradients are too gradual unless the orienting crab has approached to within close range. Horch measured acoustic velocities of the sounds in wet sand surrounding *Ocypode* choruses. Surprisingly, average velocities were 200 m/sec, much slower than the velocity of sound in air (340 m/sec), suggesting that arrival time could be an important cue in acoustic localization.

c. *Visual Signals.* The visual displays employed by crustaceans before formation fall into two general categories. On the one hand, there are those which are virtually identical to signals used in investigation or threat, but which change, either in their frequency linkages to other acts (e.g., alpheid shrimps) or by giving way to new (sexual) acts ("dances" in some grapsids; tactile contacts with the chelae and walking legs in majids: Schöne, 1968) as sex and/or reproductive state is made clear. On the other hand, there are displays which are clearly differentiated from those used in other contexts, such as those released by both visual and/or chemical stimuli in portunids (Ryan, 1966; Teytaud, 1971) or the claw waving displays of some grapsid and ocypodid crabs, which are released by females.

As stressed earlier, many crustaceans employ visual courtship signals at close range, but it is only on land that the visual channel operates at longer distances to attract (or repel) conspecifics. The amount of time spent in visual signaling, at least in *Uca,* is correlated with the ability to withstand exposure to high temperature and low humidity. For example, primitive fiddler crabs are poorly adapted to exposure and court only during early morning low tides. The more advanced species court later in the day, even during the afternoons when temperatures are high and humidity is low (Crane, 1958). Acess to water is probably another important variable permitting the evolution of diurnal display. *Ocypode gaudichaudii* and *O. kuhli* are the only ghost crabs which burrow near or in the intertidal zone. Their burrows probably descend to depths where freestanding water is available. They are also the only species of *Ocypode* which consistently exhibit diurnal activity and in which waving displays are known to occur (Wright, 1968; Vannini, 1976).

2. SPECIES IDENTIFICATION

The signals used in attraction often are also typical of the species, and therefore could serve to identify conspecifics. Hazlett (1968a) has described courtship preliminaries in hermit crabs, which are dominated by intricate, species-typical patterns of tactile (and chemotactile) displays such as stroking, jerking, and shell-rocking. Crane (1941) was the first to recognize that the waving patterns of fiddler crabs were species-specific movements, differing in the jerking pattern on the enlarged claw, embellishments by "special-steps" with the ambulatories, or even the inclusion of sonic displays with the performance of each wave. The acoustic signals of fiddler and ghost crabs differ in temporal patterning and in spectral energy distribution, though the crabs are apparently insensitive to the later (Salmon and Horch, 1976). Numerous other examples are given in Hazlett's (1975a) review.

However, there is practically no experimental evidence, comparable to the work with insects (Walker, 1957), frogs (Littlejohn, 1969), or birds (Thielcke, 1962), that these signal differences are used by crustaceans to discriminate between sympatric species. In fact, some data suggest that signal differences alone are insufficient to prevent mating attempts. For example, both Katona (1973) and Griffiths and Frost (1976) found that male calanoid copepods responded to sex pheromones from conspecific females and also to chemical stimuli emanating from related heterospecific females, though not as strongly. However, these were laboratory tests and the species involved might normally breed at different locations in the water column or at slightly different times. Also, sexual discrimination might be a prerogative of the female, not the male. The only definitive tests were those carried out by Eales (1974), discussed above.

Salmon et al. (1978) examined the mechanisms underlying reproductive isolation between two sibling species of fiddler crabs, *U. pugilator* and *U. panacea,* which were sympatric along the northwest coast of Florida. Both were reproductively active during the summer months. The males of the two species showed differences in their premating displays, which were further exaggerated by character displacement in the overlap zone. Cross mating could be induced under laboratory conditions, but larval mortality was significantly greater then in the parents. Those hybrids which survived to maturity were sterile as adults (Salmon and Hyatt, 1979). Thus, females which failed to discriminate (or allowed the males to force copulations) suffered reduced fitness when compared to those which did discriminate. Finally, laboratory-reared females of *U. pugilator* developed the ability to select conspecific males even when reared to maturity in isolation from any courting males. Thus, female discrimination, coupled with physiological incompatability, prevented gene exchange between the species. The generality of these findings to other species of fiddler crabs, or to other crustaceans, remains to be determined.

While the above study is informative, we should stress that the role of courtship signals as isolating mechanisms, and their place in the hierarchy of factors preventing interbreeding, remain virtually unexplored in other Crustacea. In some species, precopulatory signals appear to be virtually absent (Hartnoll, 1969; Hazlett, 1975a). In these species and sex discrimination may be accomplished by chemical exchanges which take place during contact, coupled with behaviors which tend to bring the conspecifics together in unique habitats.

3. MATE SELECTION AND INDIVIDUAL RECOGNITION

Evidence is beginning to accumulate that, in at least some Crustacea, conspecific mates are chosen preferentially and in some cases are recognized as individuals. The importance of such responses with respect to theories of sexual selection is obvious and will be considered in more detail below.

Manning (1975) showed that in isopods (*Asellus aquaticus* and *A. meridianus*), males select females on the basis of size. Females become attractive before their parturitional molt and are seized by males which carry them for periods of several days. During this time, a male cannot mate with other females, and so his selection of an "optimal" mate is critical. Manning showed that the males prefer large females. This response has two benefits. First, the number of eggs produced by females is proportional to female size. Secondly, the eggs of larger females mature more rapidly. This means that a male which has seized a large female can mate and move on to another female in a shorter period of time.

In fiddler crabs (*Uca rapax* and *U. pugilator*), females visit several males before selecting a mate (Greenspan, 1975; Hyatt, 1977; Christy, 1980), who is invariably larger. "Acceptance" is indicated by entry into the male's burrow, followed a few minutes later by reappearance of the male at the surface to seal his burrow entrance ("closing").

Additional evidence for mate selection comes from studies of alpheid

(Hazlett and Winn, 1962; Fishelson, 1966; Nolan and Salmon, 1970; Schein, 1975b; Knowlton, 1978), stenopid (Johnson, 1969), and gnathophyllid (Wickler and Seibt, 1970) shrimp, all of which pair for periods of several days to several months. Invariably, the males and females are "size-matched"; that is they rarely deviate by more than 5% in length from one another, a situation which cannot occur if pairing is random. Such a condition probably is the outcome of at least three factors: attempts by females to pair with the largest males, whose size reflects survival capacities; attempts by males to pair with the largest females who are capable of producing greater number of eggs; and competition within each sex which precludes extended associations with disparate-sized mates.

In *Stenopus hispidus* (Johnson, 1977), *Hymenocera picta* (Wickler and Seibt, 1970), and the desert isopod, *Hemilepistus reaumurii* (Linsenmair and Linsenmair, 1971), individual recognition of mates has been demonstrated by separating formed pairs, then measuring responses to original mates and strangers of the same sex at a later time (a form of retention test). Mate recognition is measured by reduced or absent agonistic behavior upon reintroduction. Pairs can recognize one another after separation for as long as 1 week. The responses are chemically mediated.

4. THE DEVELOPMENT AND INHERITANCE OF SEXUAL DISPLAYS

The larval stages of marine crustaceans generally disperse after hatching. The larvae possess morphological and behavioral responses designed to promote their flotation and spread to new locales. In fiddler crabs, monthly cycles of receptivity and egg development are believed to be timed by rhythms of larval release at the spring of neap tides. These adaptations may maximize their dispersion and eventual colonization of new habitats (Zucker, 1978; but see Christy, 1978, for an alternative view).

These elements of natural history suggest that crustacean mating displays and their recognition by conspecifics should develop normally, even in individuals deprived of any contact with adults after hatching. This hypothesis was recently confirmed for the fiddler crab, *U. pugilator* (Salmon et al., 1978; Salmon and Hyatt, 1979). As mentioned above, female *U. pugilator* could discriminate between conspecifics and male *U. panacea* even upon their first exposure to sexually mature males. The rapping displays of laboratory-reared males (*U. pugilator*) changed from an "immature" to "mature" pattern during the first 3–6 months of sexual maturity, mainly by the addition of a few pulses to the sounds. However, the pulse repetition rate, which is probably most important for species recognition, was virtually identical to that exhibited by conspecifics in the field.

Hybrids between *U. pugilator* and *U. panacea* exhibited temporal patterns of acoustic display which in some respects were intermediate between the species, while in others were different from each parent. These findings confirmed a long-held assumption: that species differences in crustacean sexual displays are inherited and therefore capable of modification by natural selection.

IV. EVOLUTION

A. Precursors of Signals

Very little can be said with certainty about the historical origins of crustacean visual signals except that (1) these displays invariably involve appendages which have undergone structural change for more general functions, such as feeding, cleaning, defense, or locomotion, and (2) there is no agreement concerning how the original movements used in display may have originated. Crane (1975) adheres to the "classical" notion, first proposed by Tinbergen (1952), that social signals (in fiddler crabs) are derived from "conflicts" or "displacement activities." There is no physiological evidence, and only vague behavioral data, to support these hypotheses. Wright (1968) proposed that chela movements during agonistic and sexual displays of ocypodid, grapsid, and gecarcinid crabs were derived from a primitive "lateral merus" display, which is ubiquitous in the Brachygnatha as an agonistic posture. The "chela forward" display occurs in grapsids and some ocypodids as a specialized courtship movement whose origins are more obscure. Wright's proposal, while free of any assumptions with reference to functional or neural mechanisms, is so general that it offers no testable hypotheses applicable to specific taxonomic groups.

Kittredge and Takahashi (1972) proposed that crustecdysone (20-hydroxy-ecdysone) might be the sex pheromone in crustaceans, largely because exposure to dilute solutions of this substance evoked "courtship postures" from distantly related crab genera (*Pachygrapsus* and *Cancer*). This speculation is certainly premature given that the experiments lacked critical controls and that no crustacean pheromone has been chemically identified. Eales (1974) points out a further problem. If large numbers of species employ the same pheromone, how is interference from and attraction by heterospecifics avoided in nature? The situation might be particularly disadvantageous if pheromone release by a female signaled her impending molt, when she is especially vulnerable.

The evidence suggests that social signals in crustaceans, even those which serve the same function, are independently derived even within closely related species. This is particularly obvious when sonic signals are considered. For example, the calling sounds of ghost crabs are made by rapping

(*O. laevis* and *O. ceratophthalmus*) or by stridulating (*O. cordimana:* Horch, 1975). Among the fiddler crabs, rapping with the major cheliped occurs in some subgenera (*Celuca*), while calling is associated with ambulatory leg movements in others (*Minuca*). One species (*Uca thayeri*) apparently employs both mechanisms.

B. Ritualization

While the origins of display in the Crustacea remain enigmatic, it is clear that there are species differences in the degree to which signals are "conspicuous," exhibit stereotypy, or can be used by receiving organisms to predict sender meaning. Ritualization refers to changes in signals which make their meanings less ambiguous (Huxley, 1966). Within the context of sexual behavior, such modifications might make male displays more attractive or less ambiguous with regard to species identity. Agonistic signals are considered more ritualized if they minimize injury, for example, by directing forceful components of contests to body parts which are amply protected.

There are a number of problems with this concept which make its utility questionable. Ritualization is a relative concept, most often measured by comparisons between homologous displays in related species. However, all differences between species are not necessarily due to ritualizationlike processes. Crane (1957) proposed that the courtship displays of *Uca* evolved from primitive "vertical" to more complex "lateral" waves shown by the more "advanced" forms. While it is clear that the waving displays of the latter are more complex and dramatic, there is no evidence that they are more effective as transmitters of information to females. Furthermore, the selection pressures which are responsible for the differences between vertical and lateral waving species have never been identified. The fact that most vertical waving species are found in the Indo-Pacific, whereas the lateral waving forms occur in Central, South, and North America, suggests that some correlate of locale (such as the number of sympatric species present) may be responsible. This view is further supported by the behavior of the few close relatives of vertical waving species found in Central America, which exhibit lateral waves.

A second problem is that ritualization may accurately describe certain changes, but fail to explain why they occur. For example, Hazlett (1972a) found that two species of hermit crabs (*Petrochirus diogenes* and *Clibanarius vittatus*), when held at identical group size in the laboratory, differed in mortality associated with fighting. Fights in *Petrochirus* invariably ended in injury or death due to cannibalism. This species also failed to respond to models of conspecifics presented in agonistic display postures. These findings suggest that agonistic behavior of *Petrochirus* exhibited little sign of ritualization compared to *Clibanarius,* where fights rarely resulted in injury and responses to models were invariably nonforceful. Hazlett's observations are certainly interesting and indicate profound differences between the two species. However, more fundamental issues are left unexplained. Why do these species show such divergent patterns in the first place? What are the rules that govern fight intensity and strategies of fighting?

C. Ecological and Evolutionary Considerations

1. AGONISTIC COMMUNICATION

The theoretical ideas of Brown (1964), Parker (1974), Wilson (1975), and Maynard Smith (1976) suggest alternative ways of examining the problem of why animals fight the way they do. For each species, certain resources will be in short supply (food, shelter, space, or mates), resulting in competition. The form and context of aggression will depend on the value of the resource, when and where it is available, its "defendability," what risks are involved in obtaining it, and what the benefits of successful acquisition might be. Other factors include the degree to which a species has evolved motor capacities to inflict damage and the history of its associations with competitors with whom it must contest critical resources.

The evidence in support of these predictions is fragmentary, largely because until recently the theoretical framework outlined above was largely ignored by crustacean biologists. However, some interesting correlations do exist. It is well known that most "wild" fighting occurs in predominantly aquatic species of decapods which meet infrequently, possess both ample weapons and armament for defense, and which compete only occasionally over patchy, ephemeral resources. Among the terrestrial crabs, contacts are more "formalized" (Schöne, 1968). A complex of factors may be responsible. For one, the offensive weapons (chela) remain formidable, while the body and legs are generally lightened as an adaptation for locomotion on land; hence, risk of injury is more likely. Secondly, semiterrestrial species tend to aggregate in suitable habitats for burrowing, climbing, and feeding, which increase the likelihood of social contacts. Thirdly, the resources over which fighting occurs may not be as immediately critical for survival. Food is probably not limiting, at least for omnivorous or herbivorous species. Burrows, while important, can, if lost, be reacquired by displacing other individuals or digging new ones in a short period of time.

Recent studies on the agonistic behavior of stomatopods have shown how habitat, resources, competitors, and aggressive behavior are interrelated (Caldwell and Dingle, 1975; Dingle and Caldwell, 1975). Fighting in these animals is invariably over space (shelters) and involves a pair of raptorial appendages which can be presented in a graded series of threats or rapidly extended as a "strike." The appendages are also used in prey capture. The "smashers" have heavily armored appendages which are used to stun hard-shelled crabs or break open the shells of mollusks. The "spearers" feed on soft-bodied prey which are impaled upon the distal (dactyl) segment of the appendage. Strikes of both groups are generally delivered with the dactyl closed. This is the normal mode for the smashers, whose blows are much more powerful.

Comparative studies reveal that smashers strike more often and possess a larger repertoire of agonistic acts than spearers. However, smashers also possess a heavily armored telson, to which most of the strikes are delivered. Smashers live in coral and rubble and are largely dependent on preexisting cavities for shelter. The spearers are generally found in sand or mud burrows which they defend but can construct anew if displaced. They show fewer agonistic acts, a reduced tendency to deliver their less powerful strikes, and rarely strike at the telson area. Are these differences a result of prey capture mechanisms, habitat, or other variables? Variation in ecological require-ments of both groups allows separation of correlations and causes. Those spearers which live in coral (e.g., *Pseudosquilla ciliata*) are the most aggres-sive of their group, whereas smashers which are found beneath coral "mushrooms" (e.g., *Gonodactylus platysoma*), rather than in small cavities, are the least aggressive of their group. Thus, the extent of aggressiveness is strongly coupled with the "value" of space; when it is most limiting, cavity related defense is more vigorous.

Caldwell and Dingle have also shown that agonistic behavior varies be-tween populations, depending on which other stomatopods species are present. For example, *Haptosquilla glyptocercus* is the most abundant spe-cies in Enewetok, Marshall Islands, but is the least abundant in Phuket, Thailand. However, its density in the two areas shows no absolute dif-ference, but rather a relative difference when compared to other sympatric, competing species. Enewetok individuals delivered more strikes per fight to both conspecific and heterospecific (*G. incipiens* and *G. falcatus*) comba-tants than did *H. glyptocercus* from Phuket, whereas Phuket *H. glyptocercus* gave more meral spread threats to its more dominant local competitors (*G. chiragra* and *G. viridus*). These differences could not be correlated with any variables other than different sets of competing species and were presum-ably evolved as a result of interactions with these species (Caldwell and Dingle, 1977). The correlations revealed in this and other studies (Dingle et al., 1973; Dingle and Caldwell, 1975) are the following. Where two or more stomatopod species compete for cavities or burrows, the behaviorally more aggressive form exists in greater abundance. Second, this effect occurs to the degree that the two species overlap in their ecological requirements, as well as in space. Third, geographic differences in agonistic behavior within a species are evident, their nature depending on it local competitors and their fighting strategies. Finally, competitively dominant species are more apt to escalate fights by striking, while competitively inferior species more frequently employ "display–retreat" strategies, such as meral threats in the presence of dominant competitors.

2. SEXUAL COMMUNICATION

The nature of the cues exchanged between the sexes, the degree to which courtship is symmetrical, and the consequences of sexual selection are largely determined by the type of mating system (Williams, 1966; Trivers, 1972; Halliday, 1978). In the Crustacea, females generally produce many eggs and perform most of the parental care duties (guarding the eggs, ventilating them, selecting optimal habitats for egg development, and releasing larvae at appropriate places and at the appropriate time). With few exceptions, contact between parents and young then ends. In general, males can do little to assist in these operations, which biases the crustaceans toward polygyny.

In the vertebrates and other arthropods, polygyny often results in intense intrasexual competition between males and strong male dimorphism based on structure, behavior, or both. Females, on the other hand, are highly selective, choosing as mates males whose courtship "vigor," size, or competitive ability, (with reference to other males) is superior. However, the nature of the mating system depends not only on physiological constraints, but also on the distribution of resources necessary for adult survival and for rearing young (Orians, 1969), the ability of mates to garner them (Emlen and Olring, 1977), their consequences upon dispersion, and the risks involved in searching for mates. There are few definitive studies aimed at identifying these variables. However, there is little reason to doubt that both the particular characteristics of crustacean polygyny and the few deviations toward monogamy are probably correlated with ecological variables as well as the pattern of parental investment.

a. *Polygyny.*   Many of the characteristic features of polygyny in vertebrate or insect groups are unreported from, or deviate in, the Crustacea. For the majority of species studied (from ostracodes, copepods, and branchiopods to most decapods), there is little evidence that females select particular subsets of males or that males compete with one another for access to receptive females. On the other hand, there are framentary reports indicative of optimal strategies for protection of genetic investment. For example, the deposition of male sperm plugs, in addition to postcopulatory attendance in a number of decapod species, could be viewed as adaptations to prevent sperm displacement by other males.

For many aquatic species, the absence of apparent selection by either sex, above and beyond what is necessary to identify species, could be real or an artifact. The latter is a possibility simply because so few studies have provided mates with choices or examined the specific attributes of individuals that pair. On the other hand, aquatic crustaceans are generally widely dispersed and may meet infrequently, even under ideal conditions. The search for a mate may also entail a large risk of attack from predators. These factors could select against individuals who are too choosy. The matter will only be resolved through further study.

For the terrestrial ocypodids, there is more information. Male fighting over suitable display burrows is well documented (e.g., Lighter, 1974; Zucker, 1974), and so are their efforts to aggressively interfere with copulations by other males (Crane, 1975) or to minimize the possibility of this interference from neighbors (Zucker, 1974). With regard to female selection, recent studies by Christy (1980) with *U. pugilator* are germane. Females typically select male mates which are larger than themselves, but are they doing so because large size is symbolic of greater fitness or for some other reason? Christy found that receptive females not only mate in male burrows, but also remain in the burrow while their eggs are extruded, attached to the pleopods, and incubated. This suggested that the quality of a male's burrow was an important consideration to females in selecting mates. For example, females selecting males whose burrows were flooded at high tide lost many eggs, as flooding resulted in burrow collapse before newly extruded eggs were firmly attached. Christy designed experiments in which male size and position of his display burrow in the intertidal zone were independently varied. His results showed that females consistently selected high beach burrows, regardless of the size of the males that occupied them. Since larger males have a competitive advantage over smaller males in procuring and holding these burrows, they are usually chosen as mates by females.

b. Monogamy. Monogamy is relatively rare in the Crustacea studied to date, and much basic information is needed before its character and causation are understood. For example, monogamy has been assumed to occur in those species typically found in pairs. However, unless the mates are individually marked and their fidelity documented by long-term observation, the assumption may be false. Knowlton (1978) found that some male snapping shrimp (*Alpheus armatus*) paid frequent visits to unpaired, neighboring females which were reproductively out of phase with their original mates. Such mates were therefore successful polygynists. Knowlton's study also illustrated another important point. She worked with two populations of shrimp located in study areas 0.5 km apart. In one area, predator pressure made the search for additional mates too risky, while in the other area the risks were eased and polygyny occurred. Thus, ecological conditions may vary sufficiently to alter the mating system for the species as a whole. Such flexibility in mating systems is well known in other animals (Wilson, 1975).

A second issue, also unresolved, centers on the ecological determinants of monogamy, or at least reduced polygamy, in the Crustacea. All the species involved have in common relative small size and dependence upon suitable shelters for protection from predators or environmental extremes. This situation, in turn, predisposes females within safe places to stay there and males who have found them to avoid risky searches for additional copulations.

V. CONCLUSION

We emphasized, at the outset, that ecological diversity is one of the major characteristics of crustaceans. This diversity has led to the evolution of communication systems whose variety (in terms of signal types employed), informational content, and survival value are only beginning to be understood. In that sense, the analysis of communication in the Crustacea is reflective of the field of animal communication in general: the emphasis has been upon collecting factual information, and there are few guiding principles. We expect this situation will change radically in the next 10 years.

There are two broad areas where research can be directed. The first, or proximate, area includes (1) more fine-grained studies of displays with the goal of organizing and defining functional categories; (2) experiments with models to dissect qualities of displays in relation to the behavioral state of the receiver; (3) ontogenetic studies of display development and recognition; and (4) experiments to determine sensory mechanisms involved in decoding. The second (ultimate) area is ecological and evolutionary in emphasis, and it attempts to define the place of crustacean social behavior within the framework of evolutionary theory. The thrust of such studies should be directed toward understanding (1) what selection pressures have shaped the evolution of communication in the past; (2) what factors are responsible for variation shown in presently existing populations, both within and between species; and (3) how particular constraints imposed by body plan, sensory capabilities, and physiology have affected the ability of crustaceans to adapt to their environment as compared to other vertebrate and invertebrate organisms. We expect the answers will provide two general types of conclusions. On the one hand, much of what crustaceans do will be understood within the framework of unfolding evolutionary principles, applicable to all organisms. On the other, the Crustacea will present some unique adaptations which will expand current theory to encompass their particular solutions to the problem of survival.

ACKNOWLEDGMENTS

We are grateful to John H. Christy, Dale Crusoe, Brian Hazlett, and Sherry Walker for their critical readings and helpful suggestions. Computing services were provided by the Computer Center of the University of Illinois at Chicago Circle. We also thank Christopher Drubel for his superb programming talents. This paper was written with support from the National Science Foundation (BNS 77-15902 to M.S.; BNS 76-08524 to G.W.H.).

REFERENCES

Allee, W. C., and Douglis, M. B. (1945). A dominance order in the hermit crab *Pagurus longicarpus* Say. *Ecology* **26**, 411–412.

Altmann, J. (1974). Observational study of behavior: Sampling methods. *Behaviour* **49**, 227–267.

Bardach, J., and Todd, J. (1970). Chemical communications in fish. *Adv. Chemorecept.* **1**, 205–240.

Barlow, G. (1968). Ethological units of behaviour. *In* "The Central Nervous System and Fish Behavior" (D. Ingle, ed.), pp. 217–232. Univ. of Chicago Press, Chicago, Illinois.

Barlow, G. (1977). Modal action patterns. *In* "How Animals Communicate" (T. A. Seboek, ed.), pp. 98–134. Indiana Univ. Press, Bloomington.

Barnett, C. (1977). Aspects of chemical communication with special reference to fish. *Biosci. Commun.* **3**, 331–392.

Beer, C. G. (1977). What is a display? *Am. Zool.* **17**, 155–165.

Bossert, W. H., and Wilson, E. O. (1963). The analysis of olfactory communication among animals. *J. Theor. Biol.* **5**, 443–469.

Bovbjerg, R. V. (1953). Dominance order in the crayfish *Orconectes virilis* (Hagen). *Physiol. Zool.* **26**, 173–178.

Bovbjerg, R. V. (1956). Some factors affecting aggressive behavior in crayfish. *Physiol. Zool.* **29**, 127–136.

Brown, J. L. (1964). The evolution of diversity in avian territorial systems. *Wilson Bull.* **6**, 160–169.

Caldwell, R. L., and Dingle, H. (1975). The ecology and evolution of agonistic behavior in stomatopods. *Naturwissenschaften* **62**, 214–222.

Caldwell, R. L., and Dingle, H. (1977). Variation in agonistic behavior between populations of the stomatopod, *Haptosquilla glyptocercus*. *Evolution* **31**, 221–224.

Christofferson, J. P. (1970). An electrophysiological and chemical investigation of the female sex pheromone of the crab *Portunus sanguinolentus* (Herbst). Ph.D. Thesis, Univ. of Hawaii, Honolulu.

Christy, J. H. (1978). Adaptive significance of reproductive cycles in the fiddler crab *Uca pugilator:* A hypothesis. *Science* **199**, 453–455.

Christy, J. H. (1980). The mating system of the sand fiddler crab, *Uca pugilator*. Doctoral Dissertation, Cornell Univ., Ithaca, New York.

Crane, J. (1941). Crabs of the genus *Uca* from the West Coast of Central America. *Zoologica (N. Y.)* **26**, 297–310.

Crane, J. (1957). Basic patterns of display in fiddler crabs (Ocypodidae, Genus *Uca*). *Zoologica (N. Y.)* **42**, 69–82.

Crane, J. (1958). Aspects of social behavior in fiddler crabs, with special reference to *Uca maracoani* (Latreille). *Zoologica (N. Y.)* **43**, 113–130.

Crane, J. (1975). "Fiddler Crabs of the World." Princeton, New Jersey.

Dahl, E., Emanuelsson, H., and Von Mecklenberg, C. (1970). Pheromone reception in the males of the amphipod *Gammarus dueberni* Lilljeborg. *Oikos* **21**, 42–47.

Dawkins, R., and Krebs, J. R. (1978). Animal signals: Information or manipulation? *In* "Behavioral Ecology" (J. R. Krebs and N. B. Davies, ed.), pp. 282–312. Sinauer Associates, Sunderland, Massachusetts.

Dingle, H. (1969). A statistical and information analysis of aggressive communication in the mantis shrimp *Gonodactylus bredini* Manning. *Anim. Behav.* **17**, 561–575.

Dingle, H. (1972). Aggressive behavior in stomatapods and the use of information theory in the analysis of animal communication. *In* "Behavior of Marine Animals." (H. E. Winn and B. Olla, eds.), pp. 126–156. Plenum, New York.

Dingle, H., and Caldwell, R. L. (1969). The aggressive and territorial behaviour of the mantis shrimp *Gonodactylus bredini* Manning (Crustacea: Stomatopoda). *Behaviour* **33**, 115–136.

Dingle, H., and Caldwell, R. L. (1975). Distribution, abundance, and interspecific agonistic behavior of two mudflat stomatopods. *Oecologia* **20**, 167–178.

Dingle, H., Highsmith, R. C., Evans, K. E., and Caldwell, R. L. (1973). Interspecific aggressive behavior in tropical reef stomatopods and its possible ecological significance. *Oecologia* **13**, 55–64.

Dunham, P. J. (1978). Sex pheromones in Crustacea. *Biol. Rev.* **53**, 555–583.

Eales, A. J. (1974). Sex phermones in the shore crab *Carcinus maenas* and the site of its release from females. *Mar. Behav. Physiol.* **2**, 345–355.

Emlen, S. T., and Oring, L. W. (1977). Ecology, sexual selection, and the evolution of mating systems. *Science* **197**, 215–223.

Fielder, D. R. (1965). A dominance order for shelter in the spiny lobster *Jasus lalandei*. *Behaviour* **24**, 236–245.

Fishelson, L. (1966). Observations on the littoral fauna of Israel. V. On the habitat and behaviour of *Alpheus frontalis* H. Milne Edwards (Decapoda, Alpheidae). *Crustaceana* **11**, 98–104.

Frings, H., and Frings, M. (1967). Underwater sound fields and behavior of marine invertebrates. *In* "Marine Bioacoustics" (W. N. Tavolga, ed.), Vol. II, pp. 261–282. Pergamon, New York.

Grant, W. C., Jr., and Ulmer, K. M. (1974). Shell selection and aggressive behavior in two sympatric species of hermit crabs. *Biol. Bull. (Woods Hole, Mass.)* **146**, 32–43.

Greenspan, B. (1975). Male reproductive strategy in the communal courtship system of the fiddler crab, *Uca rapax*. Ph.D. Thesis, Rockefeller University.

Griffiths, A. M., and Frost, B. W. (1976). Chemical communication in the marine planktonic copepods *Calanus pacificus* and *Pseudocalanus* sp. *Crustaceana* **30**, 1–8.

Hailman, J. P. (1977). "Optical Signals." Indiana Univ. Press, Bloomington.

Halliday, T. R. (1978). Sexual selection and mate choice. *In* "Behavioral Ecology" (J. R. Krebs and N. B. Davies, ed.), pp. 180–213. Sinauer Associates, Maine.

Hartnoll, R. G. (1969). Mating in the Brachyura. *Crustaceana* **16**, 161–181.

Hazlett, B. A. (1966a). Social behavior of the Paguridae and Diogenidae of Curacao. *Stud. Fauna Curacao.* **13**, 1–143.

Hazlett, B. A. (1966b). Factors affecting the aggressive behavior of the hermit crab *Calcinus tibicen*. *Z. Tierpsychol.* **6,** 655–671.

Hazlett, B. A. (1966c). Temporary alteration of the behavioral repertoire of the hermit crab. *Nature (London)* **210,** 1169–1170.

Hazlett, B. A. (1968a). The sexual behavior of some European hermit crabs. *Pubbl. Staz. Zool. Napoli* **36,** 238–252.

Hazlett, B. A. (1968b). Size relationship and aggressive behavior in the hermit crab *Clibanarius vittatus*. *Z. Tierpsychol.* **25,** 608–614.

Hazlett, B. A. (1972a). Ritualization in marine crustacea. *In* "Behavior of Marine Animals" (H. E. Winn and B. Olla, eds.), pp. 97–125. Plenum, New York.

Hazlett, B. A. (1972b). Stimulus characteristics of an agonistic display of the hermit crab (*Calcinus tibicens*). *Anim. Behav.* **20,** 101–107.

Hazlett, B. A. (1975a). Ethological analysis of reproductive behavior in marine Crustacea. *Pubbl. Staz. Zool. Napoli* **39** (Suppl.), 677–695.

Hazlett, B. A. (1975b). Agonistic behavior of two sympatric species of xanthid crabs, *Leptodius floridanus* and *Hexapanopus angustifrons*. *Mar. Behav. Physiol.* **4,** 107–119.

Hazlett, B. A. (1975c). Individual distances in the hermit crabs *Clibanarius tricolor* and *Clibanarius antillensis*. *Behaviour* **52,** 253–265.

Hazlett, B. A. (1978). Individual distance in Crustacea II: The mantis shrimp *Gonodactylus oesstedii*. *Mar. Behav. Physiol.* **5,** 243–254.

Hazlett, B. A. (1982). Patterns of information flow in the hermit crab *Calcinus tibicen*. *Aggressive Behav.*

Hazlett, B. A. and Bossert, W. (1965). A statistical analysis of the aggressive communications systems of some hermit crabs. *Anim. Behav.* **13,** 357–373.

Hazlett, B. A., and Bossert, W. (1966). Additional observations on the communications systems of hermit crabs. *Anim. Behav.* **14,** 546–549.

Hazlett, B. A., and Estabrook, G. (1974). Examination of agonistic behavior by character analysis. I. The spider crab *Microphrys bicomutus*. *Behaviour* **48,** 131–144.

Hazlett, B. A., and Winn, H. E. (1962). Sound production and associated behavior of Bermuda crustaceans (*Panulirus, Gonodactylus, Alpheus* and *Synalpheus*). *Crustaceana* **4,** 25–38.

Helfman, G. S. (1977). Agonistic behaviour of the coconut crab. *Birgus latro* (L.) *Z. Tierpsychol.* **43,** 425–438.

Horch, K. W. (1971). An organ for hearing and vibration sense in the ghost crab *Ocypode*. *Z. Vergl. Physiol.* **73,** 1–21.

Horch, K. W. (1975). The acoustic behavior of the ghost crab *Ocypode cordimana* Latreille, 1818 (Decapoda, Brachyura). *Crustaceana* **29,** 193–205.

Horch, K. W., and Salmon, M. (1969). Production, perception and reception of acoustic stimuli by semiterrestrial crabs (Genus *Ocypode* and *Uca,* family Ocypodidae). *Forma Functio* **1,** 1–25.

Horch, K. W., and Salmon, M. (1971). Responses of the ghost crab, *Ocypode,* to acoustic stimuli. *Z. Tierpsychol.* **30,** 1–13.

Huxley, J. (1966). Introduction: A discussion of ritualization of behaviour in animals and man. *Phil. Trans. R. Soc. London, Ser. B.* **251,** 249–271.

Hyatt, G. W. (1977). Quantitative analysis of size-dependent variation in the fiddler crab wave display (*Uca pugilator,* Brachyura, Ocypodidae). *Mar. Behav. Physiol.* **5,** 19–36.

Hyatt, G. W., and Salmon, M. (1978). Combat in the fiddler crabs *Uca pugilator* and *U. pugnax:* A quantitative analysis. *Behaviour* **65(3–4),** 182–211.

Hyatt, G. W., and Salmon, M. (1979). Comparative statistical and information analysis of combat in the fiddler crabs, *Uca pugilata* and *U. pugnax*. *Behaviour* **68,** 1–23.

Jachowski, R. L. (1974). Agonistic behavior of the blue crab, *Callinectes sapidus* Rathbun. *Behaviour* **50**, 232–253.

Jacobson, S. M. (1977). Agonistic behavior, dominance and territoriality in the American lobster, *Homorus americanus*. Ph.D. Thesis, 136 pp. Boston Univ. Graduate School, Massachusetts.

Johnson, V. R. (1969). Behavior associated with pair formation in the banded shrimp *Stenopus hispidus* (Olivier). *Pac. Sci.* **23**, 40–50.

Johnson, V. R. (1977). Individual recognition in the banded shrimp *Stenopus hispidus* (Olivier). *Anim. Behav.* **25**, 418–428.

Katona, S. K. (1973). Evidence for sex pheromones in planktonic copepods. *Limno. Oceanogr.* **81**, 574–583.

Kittredge, J. S., and Takahashi, F. T. (1972). The evolution of sex pheromone communication in the Arthropoda. *J. Theor. Biol.* **35**, 467–471.

Kleerekoper, H. (1969). "Olfaction in Fishes." Indiana Univ. Press, Bloomington.

Knowlton, N. (1978). The behavior and ecology of the commensal shrimp *Alpheus armatus,* and a model for the relationship between female choice, female synchrony and male parental care. Ph.D. Thesis, Univ. of California, Berkeley.

Lighter, F. J. (1974). A note on the behavioural spacing mechanism of the ghost crab *Ocypode ceratophthalmus* (Pallus) (Decapoda). *Crustaceana* **27**, 313–314.

Linsenmair, K. E. (1967). Konstruktion und Signalfunktion der Sandpyramide der Reiterkrabbe *Ocypode saratan* Forsk. (Decapoda Brachyura Ocypodidae). *Z. Tierpsychol.* **24**, 403–456.

Linsenmair, K. E., and Linsenmair, C. (1971). Paarbildung und Paarzusammenhalt bei der monogamen Wustenassel *Hemilepistus reamuri* (Crustacea, Isopoda, Oniscoidea). *Z. Tierpsychol.* **29**, 135–155.

Littlejohn, M. J. (1969). The systematic significance of isolating mechanisms. *Natl. Acad. Sci. Publ.* **1692**, 459–482.

Losey, G. (1978). Information theory and communication. *In* "Quantitative Ethology" (P. Colgan, ed.), pp. 43–78. Wiley, New York.

Manning, J. T. (1975). Male discrimination and investment in *Asellus aquaticus* (L.) and *A. meridianus* Racositsza (Crustacea: Isopoda). *Behaviour* **60**, 1–14.

Marler, P. (1968). Visual systems. *In* "Animal Communication" (T. A. Sebeok, ed.), pp. 103–126. Indiana Univ. Press, Bloomington.

Maxwell, A. (1961). "Analyzing Qualitative Data." Wiley, New York.

Maynard Smith, J. (1976). Evolution and the theory of games. *Am. Sci.* **64**, 41–45.

Meyer-Rochow, V. B., and Penrose, J. D. (1976). Sound production by the western rock lobster *Panulirus longipes* (Milne-Edwards). *J. Exp. Mar. Biol. Ecol.* **23**, 191–209.

Molenock, J. (1976). Agonistic interactions of the crab *Petrolisthes* (Crustacea, Anomura). *Z. Tierpsychol.* **41**, 277–294.

Moynihan, M. H. (1970). Control, suppression, decay, disappearance, and replacement of displays. *J. Theor. Biol.* **29**, 85–112.

Murton, R. K., Isaacson, A. J., and Westwood, N. J. (1966). The relationships between wood-pigeons and their clover food supply and the mechanism of population control. *J. Appl. Ecol.* **3**, 55–96.

Nolan, B. A., and Salmon, M. (1970). The behavior and ecology of snapping shrimp (Crustacea: *Alpheus heterochelis* and *A. normanni*). *Forma Functio* **2**, 289–335.

Oden, N. (1977). Partitioning dependence in nonstationary behavioral sequences. *In* "Quantitative Methods in the Study of Animal Behavior" (B. A. Hazlett, ed.), pp. 203–220. Academic Press, New York.

Orians, G. H. (1969). On the evolution of mating systems in birds and mammals. *Am. Nat.* **103**, 589–603.

Parker, G. A. (1974). Assessment strategy and the evolution of fighting behavior. *J. Theor. Biol.* **47**, 223–243.

Preston, J. L. (1978). Communication systems and social interactions in a goby shrimp symbiosis. *Anim. Behav.* **26**, 791–802.

Quastler, H. (1958). A primer on information theory. *In* "Symposium on Information Theory in Biology" (H. P. Yockey, ed.), pp. 3–49. Pergamon, New York.

Reese, E. S. (1962). Submissive posture as an adaptation to aggressive behavior in hermit crabs. *Z. Tierpschyol.* **19**, 645–651.

Reese, E. S. (1964). Ethology and marine zoology. *Oceanograph. Mar. Biol. Annu. Rev. 1964* **2**, 455–488.

Rubenstein, D., and Hazlett, B. A. (1974). Examination of the agonistic behavior of the crayfish *Orconectes virilis* by character analysis. *Behaviour* **50**, 193–216.

Ryan, E. P. (1966). Pheromone: Evidence in a decapod crustacean. *Science* **151**, 340–341.

Salmon, M. (1965). Waving display and sound production in *Uca pugilator*, with comparisons to *U. minax* and *U. pugnax*. *Zoologica* **50**, (N.Y.) 123–150.

Salmon, M., and Atsaides, S. P. (1968). Visual and acoustical signalling during courtship by fiddler crabs (Genus *Uca*). *Am Zool.* **8**, 623–639.

Salmon, M., and Horch, K. W. (1973). Vibration reception in the fiddler crab *Uca minax*. *Comp. Biochem. Physiol.* **44A**, 527–541.

Salmon, M., and Horch, K. W. (1976). Acoustic interneurons in fiddler and ghost crabs. *Physiol. Zool.* **49**, 214–226.

Salmon, M., and Hyatt, G. W. (1979). The development of acoustic display in the fiddler crab *Uca pugilator*, and its hybrids with *U. panacea*. *Mar. Behav. Physiol.* **6**, 197–209.

Salmon, M., Horch, K. W., and Hyatt, G. (1977). Barth's myochordotonal organ as a receptor for auditory and vibrational stimuli in fiddler crabs (*Uca pugilator* and *U. minax*). *Mar. Behav. Physiol.* **4**, 187–194.

Salmon, M., Hyatt, G. W., McCarthy, K., and Costlow, J. D., Jr. (1978). Display specificity and reproductive isolation in the fiddler crabs, *Uca panacea* and *U. pugilator*. *Z. Tierpsychol.* **48**, 251–276.

Schein, H. (1975a). Agonistic behavior of the big-clawed snapping shrimp *(Alpheus heterochaelis Say)* with special reference to aggressive and sexual behavior (Part I) and communicative behavior (Part II). Ph.D. Thesis, Univ. of Illinois, Urbana-Champaign.

Schein, H. (1975b). Aspects of the aggressive and sexual behaviors of *Alpheus heterochaelis* Say. *Mar. Behav. Physiol.* **3**, 83–96.

Schein, H. (1977). The role of snapping in *Alpheus heterochaelis* Say, 1818, the big-clawed snapping shrimp. *Crustaceana* **33**, 182–188.

Schöne, H. (1961). Complex behavior. *In* "The Physiology of Crustacea" (T. H. Waterman, ed.), Vol. II, pp. 465–520. Academic Press, New York.

Schöne, H. (1968). Agonistic and sexual display in aquatic and semi-terrestrial Brachyuran crabs. *Am. Zoologist* **8**, 641–654.

Shorey, H. H. (1973). Behavioral responses to insect pheromones. *Annu. Rev. Entomol.* **18**, 349–380.

Sinclair, M. (1977). Agonistic behaviour of the stone crab, *Menippe mercenaria* (Say). *Anim. Behav.* **25**, 193–207.

Steinberg, J. (1977), Information theory on an ethological tool. *In* "Quantitative Methods in the Study of Animal Behavior" (B. A. Hazlett, ed.), pp. 47–74. Academic Press, New York.

Steinberg, J., and Conant, R. (1974). An informational analysis of the inter-male behaviour of the grasshopper Chortophaga viridifasciata. Anim. Behav. **22,** 617–627.

Tamm, G. R., and Cobb, S. (1978). Behavior and the crustacean molt cycle: Changes in aggression of Homanus americanus. Science **200,** 79–81.

Taylor, J. (1976). The advantage of spacing out. J. Theor. Biol. **59,** 485–490.

Teytaud, A. R. (1971). The laboratory studies of sex recognition in the blue crab Callirectes sapidus Ratubum. Sea Grant Tech. Bull., Univ. Miami No. 15.

Thielcke, G. (1962). Versuche mit Klangettrappen zur Klärung der Verwandtschaft der Baumlaufer Certhia familiarus, C. brachydactyla und C. americana. J. Ornithol. **103,** 266–271.

Tinbergen, N. (1952). "Derived" activities; their causation, biological significance, origin, and emancipation during evolution. Q. Rev. Biol. **27,** 1–32.

Trivers, R. L. (1972). Parental investment and sexual selection. In "Sexual Selection and the Descent of Man" (C. B. Campbell, ed.), pp. 136–179. Aldine Press, Chicago.

Vannini, M. (1976). Researches on the coast of Somalia. 10. Sandy beach decapods. Monit. Zool. Ital. NS Suppl. **8,** 255–286.

Vannini, M., and Sardini, A. (1971). Aggressivity and dominance in river crab Potamon fluriatile (Herbst). Monit. Zool. Ital. **5,** 173–213.

von Hagen, H. O. (1962). Freilandstudien zur Sexual-und Fortpflanzungbiologie von Uca tangeri in Andalusien. Z. Morphol. Okol. Tiere **51,** 611–725.

Walker, T. J. (1957). Specificity in the response of female tree crickets (Orthoptera, Gryllidae, Oecanthinae) to calling songs of the males. Ann. Entomol. Soc. Am. **50,** 626–636.

Warner, G. F. (1977). "The Biology of Crabs." Van Nostrand Reinhold, Princeton, New Jersey.

Weygoldt, P. (1977). Communication in crustaceans and arachnids. In "How Animals Communicate" (T. A. Sebeok, ed.), pp. 303–333. Indiana Univ. Press, Bloomington.

Wickler, W., and Seibt, U. (1970). Das Verhalten von Hymenocera picta (Dana), einer Seesterne fressenden Garnele (Decapoda, Natantia, Gnathophyllidae). Z. Tierpsychol. **27,** 352–368.

Williams, G. C. (1966). "Adaptation and Natural Selection." Princeton Univ. Press, Princeton, New Jersey.

Williamson, D. I. (1953). Mating behaviour in the Talitridae (Amphipoda). Br. J. Anim. Behav. **1,** 83–86.

Wilson, E. O. (1970). Chemical communication within animal species. In "Chemical Ecology" (E. Sondheimer and J. B. Simeone, ed.), pp. 113–156. Academic Press, New York.

Wilson, E. O. (1975). "Sociobiology: The New Synthesis." Belknap Press, Harvard Univ., Cambridge, Massachusetts.

Winston, M. L., and Jacobson, S. (1978). Dominance and effects of strange conspecifics on aggressive interactions in the hermit crab Pagurus longicarpus (Say). Anim. Behav. **26,** 184–191.

Wright, H. O. (1968). Visual displays in brachyuran crabs: Field and laboratory studies. Am. Zool. **8,** 655–665.

Zucker, N. (1974). Shelter building as a means of reducing territory size in the fiddler crab, Uca terpsichores (Crustacea: Ocypodidae). Am. Midl. Nat. **91,** 225–236.

Zucker, N. (1978). Monthly reproductive cycles in three sympatric hood-building tropical fiddler crabs. Biol. Bull. (Woods Hole, Mass.) **155,** 410–424.

2

Movement Patterns and Orientation

WILLIAM F. HERRNKIND

I. INTRODUCTION

Orientation is basic to all motile organisms and occurs characteristically during movement patterns in support of necessary biological processes (e.g., escape, food search, migration). Contemporary definitions of spatial orientation all convey the sense that animals are able to maintain or change the position of the body, or bodily parts, with respect to the geometry of environmental space (Jander, 1975; Schöne, 1975). Past reviewers recognized that animals use oriented movements to reach and remain within habitats (Fraenkel and Gunn, 1961; Pardi and Papi, 1961; Newell, 1970) and that such movements strongly affect ecological dynamics (Kinne, 1975).

THE BIOLOGY OF CRUSTACEA, VOL. 7

However, nearly all previous symposia and reviews on orientation have focused on the physiological mechanisms of orienting processes (Carthy, 1958; Pardi and Papi, 1961; von Frisch, 1967; Adler, 1971; Galler et al., 1972; Creutzberg, 1975; Schöne, 1975; Tesch, 1975; Schmidt-Koenig and Keeton, 1978; Gauthreaux, 1980). Those pertaining to invertebrates are organized around the taxis and kinesis classification schemes of Kuhn and Loeb, as presented by Fraenkel and Gunn (1961; e.g., Carthy, 1958; Pardi and Papi, 1961). This scheme is inappropriate for the ecological perspective because it relates narrowly to isolated stimulus–response phenomena. We now know that much orientation occurs in complex and variable conditions with strong central control (see especially the critique by Schöne, 1975). Recent reviews are organized around the classes of orienting stimuli (i.e., guideposts such as sun, polarized light, visual objects, water currents, and chemicals: Creutzberg, 1975), some functional categorization (e.g., homing and migration: Herrnkind, 1980), or simply by taxon. These schemes do not provide a satisfactorily unifying conceptual theme.

Recently, Jander (1975) suggested an integrative theme based on an adaptive framework. Irrespective of phyletic level or species-specific life-style, animals that minimize their effective distance from primary resources (e.g., food, shelter, mate, and host) and maximize their effective distance from sources of stress (e.g., predators, thermal extremes, and desiccation) should have greater relative fitness than those that do not. This evolutionary perspective leads one to seek ". . . to know how and how well the orientation systems of organisms match the spatio-temporal structure of their niches." It becomes necessary to know how movements with different biological functions relate to the location of life-supporting resources, stresses, and the available guideposts. Such factors as the releasing conditions, temporal pattern of movements, and the components of path become significant, while resultant direction of movement per se and neurosensory mechanisms take their place as essential, but only component, factors in the overall process.

This chapter examines the orientation of crustaceans in relation to the resources and stress of certain biotopes. Movement patterns are characterized according to their biological function, while the treatment of orientational mechanisms emphasizes their operation under natural conditions.

II. ADAPTIVE COMPONENTS OF ORIENTATION

Adaptive radiation of the Crustacea into diverse niches is reflected in their form, modes of locomotion, motility, life histories, and sensorial capacities. The adaptive significance of the diverse movements of the crustaceans can be better understood by examining the following: life history stage, spatiotemporal organization of the biotope, factors releasing movement patterns, guideposts and orientation mechanisms, spatiotemporal features of the movement patterns, and biological function.

Many crustaceans exploit more than one biotope during a lifetime. Obvious examples are the macrurans (lobsters) and brachyurans (crabs), benthic as adults but with planktonic larvae. The spatial orientation of the late stages bears little functional relationship to that of the larvae, yet both contribute to a successful life history. The pelagic environment demands responses to photic and pressure guideposts, while the directional guideposts of the benthic zone are often hydrodynamic and chemical. A study of orientation of all life stages of a fiddler crab is much like the study of several different, apparently unrelated, organisms (Herrnkind, 1968b, 1972). Orientational changes underlie the transitions between biotopes and life-styles (Forward and Costlow, 1974; Forward, 1976; Forward and Cronin, 1979, 1980).

Each organism moves about within a biotope in which certain objects and events are crucial resources for existence and others constitute stresses or detriments. Just as significantly, the environment presents an array of stimuli providing spatial information (i.e., informational resources: Jander, 1975). Understanding the biotope, both spatially and temporally, is essential to avoiding the misconception that, because an animal can orient by one particular guidepost, it uses no other. The development of diverse capacities of organisms to sense, select, and integrate particular components of a spectrum as guideposts is one of the most interesting games played by evolution.

The array of oriented movements made by an individual animal occurs with some probabilistic or predictable sequence and duration, often within daily, tidal, lunar, or seasonal cycles. Each activity occurs as a product of internal conditions and specific external events. Hence, decreased humidity initiates downshore orientation of talitrid amphipods, whereas increased hydrodynamics triggers mass migration of spiny lobsters. Temperature, salinity, hydrostatic pressure, turbulence, desiccation, pheromones, and other chemicals serve as releasing factors; such stimuli do not ordinarily serve as guideposts (Creutzberg, 1975). Releasing stimuli variously influence the recipient organism according to its physiological state, which is often difficult to specify. Commonly, nutritive state, phase of the circadian or tidal rhythm, and gonadal stage are inferred to act via neurohormonal pathways, predisposing the animal to respond to releasing stimuli.

Perusal of the literature since 1960 might lead one to assume that the only significant feature of oriented movements is direction. Clearly, the essential nature of many movements can be reduced to the question of whether the animal is, or is not, progressing in the direction of a goal (e.g., home, food). However, other path features are consequential although more difficult to analyze and compare statistically. For example, describing even two-di-mensional paths includes specifying starting point, locomotory rate, turning angles and frequencies, frequencies of stops and accelerations, lengths of straight components, total path length, resultant length, resultant direction, randomness/straightness, and end-point location (Siniff and Jessen, 1969). Such components must be known to assess how oriented movements serve food search, home relocation, migration, and other life-supporting activities.

Endogenous, central neural processes direct certain movement patterns, suggesting that animals have evolved innate knowledge of redundant or simple environmental properties (Jander, 1975). For example, animals usu-ally exhibit two phases of search: ranging and approach (Jander, 1975). The path is initially straight, interspersed with frequent turning; a tactic well adapted to patchily distributed resources (Pyke *et al.*, 1977). When a re-source is sensed, but only imprecisely, ranging ceases and local search begins, usually involving convoluted turning, e.g., blue crabs (*Callinectes sapidus*) foraging for food over the submerged beach at flood tide (Nish-imoto, 1980). However, migration and homing over long distances under variable conditions cannot be explained by simply generalized mechanisms.

The greatest increase in general orientational knowledge since 1960 has come from discovering and specifying environmental stimuli providing ori-entational information (i.e., guideposts), and the neurosensory basis for their perception and integration (Adler, 1971; Galler *et al.*, 1972; Schmidt-Koenig and Keeton, 1978; Able, 1980). Guideposts vary in availability and influence within the biotopes of the various crustaceans and over the stages in their life-cycles. For example, celestial cues are available nearly con-stantly to a diurnal ocypodid crab, but are unavailable to a nocturnal pal-inurid lobster. Furthermore, orientation upcurrent or downcurrent may well serve the stream dweller or estuarine invader, but in many places, for vary-ing durations, current is nil or variable in direction; which cues are then used? Although a number of animals, such as homing pigeons and hon-eybees, exhibit a hierachy or redundance of responsiveness to multiple guideposts, one cannot assume this to be true of all animals. One must ascertain experimentally the role of specific cues to comprehend validly how orientation operates and to discover previously unrecognized mecha-nisms. The most effective way to approach this problem is to know well the physics and geometry of a given biotope.

The biological function of an oriented activity may appear obvious, e.g., escape, otherwise it is often either assumed or ignored. This approach is unacceptable from the adaptive viewpoint because we must infer that the "strategy" of orientation, as well as the mechanism, is inextricably tied to the efficiency of performance. The marine and aquatic environments of the majority of crustaceans prevent the direct observation of movement patterns in many species, therefore obscuring their function. Demonstrating a consistent oriented response to laboratory stimulation is satisfactory physiologically, but provides little suggestion of its value in the organism's life cycle.

III. ORIENTATION IN THE BIOTOPE

This section examines and generalizes the environmental contexts in which crustaceans operate. It considers the stresses and resources, both primary and informational, of each in relation to environmental geometry and temporal factors; oriented movements of representative crustaceans within or across these zones are examined. Emphasis is placed on the patterns of movement in relation to the biological functions performed and the orientation mechanisms employed.

The crustacean biotope can be usefully divided into regions, based on distinct differences in physical milieu: e.g., terrestrial, exposed intertidal, submerged intertidal, subtidal, fresh water, and estuarine. Crustaceans inhabiting a given biotope are rarely equipped to reside indefinitely in another. Hence, shifting boundaries between regions require oriented movements to sustain appropriate zonation (Newell, 1970) or to permit traversing or temporarily invading another zone (Nishimoto and Herrnkind, 1978).

Discussion begins with the exposed intertidal zone primarily because most knowledge of crustacean orientation concerns semiterrestrial species (decapods, Ocypodidae; amphipods, Talitridae; isopods, Tylidae). Despite the emphasis on intertidal forms, they should not be assumed to represent general models for crustacean orientation.

A. Exposed Intertidal Region

The semiterrestrial crustaceans collectively occupy the aerially exposed intertidal shore from the lower tidal limits to the landward fringe, usually demarcated by terrestrial vegetation, where immersion from seasonal or aperiodic storm tides occurs only occasionally. The crustaceans of this biotope typically share the fundamental requirements of (1) access to seawater, saturated substrate, or air, associated with proximity to the sea; and (2) movement to sources of nutrition located some distance from refuge.

The stresses of this region encompass a formidable spectrum and include abrasion (from substrate driven by wind and water), physical displacement (by wind, and waves), burial (sedimentation), insolation and drying (from stranding), osmotic imbalance (salinity changes in tide pools), and anoxia and poisoning (from high substrate concentrations of hydrogen sulfide). A variety of predators occur, including other crustaceans of the same biotope (e.g., *Ocypode*), tidal invaders (fishes and crustaceans), arachnids, shorebirds, and mammals. Despite stresses, the intertidal zone presents rich nutritional resources, including terrestrial vegetation and organic debris, tidally deposited organic material, and rich microflora on substrate exposed at low tide, resident infauna, epifauna and epiflora, and stranded macroorganisms. Tide is the commanding factor in both the behavioral and the ecological sense.

The shore-dwelling crustaceans show movements patterned along $X–Y$ coordinates (*sensu:*Ferguson, 1967); substrate features tend to be uniform along a particular horizonal level, usually running parallel to shore. The animal moves along this X-axis to maintain contact with such resources as moisture, food, conspecifics, and habitation. However, the animal commonly resides at another level, upshore or downshore, to and from which it travels along the Y-axis during each tidal period. Daily intratidal movements of resident amphipods, isopods, and decapods represent what Jander (1975) termed zonal orientation and Baker (1978) described as return migration in a stratified environment.

The availability and utility of guideposts in this region vary depending on weather, cyclic factors, and the instantaneous location of the organism. Visual cues are especially available to the aerially active forms, since the optical transmission quality of air permits visualization over long distances with little distortion of form, relative position, or wavelength. Celestial cues (sun, moon, polarized skylight), unobscured by turbid water or surface ripples, are consistently available over lengthy stretches of open beach. Visual landmarks include nearby objects serving directly as a source of refuge and distant objects such as vegetation, or topographic features typically indicating the landward direction (Williamson, 1954; Herrnkind, 1972; Vannini, 1975a). The angular or anisotropic radiance distribution (ARD: Verheijen, 1978) resulting from the sun (and skylight) and the differential reflection from sea and land, as well as the scattering of light by water droplets in the air above a breaking surf, represent potential guideposts.

The nonvisual guidepost most often suggested, if not demonstrated, as significant to shore dwellers is gravitational information associated with beach slope (Hamner et al., 1968; Young and Ambrose, 1978). The detectability and reliability of slope as a Y-axis guide vary within and among beaches. The length and steepness are a function of the substrate and water dynamics; a low-energy marsh seldom has slopes exceeding 2° (Nishimoto and Herrnkind, 1978), while the foreshores of surf beaches may slope as much as 7° (Shepard, 1963). Vertical relief is characteristic of hard substrate where topographic variation complicates direct slope orientation, but makes available topographic pathways.

Sea breezes represent a potentially useful guidepost in many areas where trade winds or monsoon conditions provide constant velocities and directions for long periods; airborne chemical information is likewise available. Acoustic, low frequency cues from a breaking surf are often suggested, but have not been proven to be a source of Y-axis information to crustaceans, although ocypodids orient over short distances to substrate vibration in social situations and in response to predators (Salmon and Horch, 1972).

The persistent physiological and predatory stresses in this region hypothetically select strongly for rapid and direct orientation. Indirect location of refuge by circuitous search patterns or by kineses (*sensu:* Fraenkel and Gunn, 1961) is unlikely except in certain situations.

Daily intratidal movements of semiterrestrial crustaceans often can be described as Y-axis orientation, although spatial patterns vary among species. Movement downshore from refuge near the high tide level to feed, with subsequent return upshore to the original zone, is characteristic of most ocypodid crabs, talitrid amphipods, tylid isopods, and some grapsids (e.g., *Sesarma*), all of which are marsh, sand, or mud-shore dwellers. Those species on hard substrate, such as the grapsids *Pachygrapsus, Grapsus,* and *Hemigrapsus* (van Tets, 1956), the isopod, *Ligia* (Barnes, 1932), and the mangrove crab, *Goniopsis* (Schöne and Schöne, 1963) exhibit shorter movements, usually with resultant movement along a Y-axis. Y-Axis movement is a component of other activities, including spawning (most decapods), escaping from predators, seeking shelter from desiccation and storm flooding, and, in the case of the land hermit crabs, seeking gastropod shells. The seasonal spawning migrations of the terrestrial decapod genera *Coenobita, Cardisoma,* and *Gecarcinus* cover up to several kilometers and may not be based on the same mechanisms operating in strictly intertidal forms.

The general pattern of activity reflects underlying rhythmic processes (Enright, 1978) by which a given species may emerge on ebbing day tides (*Uca* spp.), night tides (various talitrids), or both (*Uca pugilator, U. tangeri*). Tidal influence is also apparent for *Coenobita, Ocypode,* and *Sesarma,* which often become active at dusk and retire at or near dawn. The animals move downshore directly, either individually (e.g., *Ocypode* spp.) or as coherent groups (e.g., *Uca* and *Mictyris*). Locomotory rates of semiterrestrial forms are sufficiently rapid that the trans-beach crossings do not infringe significantly on foraging time. Y-Axis distances may exceed 100 m in *Uca, Ocypode, Coenobita,* and talitrids.

1. FIDDLER CRABS

Low energy beaches of estuaries and embayments are the domain of fiddler crabs (Crane, 1975). Broad mud or sand flats are typical for *Uca pugilator* and *U. tangeri*, whereas *U. rapax, U. virens*, and *U. thayeri* inhabit heavily vegetated marshes or mangrove areas. The former condition demands rapid travel across long arid stretches where refuge is sparse. The latter biotope restricts the range of visual reception and obstructs long distance travel. Interspecific and interpopulational differences in Y-axis astronomical orientation often reflect the physiography of the habitat and consistency of its shore—water axis (Herrnkind, 1972).

The activity repertoire of fiddler crabs is extensive and complex in comparison to that of the intertidal amphipods and isopods. A number of acoustic species, such as *U. pugilator, U. tangeri*, and *U. panacea*, carry on social activity and long-distance feeding forays on nocturnal low tides (Altevogt, 1965; Salmon and Atsaides, 1968). To such species able to orient under reduced vision, night probably offers reduction in both predation by shorebirds and physiological stress from insolation. Heavy overcast and rain limits activity in general and long distance forays in particular (Crane, 1975). Fiddler crabs on open beaches typically do not emerge when the beach is innundated, but may enter or traverse shallow waters to feed.

Fiddlers undertake Y-axis orientation both tidally and when induced to flee by a predator or by manually displacing them upshore, downshore, or under water (Altevogt and von Hagen, 1964). This response has been used to test the guideposts and the sensory basis of orientation in fiddlers (Altevogt and von Hagen, 1964; Altevogt, 1965; Herrnkind, 1968a, 1972; Langdon, 1971; Young and Ambrose, 1978). Y-Axis orientation in the home beach, upshore direction is shown by crabs in pan experiments where only the sun and/or polarized skylight is visible. Because the directional choice is shown by inactive crabs dug from burrows and also appears in crabs held for various periods indoors, the mechanism is inferred to be time-compensating (Altevogt and von Hagen, 1964; Herrnkind, 1972).

Mirror experiments and the interposing of polaroid sheets, both of which modify the perceived position of celestial cues, suggest that:

1. Either the sun or polarized skylight is sufficient for orientation.
2. The sun, as a discrete light source, and the polarization pattern are processed separately in the peripheral visual system; e.g., in *U. tangeri*, only that polarized light impinging from above on the apical ommatidia induces orientation (Korte, 1966).
3. Gross shifts in the angle of polarized skylight do not cause disorientation if the solar disk is visible.
4. Large azimuthal changes of the sun (by mirror) induce a corresponding shift in the orientation of the crabs.

Fiddlers orient by diurnal celestial guideposts both in air (Altevogt, 1965) and under water (Young and Ambrose, 1978). The features of sky polarization essential for orientation, as are known for honeybees (von Frisch, 1967), are not known for fiddlers. Zenithal, rather than peripheral, sky polarization is suspected to be most influential since it is the most linearly polarized and encompasses the most visible section of sky (except when the sun is directly overhead). The physioanatomical basis for polarized light sensing in decapod crustaceans is well studied by virtue of the extensive work by Waterman and colleagues (Waterman, 1966, 1972, 1975; Waterman and Horch, 1966; also see Horridge, 1967).

The attributes of the fiddler crab astronomical mechanism correspond well with the Y-axis orientational tasks the crabs perform. The guideposts are nearly always available during the day when activity is greatest. Conditions such as haze, scattered clouds, shallow submersion, dense marsh grass, and contradirectional visual objects are usually insufficient to obscure minimally effective guidance information. Because Y-axis directionality is not as precise as homing to an ant nest or beehive, slight distortions or reductions of information probably do not prevent the essential orientation. The functional limits and precise specifications of visual and temporal features of fiddler crab orientation are yet to be thoroughly examined.

Strong Y-axis astronomical orientation is shown particularly by adult crabs living on shorelines where the shore–water axis is consistent (Herrnkind, 1972). The origin of this directionality is largely experiential; for example, groups of crabs from swamps show no persistent directional preference, as do groups of conspecifics from straighter portions of adjacent beaches (Herrnkind, 1972). Altevogt (1965) trained *U. tangeri* to a new Y-axis by forcing them to flee repeatedly to a simulated burrow in a large circular chamber providing a view of only the sun and sky. The crabs most easily trained had no spontaneous preference; crabs lost directional specificity when held for several days in barren containers. Young *U. pugilator,* either wild or laboratory-reared, show no strong directional response to astronomical cues (Herrnkind, 1972). However, upshore orientation appears at the age of sexual maturity when the young crabs enter droves and begin long-distance Y-axis movement patterns. Training experiments using polarization cues also suggest a learned preference (Herrnkind, 1968a), but inadequate behavioral criteria and optical artifacts in those studies make them inconclusive.

A mechanism yielding labile directional preferences seems appropriate to fiddler crabs. Many larvae settle on shorelines greatly different in configuration from that of the parent, so any innate directional tendency would be useless or even maladaptive. Also, along shore migration and beach erosion may yield differing Y-axes over a crab's lifetime. The ability to adopt new directionality as rapidly as possible seems essential.

Natural and artifactual objects, including two-dimensional shapes, can influence or guide directional orientation in specific situations (Herrnkind, 1968a, 1972; Langdon, 1971). Crabs released anywhere on a beach move directly to nearby natural objects, such as mangrove trees or beach grass clumps, from relatively great, but indeterminate, distances when heavy overcast occludes astronomical cues. Crabs approach such objects under clear sky, but only when nearby (within a few meters) or pursued closely; the object then serves as refuge. The minimally essential feature of a landmark is sharp optical contrast with the background since crabs approach both black-on-white and white-on-black rectangular screens (Herrnkind, 1968a, 1972). Shape may be a significant attribute, since strongly vertical objects with a high perimeter squared to area (P^2:A) ratio elicit strong spontaneous approaches by fiddlers (Fig. 1a,b; Langdon, 1971).

Vertically oriented rectangular screens were significantly more effective in eliciting approach than the same shape placed horizontally. Curved shapes, including spheres, characteristically elicited avoidance or a deflection away from them. Two-choice pan experiments using sets of a variety of geometric shapes (controlled for area) suggest that fiddlers differentially approach or avoid forms according to specifiable features—i.e., they demonstrate form discrimination. Field studies show oriented responses ranging from approach, deflection away, and movement directly away to no obvious response. Trees, shrubbery, and beach grass that provide both strong contrast and verticality elicit approach by fiddlers. On wide beaches, vegetation is usually to landward of feeding sites near the water's edge and hence, serves as an additional long-distance guidepost for upshore movement. Avoiding circular forms is likely related to the strong aversion of fiddlers to shorebirds, which vary from circular in silhouette (head-on) to oblong with curved features (e.g., a lateral aspect). Individual fiddlers deflect their upshore path by 4–6 m when detouring around a stationary bird form. Fiddlers are generally attracted to forms with high P^2:A ratios elongated vertically and avoid those with highly curved outlines and low P^2:A ratios.

Responses to visual objects and astronomical cues change during ontogeny and appear susceptible to experiential modification. Very young crabs orient toward a black rectangular screen on a white background, independent of its location in a circular pan, whether under the natural sky or diffuse artificial illumination. This behavior, identical in wild and laboratory-reared crabs, is maintained until approximately the age of sexual maturity, when wild crabs first show astronomically guided Y-axis orientation. Adult crabs preferentially orient in accord with the astronomic mechanism when a screen is presented in opposition, although the screen still elicits approach when the sky is obscured. Some *U. tangeri* are not spontaneously attracted to screens, but can be "trained" to do so if they are repeatedly forced to approach a screen in order to enter a simulated burrow (Altevogt, 1965). Langdon (1971) found that the spontaneous preference for vertical shapes by wild crabs waned when crabs were maintained in a tank with uniform surrounds. New preferences developed in crabs held several weeks with different shapes (e.g., a horizontal rectangle) on the tank wall above the burrow area (Fig. 1c).

For fiddlers, the adaptive consequence of orientation to visual objects often is the attainment of refuge. The direct movement by early instars toward a dark object or area of contrast often translates into reaching a crevice, stalks of beach grass, or mangrove root. This is of special value to the newly metamorphosed first stage, which must secure refuge soon after terrestrial emergence. Astronomic orientation and the ability to perceive objects of different shapes seems essential to the ability to learn or adjust responses later in life. Generally, landmark orientation provides the ability to locate refuge in the absence of astronomic cues and when they provide no useful information.

Fiddler crabs exhibit an ability to discriminate both intensity and certain wavelengths of light by directional orientation (Herrnkind, 1968b, 1972; Hyatt, 1974, 1975). Hyatt (1974, 1975) showed that several species typically oriented toward the dimmer of two white lights. This negative "phototaxis" conflicted with previous literature, which indicated movement toward sources of light. Palmer (1964) gave evidence of a daily rhythmic change in responsiveness, while Hyatt (1974) found tidal variations in responsiveness. The conflicting results may actually result from responses not to the strength of illumination, but rather to the degree of anisotropy of the radiance field (ARD) in the test chambers (Verheijen, 1978). This photic feature is seldom controlled in artificially lighted apparatuses and often is ignored in field situations.

Nonvisual Y-axis orientation in fiddlers incorporates cues from beach slope (Young and Ambrose, 1978) and possibly wind (W. Herrnkind, unpublished data; Vannini, 1975a), whereas short distance movements to and from the home burrow are oriented idiothetically (Fig. 1d; von Hagen, 1967; Hueftle, 1977), and location of conspecifics at night is mediated acoustically (Salmon and Horch, 1972). Fiddlers blinded by opaque eyecaps show no Y-axis directionality on the level beach (Herrnkind, 1972), but orient up inclines of 2–5° (Fig. 1f). Clear upslope orientation by unblinded crabs occurs in a chamber where visual and hydrostatic pressure cues are obviated (Young and Ambrose, 1978), but at an atypically steep slope of 9°. Slope orientation was superceded by responses to astronomic cues and visual objects in this latter study. However, a geotactic mechanism may well come into play as crabs move through dense vegetation or move about at night. Evidence of wind-influenced orientation is lacking, although I have

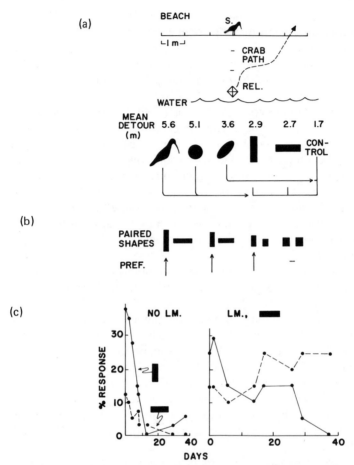

Fig. 1. (a) Paths of fiddler crabs, *Uca pugilator*, are deflected by bird-shaped and rounded models lying upshore, but significantly less so by rectangular objects (after Langdon, 1971). (b) *Uca pugilator* in a circular arena discriminate rectangular landmarks by more frequently approaching the more vertical one (shapes equal in area) (after Langdon, 1971). (c) Strong preference for vertical versus horizontal landmarks shown by wild *U. pugilator* declines after several weeks in a visually uniform habitat (left). Reversal of preference occurs in crabs held in habitats with horizontal shape adjacent to burrowing sites (after Langdon, 1971). (d) Short-distance idiothetic return to a burrow in *Uca rapax*; crabs show zigzag search at appropriate location when entrance is blocked (after von Hagen, 1967). (e) Idiothetic detour integration by *U. crenulata*, which return directly to the location of a burrow entrance (blocked) after a differently oriented outbound path (after Hueftle, 1977). (f) Eyecapped *U. pugilator* orient upslope at angles of 2–5° (upslope 5° = 295°; 2° = 240°: G. Gittschlag, K. Hueftle, and W. Herrnkind, original data). (g) Composite summary hypothesizing guideposts used by fiddler crabs during Y-axis movements at various locations in the biotope. (1) Suncompass, (2) polarized skylight, (3) landmarks, (4) visualized conspecifics, (5) visualized avian predator, (6) idiothetic, (7) acoustic signals (conspecific), (8) beach slope, (9) wave surge, and (10) light intensity. B, Burrow area; G, beach grass; S, sand; W, water.

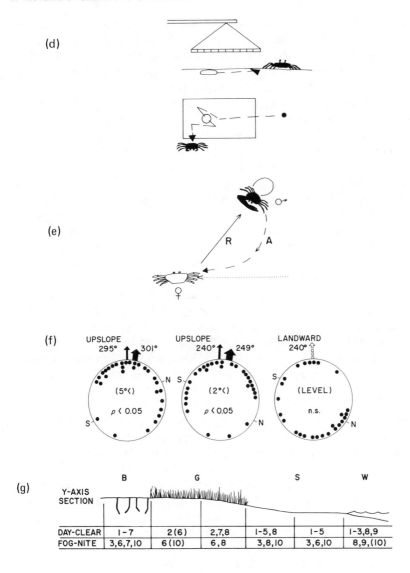

observed movement in the downwind direction [approximately in the Y-axis direction during beach releases of eyecapped crabs on level substrate during 20 km/hr sea breezes (W. Herrnkind, unpublished data)].

Vibrations emanating from a rapping male and carried through the substrate are sensed by fiddler crabs up to 1 m away (Salmon and Horch, 1972) and perhaps other sounds as far away as 10 m (Horch and Salmon, 1972). Field and laboratory observations suggest that individual crabs orient to these signals, especially during nocturnal social activities. The range of detected frequencies is mainly below 2000 Hz, including other acoustic phenomena, such as breaking waves and footfalls of large predators. It is not known yet whether fiddlers orient acoustically outside the social milieu.

Idiothetic orientation independent of (and in spite of) visual cues allows fiddlers to relocate their home burrows rapidly and accurately from distances of about 1 m (Altevogt and von Hagen, 1964). The ability to move out from a burrow and then return in the proper direction for the proper distance occurs in *U. rapax* and probably all fiddlers (van Hagen, 1967). *Uca crenulata* integrates detours into the return (Hueftle, 1977), an ability shared with the spider, *Agelena* (Gorner, 1958).

Anyone witnessing a colony of fiddler crabs disappear almost instantaneously will attest to the potential adaptive significance of their idiothetic orientation. The crab is assured of reaching a burrow of proper aperture size without blockage; especially significant since crab densities may exceed 100/m². While a crab seldom misses its burrow, those fooled by plugging of the burrow or by placement of a flat plate over it zigzagged about over the immediate area, a characteristic search tactic (van Hagen, 1967). The accuracy of idiothetic orientation in *Uca* suggests that the same mechanism mediates straight-line orientation and detour integration when instantaneous guideposts are briefly obscured. For example, a crab can maintain direction over several meters through vegetation along a course set originally by sun compass. A hypothetical summary of orientation by redundant and synergistic guideposts is summarized in Fig. 1g.

2. TALITRID AMPHIPODS

Talitrid amphipods, or beach hoppers, are ubiquitous over the world's temperate and tropical sandy beaches (Hurley, 1959, 1968). They are characteristically nocturnal, remaining burrowed by day in damp sand near or above the high tide line or among vegetational debris deposited high on the beach. Despite their prominence in this biotope, they are physiologically susceptible to death by drying, heating by insolation, and long-term immersion in either fresh or saltwater (Williamson, 1951). Predation by ghost crabs, shorebirds, and surf fishes is a formidable problem. The distribution and abundance of each species varies over the beach, dependent primarily on grain size and moisture content (Bowers, 1964).

Species such as *Orchestoidea californiana* burrow deeply in dry landward zones of fine sand beaches, while the often sympatric *O. corniculata* makes shallow burrows with an air pocket in less well sorted sand on the steeper areas (Bowers, 1964). Habitat selection for each species is also influenced by the physical character of the substrate controlling the mechanical ability to dig burrows effectively.

Talitrid movement patterns center on nocturnal migrations from the burrow region to feeding zones, either downshore (*Orchestoidea corniculata*) or inland *(Talitrus saltator:* Gepetti and Tongiorgi, 1967a,b). Tidal state and surf energetics control the distance and timing of downshore movements (Craig, 1971, 1973a), but may not restrict the activities of inland migrants such as *T. saltator* (Gepetti and Tongiorgi, 1967a,b). The latter emerge from diurnal habitation at dusk and reach peak inland movement several hours after midnight (independent of moon phase). Movement distances vary, but 30–50 m is common, and may approach 100 m. Return occurs just before dawn (Fig. 2a). Shorter movements of 5–10 m along the Y-axis are recorded for the seaward moving *O. corniculata* (Craig, 1971) and *Talorchestia* sp. (W. Herrnkind, personal observation) (Fig. 2f). Talitrids both crawl and hop. Hopping amphipods move most rapidly but often land facing in a different direction from that of takeoff or from the axis of travel, necessitating reorientation before the next hop. Paths are usually straight, despite the disorienting effect of air turbulence and uneven substrate (Williamson, 1951).

The best known orientational response is the escape behavior directed along the Y-axis by talitrids exposed during the day. Amphipods displaced seaward onto wet substrate hop upshore before burrowing, while those released on dry substrate inland move back toward the sea (Pardi and Papi, 1952), although some may burrow. (No carcinologist should leave a beach before witnessing this behavior!) Amphipods behave similarly when placed in a wet or dry circular pan. Since the amphipods repeatedly hop against the pan side, their preferred orientation can be quantified conveniently.

Immersion and wet substrate induce *T. saltator* to landward movement oriented by astronomical cues. However, dilutions to 3.5 ‰ salinity or less cause seaward movement (Scapini, 1979). This reversal does not depend on osmotic concentration per se, since seaward tendencies did not appear in seawater-equivalent solutions of mannitol or NaCl unless Mg^{2+} ions were added.

Various species of talitrids show strong Y-axis directions in pan experiments when the sun or blue sky are visible. The amphipod sun compass exhibits attributes similar to those reported for other animals from ants to pigeons (von Frisch, 1967). These attributes include time compensation for apparent changes in solar azimuth, such that the directional choice varies only slightly over the day. Directionality can be predictably altered by phase shifting the light–dark cycle for 4 or more days (Pardi and Grassi, 1955). Solar altitude is not used as a timer because phase-shifted hoppers adopt the Y-axis direction appropriate to the shifted time base, even though, for them, the sun is at the wrong height. Mirror experiments cause corresponding reversals or other large azimuthal shifts in direction within minutes. Equivalent orientation under blue sky (sun not visible) suggests polarized light as an alternative guidepost; a polaroid sheet induces orientation on overcast days.

Although the sun compass has been confirmed by several workers, its

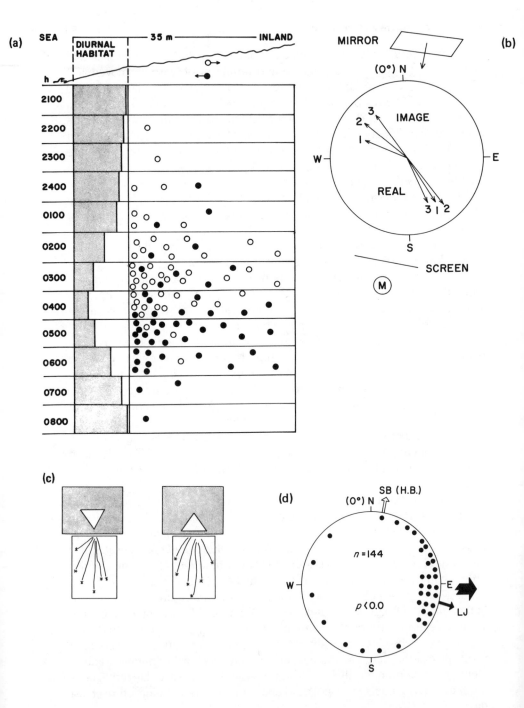

(a)

SEA — 35 m — INLAND

DIURNAL HABITAT

(b)

MIRROR

(0°) N

IMAGE

W — E

REAL

S

SCREEN

M

(c)

(d)

SB (H.B.)

(0°) N

$n = 144$

W — E

$p < 0.0$

LJ

S

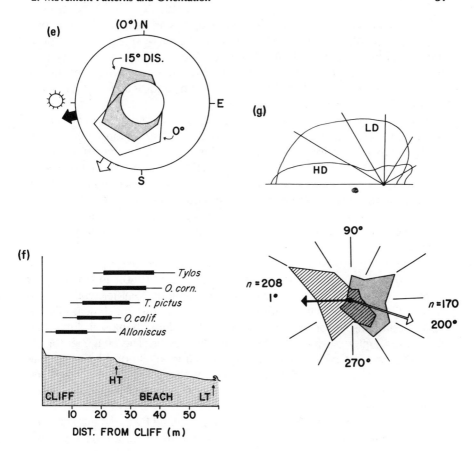

Fig. 2. (a) Nocturnal Y-axis migration of amphipods, *Talitrus saltator* (after Gepetti and Tongiorgi, 1967b). Open circles, positions during the inland movement; filled circles, positions during the seaward return movement. (b) Lunar orientation by amphipods, *Orchestoidea corniculata:* vector arrows show 180° change in distribution of amphipods in a chamber resulting from a 180° shift by mirror of the apparent moon position (after Enright, 1961). (c) Paths of *T. saltator* released at X in view of contrasting, two-dimensional landmarks (after Williamson, 1951). (d) *Orchestoidea corniculata* apparently orient toward inland landmarks to the east (arrow) when displaced to a new beach differing in compass configuration from their home beach, where inland was to the north (arrow) (after Hartwick, 1976a). (e) Seaward orientation by sun compass occurs in *T. saltator* in a level arena (open arrow). Deflection occurs in an arena tilted (15°) in the discordant reference (solid arrow), i.e., upward in the seaward direction (after Ercolini and Scapini, 1974). (f) Intrazonal migratory ranges differ among species of beach-dwelling isopods (*Tylos* and *Alloniscus*), amphipods *(O. corniculata),* and insects *(Thinopinus pictus)* (after Craig, 1973a). (g) Changing the light radiance field (top) from low (ld) to high (hd) directivity yields reversal in escape directions of talitrids (bottom). Hatched, (hd); shaded, (ld).The radiance field is actually symmetrical about the horizontal axis (after Verheijen, 1978).

functional significance has been questioned. It is argued that the sun compass cannot explain nocturnal migrations, and even by day other guideposts are available coincidentally (Fig. 2c,d,g). These arguments have recently been attended by experimental tests of hypotheses as discussed below (Craig, 1971, 1973a,b; Enright, 1972; Ercolini and Scapini, 1974; Hartwick, 1976a,b; Verheijen, 1978).

At least three species of amphipods (*Talitrus saltator, Talorchestia deshayesii, O. corniculata*) with similar ecological distributions show appropriate Y-axis orientation when the moon is visible from the test chamber (Fig. 2b; Pardi and Papi, 1953; Enright, 1961, 1972). The orientation seems temporally compensated since the Y-axis direction is sustained through the night (Enright, 1972). The moon compass requires timing different from that of the sun compass, since moonrise occurs approximately 50 min later on each succeeding night; the angle of orientation to the moon must shift nightly.

A moon-compass hypothesis matching the above description was put forth in part by Papi and Pardi (1959, 1963), by Papi (1960), and in modified form, after lengthy experimentation and evaluation of earlier work, by Enright (1972). It states that talitrids orient to the lunar disc (or some portion of it) and compensate for overnight and night-to-night changes in lunar azimuth via an internal rhythm of approximately 25 hr. The Zeitgeber for the "clock" is associated with lunar phenomena (moonlight or tides), such that after dark periods longer than a few days without entrainment, the lunar clock no longer exerts compensatory influence, and the orientation deteriorates to a fixed-angle response (Enright, 1972).

Data consistent with this hypothesis are presented by Enright (1972) with the cautionary interpretation of other nonsupportive data (Craig, 1971; Enright, 1972). At present, it is certain that the moon influences optical orientation in talitrids, but that influence varies from population to population or from time to time. Enright presents internally consistent data demonstrating all components of the hypothesis, but he points out that the responses are neither universal, species specific, nor apparent under all test conditions. Some inconsistencies can be laid to artifacts and protocol differences, but others may reflect behavioral differences among populations and individuals. For example, the short 5- to 10-m movements of *Orchestoidea corniculata* on short, sloped downshores occur in the absence of lunar cues and independent of them (Craig, 1971, 1973a,b).

Talitrus saltator on the beach will approach an object, depending on its size and proximity (Williamson, 1951); for example, a human more than 3 m distant caused no obvious deflection in the paths of moving amphipods. Experiments using silhouettes (approximately 0.4 m²) projected on a screen yielded movements by amphipods of approximately 1 m toward dark–light boundaries at the screen's base (Fig. 2c). Williamson (1951) suggested that the amphipods oriented toward natural landmarks in the same fashion, especially for landward orientation from the lower beach toward dunes or terrestrial vegetation. Later field studies (Hartwick, 1976a) confirmed that amphipods placed on damp sand on a novel beach orient toward high visual landmarks independent of the home beach Y-axis or availability of astronomical cues (Fig. 2d). In such beach tests, the landmarks were apparently sufficiently distant or of such a nature that the landward direction was indistinguishable from orientation toward a particular point of visual reference. Results of screen experiments predict a concentration of bearings toward vertical edges of contrast (Williamson, 1951). Of course in nature, the amphipods approach a distant object only until reaching appropriate substratum. The simple anatomy of the talitrid eye (few facets) suggests that amphipods detect only the gross features of contrasting forms.

None of the reviewed studies show evidence for seaward orientation over dry substrate by moving away from landmarks; either astronomical or slope cues were also available and were the probable guidepost to seaward movement (Hartwick, 1976a). It is unknown how an amphipod orients seaward on a level beach in the absence of astronomical cues.

The talitrid, *Orchestoidea corniculata,* from the foreshore of high energy sand beaches, shows unequivocal orientation to slope in the absence of visual stimuli (Craig, 1973a). Upslope orientation occurs on wet sand from a 3 to 9° slope, more amphipods responding with increasing slope. Slopes of 3–9° on dry substrate induce downslope movement. The degree of responsiveness varies with the apparent motivational state of the hoppers in accord with their original location on the beach and nutritive state. Those unfed for 4 days orient downslope even on wet sand. Craig's most parsimonious hypothesis states that slope is the only necessary directive guidepost for short nocturnal movements by *O. corniculata* on steep beaches. The backshore dwelling *Talitrus saltator* show responses similar to those of *O. corniculata,* but at slopes of 5° (minimum) to 10° (Ercolini and Scapini, 1974). The minimum slope for clear responsiveness exceeds the average slope of the beach (4°30') from which the experimental animals were collected. Talitrids seem likely to use slope cues on steep beach areas. However, the narrow, 2-m long two-choice runways with pitfall traps at each end, used by both Craig and Ercolini and Scapini, provide little indication of the ultimate path and resultant direction which might be taken by a freely moving talitrid.

Other nonvisual factors include wind, under certain conditions (Papi and Pardi, 1953), and an unconfirmed and unidentified geophysical phenomenon reported by van den Bercken et al. (1967). The former has not been shown to operate in a natural situation, while attempts to replicate the latter result have been unsuccessful (Enright, 1972; Ercolini and Scapini, 1972; Scapini and Ercolini, 1973).

Experiments in both the laboratory and field show the influence of the various talitrid guideposts to be hierarchically organized (Craig, 1973a; Ercolini and Scapini, 1974; Hartwick, 1976a; Enright, 1978). The order of the hierarchy varies among studies, if not between species or biotopes. Hartwick (1976a) showed that a sun compass orients (or dominates) daytime escape responses by O. corniculata when high land forms are absent. However, high cliffs on the backshore are approached even after hoppers are displaced to another beach under a clear sky (Hartwick, 1976a). However, T. saltator from certain beaches always prefer the sun-compass direction, although others show "compromise" directions when displaced (Ercolini and Scapini, 1974). Talitrids in the latter study showed a sun-compass direction when matched against discordant slope features under a clear sky; for example, hoppers moved seaward even when it was upslope on dry substrate (Fig. 2e). The strongest appropriate directional responses occurred when the cues were concordant; when downslope was to seaward on dry substrate. Although escape responses of T. saltator are dominated by photic cues, some synergistic effects appear, and differences exist among individuals from populations on physically different beaches.

Certain requisite conflict experiments have not been performed on O. corniculata to resolve the hierarchy with respect either to differing habitats or to influences of specific cues. Thus, it is not possible to compare between species for, as Enright (1972) suggests, it is irrelevant to compare species when such great interpopulational intraspecies differences exist. Available information yields the following: (1) evidence incontrovertibly supports a sun-compass mechanism greatly influencing daytime escape directions, especially when landforms and slope are absent; (2) evidence is adequate to support a moon-compass hypothesis for some but not all populations; (3) by day, landmarks influence landward movement more strongly in some populations than in others; (4) slopes over 3–5° serve as guideposts in the absence of visual cues; and (5) talitrids in a single population may show ordered responsiveness to all these cues depending on as yet unspecified conditions of physiography, previous experience, and motivation.

The origin and modification of preferred flight direction in talitrids have not been adequately examined. The T. saltator young tested by Pardi (1960) and colleagues show so-called innate home-shore orientation by astronomic cues, matching the direction of the parents, but with somewhat greater scatter. This result was replicated and even demonstrated in second generation amphipods by Scapini and Pardi (1979). The scatter by hatchlings is ascribed to a conflict of sun compass and phototactic tendencies which vary over the course of the day. Conflict from coincident photic cues was detailed by Ercolini and Scapini (1976) for older talitrids. How the astronomically oriented home beach response improves by the time of adulthood is not explained. Improvement via experience (learning) or neurosensory maturation was suggested but not tested. Clear modification of direction results from phase-shifting (Pardi and Grassi, 1955; Marchionni, 1962), and reversion to phototactic responses occurs under long-term constant conditions (Wiliamson, 1954). However, clock modification theoretically does not change the directional preference, merely the temporal control of compensation.

The most common interpretation is that direct development of young permits hereditary home beach orientation, such that the precocial young orient appropriately upon leaving the brood pouch (Marler and Hamilton, 1966; Herrnkind, 1972). The rationale for further study of ontogeny in talitrids is twofold. First, the adaptive value of an innate compass is difficult to explain. Second, the reason for variation in responses within and among species cannot be discovered by continued testing of arbitrarily selected subadult or adult hoppers whose responses are the product of differing but unspecified genetic, maturational, and experiential histories.

The former problem arises for the following reason: we must assume strong selection against genes for inaccurate home-beach, sun-compass orientation since arbitrary sampling shows predominately parental beach directionality of newborn. Now, suppose the structure of the beach changes seasonally or adults are transported by chance to a new beach facing in some different direction (Hartwick, 1976b). According to the literature, adults could modify or readjust their orientation appropriate to the new situation. However, their offspring would be genetically programmed to the previous home beach, resulting, perhaps, in considerable mortality due to inappropriate orientation. If true, one would expect strong selection against an inflexible home-beach directionality by newborn. Further study using selective breeding and ontogenetic studies of wild and captive hoppers is urged.

B. Terrestrial Region

This region lies beyond the influence of the tides. The absence of tidal immersion profoundly affects substrate stability and vegetation. Physiological stresses for terrestrial crustaceans include both loss of water by evaporation and, for some, elimination of excess water (Bames, 1935; Paris, 1963; Edney, 1968). Control of evaporative water loss is complicated because water vapor pressure and temperature vary widely and rapidly over short distances. Water cannot be replaced by absorption through gills or integument because of the low water potential of air. Long-term immersion in fresh water causes problems for osmotic and ionic regulation. Clearly, the climatic pattern has an accentuated impact on terrestrial crustaceans with respect to both thermal- and water-related phenomena. Movements and selection of a home or refuge reflect the influence of physical factors affecting water balance (Edney, 1968; Warburg, 1968).

As in the intertidal zone, predation is a significant factor, although predators and foraging methods differ. The small isopods and amphipods must contend with insect, arachnid, amphibian, and reptile predators uncommon in the intertidal. Birds and other predators that hunt insects also hunt small crustaceans. The larger decapods are prey to birds, raccoons, rodents, and other terrestrial vertebrates. Plant food, live, dead, and decomposing, is potentially abundant, and most terrestrial crustaceans are herbivores, detritivores or scavengers, but rarely predators.

The diversity of lifestyles among the major terrestrial groups is a consequence of the diversity of adaptations to the above stresses. The most significant are the large size and mobility of the decapods and life-cycles independent of the sea for amphipods and isopods. The ocypodids, such as *Cardisoma* and *Gecarcinus,* and the anomurans *Coenobita* and *Birgus,* are robust, well-armored, and capable of powerful, long-range locomotion. Yet, they must return to the sea to release larvae, and the young crabs must attain sufficient size to move long distances inland. Direct development in a brood-pouch frees the terrestrial amphipods and isopods from seaward migrations. Most species are less than 1 cm in length; they tend to behave cryptically.

The spatial configuration of habitable terrestrial regions, like that of the subtidal, has no $X-Y$ axial organization (excepting stream banks). The path to the sea obviously remains consequential to those releasing planktonic larvae. Otherwise, the crabs, as well as isopods and amphipods, should theoretically demonstrate patterns of movement in accord with the spacing and temporal availability of resources, including food, mates, water, and refuge from biotic and abiotic stresses. While only a few crustacean taxa occupy this region, and physiological resistance seems poorly developed compared to that of the insects and arachnids, it is a fact that the isopods are commonly present, even at great mountain altitudes and in many desert habitats. The number of terrestrial isopods is also relatively large, over 500 species within the Porcellionidae alone.

1. DECAPODS

The terrestrial decapods include members of the families Ocypodidae, Grapsidae, and Coenobitidae (Bliss, 1979). Something is known of the daily movement patterns of the supralittoral ghost crabs, *Ocypode* spp., the more terrestrial *Gecarcinus lateralis,* and the land hermit crabs *Coenobita* spp.

Extensive information on orientation is available only on *Coenobita rugosus* (Vannini, 1975 a,b, 1976a; Vannini and Chelazzi, 1981).

Movements in the Y-axis are characteristic of most species of *Ocypode* (Vannini, 1975a; Wolcott, 1978) which burrow on the upper beach or dunes and, under certain conditions, some species of *Coenobita* (Vannini, 1975a). The latter show movements directed inland as well as seaward from the dune region. Feeding, water balance, and shell supply or condition influence or even restrict activity to the beach region (de Wilde, 1973). Land crabs, *Gecarcinus lateralis* as well as *Coenobita* in optimal shells, may be found inland up to 30 km from the sea at altitudes up to 450 m (Bright, 1966). There, daily movements and their timing seem adjusted to predator activity, food location, and water conservation. Seasonal reproductive migrations are known for several species of *Coenobita,* including *C. clypeatus* (de Wilde, 1973), *Cardisoma guanhumi* (Gifford, 1962), and *Gecarcinus lateralis* (Bliss, 1979), and are assumed to occur in the other species; only the grapsids, *Sesarma jarvisi* and *S. cookei,* are known to pass the larval stage out of the sea (Abele and Means, 1977).

The ghost crabs include the fastest running crustaceans (up to 3.4 m/sec: W. Herrnkind, unpublished data), capable of covering long distances during forays. They are known from resight (Roe, 1980) and radio tracking (Wolcott, 1978) studies to move as much as 300 m during an activity period. Ghost crabs collectively exhibit many feeding tactics and activity cycles. The former includes scavenging in tidal debris, digging clams (*Donax*), ambushing mole crabs (*Emerita*), and pouncing on insects or amphipods. Large items, scavenged or killed, are sometimes transported back to a home burrow. At other times, feeding occurs at the point of capture, and the crab locates another burrow from which it may try to oust any inhabitant. Such variation invites questions regarding optimal foraging (Schoener, 1971), and centrist versus nomadic life styles (Covich, 1976). Periods of movement about the beach and feeding vary from diurnal to nocturnal, with the more littoral populations showing tidal influence, usually with activity peaking at low tide and the crabs retreating down burrows at high tide (Vannini, 1976b). The great variation in available guideposts between inland and beach areas and between night and day demands an exceptional orientational repertoire.

Anecdotal reports imply acute vision for ghost crabs (e.g., Cowles, 1908; Hughes, 1966; Vannini, 1976b), but few experimental studies detail visual orientation. Polarized skylight can serve directed movements (Daumer et al., 1963), while objects near the burrows may serve as landmarks for home recognition (Hughes, 1966; Linsenmair, 1967). The mechanisms shown in *Uca* may well serve *Ocypode,* although the latter likely possess even more sophisticated visual capabilities matching their more complex movement patterns. Study of acuity, low light vision, and form discrimination seems most appropriate. Preliminary studies might begin with assessing the effects on homing and forays of occluding various parts of the large, many-faceted eyes.

Land hermit crabs, *Coenobita rugosus,* in Somalia are supralittoral during the dry season and terrestrial during the rainy season (Vannini, 1975a, 1976a). The crabs during either season reside, when inactive by day, buried in damp sand inland of the dune under bushes or beneath tidal debris seaward of the dune. Nocturnal dune–beach–dune movements occur each day, with peaks of population activity adjusted to tidal level and its day-to-day progression (Fig. 3a). Wet season dune–bush–dune movements are nocturnal without tidal influence. The Somalia crabs show local residency with little movement along the X-axis over periods of weeks. Elsewhere, other species exhibit high emigration rates with population shifting alongshore. The Somalia crabs appear dependent on the beach for biological (egg deposition, shell source) and physiological (water and salt) needs (Vannini, 1976a). Food and, perhaps, reduced interspecific competition and predation may encourage inland migration when water balance is not problematic. In Curaçao, *C. clypeatus* in well-fitted shells move about nocturnally far inland, but those in cracked or poor-fitting shells remain near the beach (de Wilde, 1973). The patterns of movement cannot be generalized for land hermits except with respect to universal reproductive migrations. Otherwise, movements apparently reflect adjustments to local ecological conditions.

The release of oriented movements involves endogenous rhythms, as well as physiological conditions arising from the interaction of metabolic and maturational (ontogenetic, reproductive) events. In addition, movement patterns are modified by ambient conditions, varying tidally, daily, and seasonally. Directionality is also specified in some way by these interactions; i.e., inland, seaward, or along shore. Guideposts include visual landmarks, ground slope, and wind direction, all experimentally demonstrated, while airborne chemicals and celestial cues are inferred (Vannini, 1975a).

Tracks left by hermit crabs on sand provide excellent directional information without observer artifact. Path analysis shows initial downshore and inland paths to be strongly oriented in the Y-axis (Fig. 3b; Vannini, 1975a, 1976b). Later the paths show circuitous "random" movement (foraging) followed by strongly directed return paths in the reverse vector. Those crabs moving from beach to dune orient toward conspicuous landmarks (Fig. 3d), while those moving from bush to dune apparently ignore landforms, thus suggesting some celestial cue. Orientation occurs under clear sky, overcast, new and full moon.

Nonvisual guideposts, including wind and ground slope, influence orientation in certain situations. Crabs moving seaward shift from Y-axis to up-wind where visual cues and inclination are missing (Fig. 3c). This shift ordinarily lasts only until the latter cues again become available, e.g., when the dune top is surmounted. More radical upwind orientation occurs for brief periods (2 min) in crabs displaced by hand; again, the paths shift to the Y-axis thereafter. Vannini hypothesizes that orientation to the seabreeze serves as a brief or supplemental guidepost leading the animals roughly seaward. Short-term upwind movements by *Coenobita scaevola* in response to airborne scents may not be related to the Y-axis circuit (Magnus, 1960).

2. ISOPODS

Edney (1960, 1968) summarizes the surprisingly long-lived notion that the movements of terrestrial isopods simply reflect kinetic and tactic responses, tending to maintain the animals within microhabitats preventing water and thermal stress. He points out the inadequacy of this explanation—its inflex-ibility would actually confine the isopods to crevices, whereas in nature they are much more vagile. His excellent synthetic review (1968) estab-lished the labile nature of isopod orientational responses and linked them to specific habitat conditions and physiological capabilities. Cloudsley-Thompson (1956), den Boer (1961), Southwood (1962), Paris (1963), War-burg (1964, 1965a,b, 1968), and Friedlander (1965), provide both useful information and the bases for posing new hypotheses. In particular, they show that carefully documenting daily and seasonal movements, and con-comitantly measuring influential biotic and abiotic factors, is requisite to establishing testable orientational hypotheses.

None of the terrestrial isopods moves over a wide range of terrestrial habitats; rather, most movement occurs within narrow biotopes. Therefore, perspective is gained by comparing movement patterns in a variety of spe-cies (Warburg, 1968). Examination begins with coastal species and pro-gresses inland to inhabitants of areas less directly influenced by the marine environment.

The daily and seasonal movements of 12 coastal isopod species on the tropical coast of Somalia were detailed by Chelazzi and Ferrara (1978). They examined habitats ranging from the open sand beach through the dunes and retrodunal plain to the coastal hills approximately 1-km inland. The species of the beach–dune region, *Littorophiloscia compar, Alloniscus robustus, Periscyphis civilis,* and *Tura candida,* are halophilic. The most littoral species, *L. compar,* emerges from burrows or debris and migrates 10–30 m seaward for several hours after sunset, returning upshore by short-ly after sunrise. Lunar tidal phases (spring or neap) influence the timing and degree of seaward movement; rain tends to extend the activity period. The other, less littoral species are less nocturnal, some peaking at dawn (*A.*

(a)

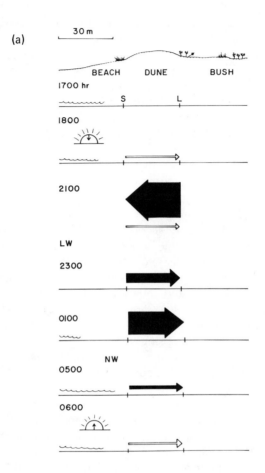

(b)

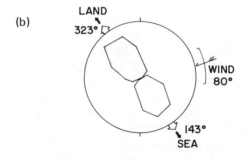

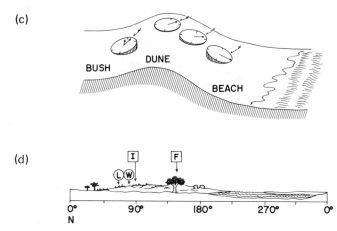

Fig. 3. (a) Tidal and nocturnal dune–beach–dune movements of land hermit crabs, *Coenobita rugosus* (after Vannini, 1976a). Arrow thickness indicates the relative number of crabs migrating in that direction. (b) Sand track directions of *C. rugosus* at start of dune–beach movement (open) and at the return beach–dune movement (shaded) (Vannini, 1975a). (c) Dune–beach paths of *C. rugosus* show deflection from Y-axis in accord with wind direction at certain locations (after Vannini, 1975a). (d) Landward orientation of *C. rugosus* is apparently influenced by prominent landmarks in the visual panorama (after Vannini, 1975a). L, landward; W, wind direction; F, final direction; I, initial direction.

robustus, T. candida) or diurnally (*P. civilis*). The latter species move about within the extralittoral zone and dune region. All move inland in the wet season, coincident with development of vegetation, and then shift seaward again at the dry season. *Periscyphis civilis* shows an especially terrestrial pattern of diurnal activity in the back dune area with midday retreat to refuge. *Periscyphis rubroantennatus,* inhabitant of the retrodunal plain, lives exclusively on vegetation, feeding on shrub soft parts and retreating to tunnels in the stem when stressed or inactive. Other essentially terrestrial species, including *Periscyphis ruficauda,* inhabit the retrodunal bush even in dry periods.

In overview, a general transition from nocturnal to diurnal activity coincides with increased distance from the littoral, perhaps facilitated by improved water conservation mechanisms in the more terrestrial species. Similarly, tidal components and daily transzonal migrations disappear and are replaced by specialization to vegetation, mammal lairs, or rock crevices in the terrestrial zones. The inland shift by the halophilic species is presumably permitted by the increased moisture and vegetative cover during the wet season. Alternatively, increased food and water supply inland may serve to attract and maintain habitation. Regardless, this opportunistic response by forms poorly equipped for terrestrial life may hint at evolutionary pathways. Oriented movements are clearly necessary to temporary invasions and possibly preceded the evolution of both physiological stress resistance and other specializations emancipating isopods from daily transzonal migration.

Warburg (1964, 1968) sought general trends in the orientation mechanisms and movements inducing stress by examining isopods over the whole range of terrestrialness. Behavioral adjustments occur with respect to both orienting and releasing stimuli, including conspecifics, contact, light, temperature, and humidity. However, the variation of measures, criteria, and controls used by different investigators calls for cautious interpretations, both in making comparisons and in applying results to nature (Warburg, 1968). For example, a thermal preferendum obtained by spot checking of the distribution of several isopods in a small chamber cannot be adequately interpreted unless one controls confounding factors such as biorhythms, humidity, conspecific attractiveness, and initial water state. Furthermore, the preferenda themselves give no evidence that thermal gradients serve as the guidepost in nature where very abrupt thermal transitions occur.

Isopods respond to conspecifics via distant olfactory (chemical) cues, contact chemoreception, and tactile cues (Kuenen and Nooteboom, 1963; Farr, 1978). The resultant aggregations reduce water loss, while the response varies in accord with temperature and humidity. Chemoreception, potentially facilitates locating both mates and favorable microhabits (i.e., a conspecific's presence ensures suitability: Farr, 1978).

Contact responses hypothetically serve to maintain the animals in physical shelter and reduce their probability of being stranded in the open (Friedlander, 1965). The response of terrestrial *Oniscus asellus* and *Porcellio scaber* to dorsal contact is reduction of movement, but the strength of the response varies with the phase of the activity cycle, relative humidity, and the species. Humidity and moistness influence both the initiation and degree of movements as well as responsiveness to other stimuli (Edney, 1968). Littoral species, such as *Ligia italica,* reduce activity, remain quiescent, or aggregate at high humidity. Such species either locomote faster, turn more frequently, or rest for shorter periods in lower humidities; the critical humidity levels differ among species and correlate with microhabitat differences. For example, *Oniscus* rested at 80% relative humidity (RH), while *Porcellio* remained active until 90% RH (Warburg, 1968). Species in xeric conditions show the least pronounced responses, while those from mesic areas are intermediate; dehydration strengthens the response in the latter.

Light responses, both directed orientation and kinesis, are interdependent on humidity, temperature, and internal water state (Perttunen, 1963). Littoral *Tylos latreillei* show Y-axis astronomic orientation similar to talitrid amphipods in the same habitat (Pardi, 1954), whereas *T. punctatus* orients to shore slope (Hamner et al., 1968). Other littoral and terrestrial forms have been examined only with respect to bidirectional orientation or turning rates, walking speed, and residence time in various photic conditions (Warburg, 1968). Overall, terrestrial isopods are more often photonegative than littoral and aquatic species. Hence, the latter tend to move faster and spend less time in lighted areas as temperature rises. However, the mesic, grassland species possessing good water regulation show positive reactions. Xeric forms, with exception, are mainly photonegative. High temperatures may induce a change from photonegative to photopositive, which facilitates evaporative cooling (Warburg, 1968).

Excessive water during rainy periods is potentially deleterious, because drowning can occur (Paris, 1963). In this situation, isopods remain exposed, apparently losing water by transpiration. Evaporative cooling may also serve survival through short-term thermal homeostasis when environmental termperatures rise abruptly (Edney, 1968).

Present knowledge suggests that movement patterns of land isopods are regulated partly by endogenous rhythms, perhaps with precise periodicity when environmental conditions are benign. The directionality of movement, especially under thermal or water stress, is dictated by humidity and ambient light; movements may be either horizontal or vertical. Favorable refuge is determined by the presence of conspecifics or by combinations of contact, high humidity, and temperature; light level is characteristically low since the above conditions occur in dark crevices, debris, or vegetation. Refuge for some is self-dug burrows in the ground or in plants. The xeric *Hemilepistus reaumurii*, a desert burrow-digger, homes to burrows from some distance along the ground surface (Linsenmair, 1972).

C. Submerged Intertidal Region

The submerged intertidal shore at flood tide is occupied or invaded by a variety of crustaceans with movement patterns and activities characteristically different from those of the semiterrestrial zone. The demarcation of crustacean faunas is especially abrupt, although some grapsids, ocypodids, and xanthids move about through the land–sea interface; the semiterrestrial forms briefly enter water and the marine forms briefly emerge onto land. The anomuran mole crabs *(Emerita)* and certain isopods and amphipods are strict residents of the wave wash zone.

The primary resources of the submerged zone include many noted for the exposed intertidal, but more trophic interaction occurs among the zone's occupants. Again, the environmental geometry is organized along X–Y axes, but the significant physical conditions are hydrodynamic, especially waves, turbulence, and alongshore currents, depth, substrate and turbidity (Fig. 4a). Turbulence varies along the Y-axis and regulates movement and type of activity; for example, hermit crabs cannot effectively feed or loco- mote in the wave wash zone, whereas mole crabs can only feed effectively there. The horizontal distance of the turbulent zone varies with depth as a product of bottom steepness and wave size. Steeply sloped sand beaches possess lengthier zones of violent water motion. For all shores, the turbulent region progresses back and forth along the Y-axis with each tide. The effect of offshore hydrodynamics and physiography on species distribution and abundance is well-known (Pearse et al., 1942).

The directional guideposts include hydrodynamic stimuli, waterborne chemicals, slope, and visual cues. Except in clear, low-energy waters, visu- al information is severely limited or modified from the aerial condition, especially in the wave zones. The popular portrayal of a fish-eye view of the optical field above water, with concomitant displacement of altitude and relative position of objects due to the change in index of refraction (Hasler, 1966), is appropriate to relatively few natural shores. Waves, surface rip- ples, and turbidity certainly reduce the utility of optical features, thus re- stricting visual orientation.

Hydrodynamic cues can be classified as monodirectional water flow (cur- rent) and wave surge oscillation (alternating, bidirectional water movement near the substrate). The two often possess different directional attributes and relative availabilities in the nearshore region. Shoreward of the surf, wave wash currents displace water alternately shoreward and seaward, with net transport alongshore as well, depending primarily on the angle of attack of the wave front (Shepard, 1963). Currents seaward of the surf show more consistent velocity along shore but change direction as a result of tides in low-energy areas, and/or wave–tide interaction in high-energy areas. Where current is consistent over time, waterborne chemical information provides orientation to food (Pearson and Olla, 1977; Ache and MacMillan, 1980), mates (Ryan, 1966; Atema and Engstrom, 1971), enemies (Snyder and Snyder, 1970), hosts (Ache and Case, 1969), and other resources (McLean, 1974; Rittschof, 1980).

Probably the most prominent and conservative Y-axis cue is wave surge (Fig. 4b,c; Herrnkind and McLean, 1971; Rudloe and Herrnkind, 1976, 1980; Gendron, 1977; Walton and Herrnkind, 1977; Nishimoto and Herrn- kind, 1978). Waves generated by local winds or oceanic swells refract shoreward, approaching the Y-axis before breaking. Net water transport is small during the horizontal oscillation, thus differing markedly from mono- directional currents. The axis of oscillation is essentially perpendicular to the wave front, the displacement within each surge varying with wavelength and velocity. Waves create circular water motion at the surface, which becomes more elliptical toward bottom, flattening to horizontal to-and-fro oscillations at the substrate. The surge spectrum ranges from short, rapid oscillations in very shallow water to displacements of several meters at varying maximum velocities seaward of a breaking surf (Hunt, 1961). Wind gusts cause the former condition on shallow sandflats, while the latter is characteristic of high-energy beaches subject to oceanic swells. Wave action is nearly omnipresent in the marine environment and a frequent feature on shores of large bodies of fresh water. Tidal fluctuations and bottom irregularities have little influence on surge direction, making it a more conservative guidepost than current for Y-axis orientation.

Bottom contour, especially slope, is another relatively conservative Y-axis cue. However, it is seldom persistently directional over the entire region, being dependent on submarine topography, substrate, and hydrodynamics. Calm embayments, the wave wash region of the surf, and some sand areas seaward of the surf may show relatively long stretches of uniform Y-axis slope. The low-energy areas and back surf may slope only a few degrees, while the wave wash area of a surf beach and the drop-off into estuarine channels show the steepest slopes (up to 9°: e.g., Shepard, 1963). Hard substrate areas possess varying slopes, as well as features such as surge channels in spur and groove reefs yielding both topographic pathways and directional water flow in the Y-axis (Herrnkind and McLean, 1971).

1. BLUE CRABS

Numerous species of *Callinectes,* bottom-dwelling, swimming crabs of the family Portunidae, inhabit the submerged littoral zone of temperate and tropical seas (Williams, 1974). Much of the crab stage is spent in nearshore embayments, with some species commonly found in estuaries and into essentially fresh water (Norse, 1978). The blue crab, *C. sapidus,* is especially abundant along the Atlantic coast of the United States and the Gulf of Mexico, where it is found in estuaries as well as at the land–sea interface of marshes, low-energy beaches, and tide pools. Blue crabs regularly invade the littoral region at flood tide where they primarily forage and feed, but also seek mates, molt, and then retreat offshore at ebb (Nishimoto and Herrnkind, 1978; Nishimoto, 1980). Tidal invasions occur both day and night: individuals in a population show a variety of activity cycles, including nocturnal, diurnal, crepuscular, and even aperiodic activity.

The feeding versatility of blue crabs is truly remarkable and reflects the potential to move about effectively under a greater variety of physical conditions than nearly any other crustacean. Blue crabs are known to "swarm" over oyster reefs, breaking open even large oysters (Menzel and Nichy, 1958). On particulate substrate they probe, dig, and even excavate to capture infaunal invertebrates (Nishimoto, 1980). Larger crabs prey heavily on

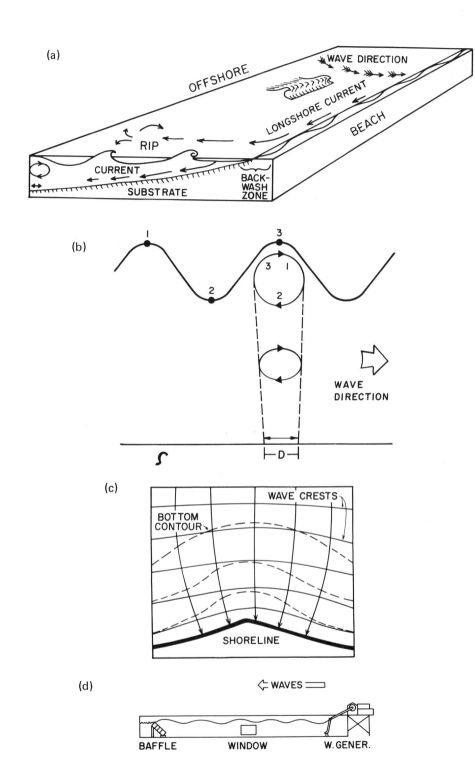

(a)

WAVE DIRECTION

OFFSHORE

LONGSHORE CURRENT

BEACH

RIP

CURRENT

SUBSTRATE

BACK-
WASH
ZONE

(b)

1

3

3   1

2

2

WAVE
DIRECTION

ʃ

D

(c)

WAVE CRESTS

BOTTOM
CONTOUR

SHORELINE

(d)

WAVES

BAFFLE   WINDOW   W. GENER.

smaller conspecifics in certain areas (Laughlin, 1979). Semiterrestrial organisms are not immune to blue crabs, which reach up to 7 cm above the water surface to pluck marsh periwinkles (*Littorina irrorata*) from *Spartina* grass blades (Hamilton, 1976). Even more spectacular behavior occurs when blue crabs rush from the water onto land to capture fiddler crabs (Nishimoto, 1980). The closely related form *Arenaeus cribrarius* moves into the surf wash zone at night, where it is exposed as the waves recede, ambushing surf clams (*Donax*) and mole crabs (*Emerita*: Leber, 1982). Each of these activities occurs during the cyclical Y-axis movements characterizing the inshore life of these crabs.

Blue crabs remain in sublittoral channels, seagrass beds, etc., during ebb and then move inshore as the tide floods, often burying in the mud-sand. They may forage, ambush, or pursue prey even in a few centimeters depth right at the waters edge. The presence of chemical stimuli (sapid materals or pheromones) causes upcurrent orientation or searching (Pearson and Olla, 1977). As the tide recedes, or in advance of it, crabs move offshore. Some individuals are transient, spending only one tidal cycle in an area before moving alongshore an indeterminate distance; others remain for up to 2 weeks near oyster bars or other concentrated resources (Nishimoto, 1980).

Blue crabs are easily induced to flee offshore, swimming rapidly near the substrate, by the approach of potential predators. This escape response is reliably obtained by a person walking slowly along the shore or by hand releasing a captured crab. A series of field and laboratory experiments using these techniques suggests how the highly variable directional cues of this zone serve Y-axis orientation (Nishimoto and Herrnkind, 1978).

Strong offshore directionality of fleeing crabs appears over a wide range of turbidity, wave action, and visual conditions (Fig. 5a–f). Under low visibility, characteristic of turbid estuarine waters, both wave surge and substrate slope serve offshore orientation (Nishimoto and Herrnkind, 1978). Crabs noninjuriously blinded by wide rubberbands consistently and repeatedly move offshore in the presence of waves, Y-axis slope, or both (Fig. 5b). The eyecapped crabs showed nondirectional scattering, slower locomotion, and circuitous paths when tested on level substrate in calm water.

Wave tank experiments confirmed the orienting influence of wave surge

---

Fig. 4.   (a) Hydrodynamics of the littoral region of a beach (after Shepard, 1963). (b) Wave cross section showing water movement changing from orbital at the surface to a horizontal oscillation with decreased displacement near bottom (after Shepard, 1963). D, displacement distance. (c) Wave fronts refract in accord with bottom depth and shoreline contour (after Shepard, 1963). (d) Schematic sideview of 42 m × 4 m × 1 m wave tank for testing responses to wave surge of blue crabs, spiny lobsters, and horseshoe crabs (after Nishimoto and Herrnkind, 1978).

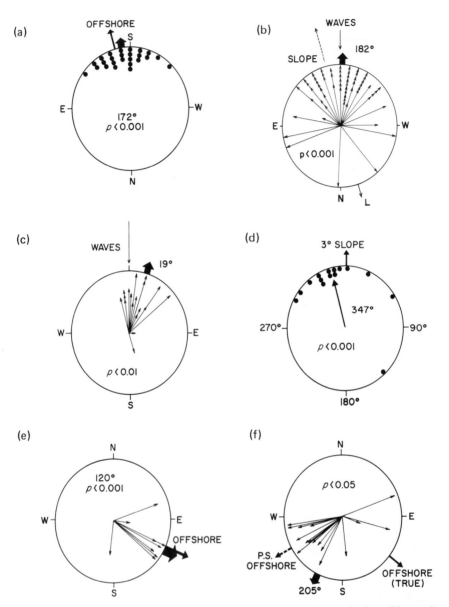

Fig. 5. (a) Escape directions of blue crabs, *Callinectes sapidus;* crabs flee offshore (after Nishimoto and Herrnkind, 1978). Thick arrow is mean direction of group; each dot represents five crabs. (b) Mean vectors of individual, eyecapped blue crabs tested in semifield arenas with waves approaching as shown by thin solid arrow, down slope as dashed arrow (heavy arrow, mean direction of group) (after Nishimoto and Herrnkind, 1978). (c) Mean vectors of blue crabs tested in wave tank (Fig. 4d); wave approach indicated by thin arrow at top; thick arrow group

under conditions precluding orientation to other cues possibly present in field tests. A large wave tank 42 m (length) × 4 m (width) × 1 m (depth) allowed observations on freely moving animals, facilitating direct comparison between the performances in field and controlled conditions (Fig. 4d). Surge levels tested also matched those of the field and were additionally adjusted with respect to two quasiindependent wave features: maximum water velocity (midway through each oscilliation) and actual displacement distance of an oscillation.

Blue crabs in the tank oriented "upwave," analogous to offshore, at long (14 cm) but not short (7 cm) surge displacements (Fig. 5c). However, even at short displacements, some individuals walked straighter and faster than in waveless control runs; clearly the blue crabs responded to the wave surge in both treatments. Offshore orientation seems to be triggered only by larger waves moving inshore from deeper water and thus correctly indicating the offshore axis, whereas crabs ignore short displacement waves that are caused by winds over the shallows and are likely to vary from the Y-axis. This interpretation should be framed as a hypothesis and tested.

Tests in field arenas and circular 1.2-m pools showed that eyecapped blue crabs consistently move downslope at inclinations of 3° and above (Fig. 5d; 1–9° tested: Nishimoto, 1980). Field performance suggests 2° as a lower limit for blinded crabs over short distances (2.5 m); a slope representative of low-energy beaches. Greater slopes occur at channel edges and on higher energy beaches with coarser substrate. Tests in air gave similar results, suggesting that hydrostatic pressure changes with depth are not necessary for the observed performance.

The visual orientation of C. sapidus is not yet well-studied, although it is evident that vision mediates many behaviors; e.g., escape from predators and prey capture. A sun compass was demonstrated in experiments using conical pans in which crabs swam actively at the edge in the direction corresponding to offshore at the place of capture (Fig. 5e; Nishimoto and Herrnkind, 1982). Gross changes (120°) in the sun angle caused by mirrors yield a corresponding change in swimming direction, while interposing a translucent cover caused dispersed headings. Convincing evidence for time compensation came from crabs phase delayed by 8 hr. Mean vectors of two groups included the predicted change (120° clockwise, Fig. 5f).

mean direction (after Nishimoto and Herrnkind, 1978). (d) Submerged blue crabs orient downslope at angles as small as 3° (R. Nishimoto, original data). (e) Sun compass orientation in blue crabs; direction corresponds to offshore (narrow arrow) at their capture site (after Nishimoto and Herrnkind, 1982). (f) Blue crabs, phase-shifted 8 hr, orient in the predicted bearing; thin solid arrow, original offshore; broken arrow, predicted direction; thick arrow, group mean direction (after Nishimoto and Herrnkind, 1982).

The blue crab sun compass is intriguing for several reasons. First, crabs are highly transient and move over highly sinuous shorelines. They apparently learn the Y-axis reference of each beach *de novo*. This condition is antithetical to that known for fiddler crabs and, especially, for talitrid amphipods, in which learning a new Y-axis is assumed to be infrequent. Second, unlike most other brachyurans, blue crabs are not confined to benthic and/ or nocturnal movements; they are commonly seen surface swimming across open water (as much as 20 km offshore) by day. Daytime steerage by sun compass is conceivable during long-distance migration (Oesterling, 1976).

Blue crabs commonly reside in flowing estuaries where they must either maintain position or migrate in the appropriate up-estuary or seaward direction. Juveniles from an estuary placed in a cylindical current apparatus (designed after Creutzberg, 1961, 1963) show tidally regulated activity including both up- and downcurrent (passive) swimming (Sulkin, 1975). By altering the activity period or swimming direction, a crab could either move about the estuary with little total displacement each tidal cycle or move further into or out of the estuary. The former behavior is known from ultrasonically tracked portunids, *Scylla serrata*, in an estuary, although the method of orientation was not examined (Hill, 1978).

2. WASH ZONE CRUSTACEANS

Inhabitants of the wave wash zone of ocean beaches, including the mole crab *Emerita*, the amphipod *Synchelidium,* and the isopod *Excirolana*, move in a manner retaining them in a dynamic, variable, and constantly shifting microhabitat (Enright, 1978). The conditions there include a rapid uprush from turbulent breaking waves constantly alternating with the gradual backwash, a sheet of water accelerating slowly as a laminar flow. Prior to the ensuing uprush, the beach is briefly exposed. The width of the wash zone varies with beach slope and wave size, while its level on theY-axis is regulated by tidal phase. The physical stresses of wave wash include abrasion and upshore stranding. Predation is intense at the offshore edge and sometimes into the lower wash region, where fishes such as drums, pompanos, flounder, and striped bass feed heavily (Pearse et al., 1942; Leber, 1982). Both ghost crabs and portunids (e.g., *Arenaeus*) anchor themselves in the shoreward reaches of the wash zone, where they ambush mole crabs and the motile surf clam, *Donax* (Hughes, 1966; Vannini, 1976b).

Food is abundant in the form of small organisms and detrital particles suspended and transported in the surf zone. Feeding occurs during the recurrent backwashes when *Emerita* actively filter. The wash zone supports enormous concentrations of amphipods, *Donax,* and mole crabs (thousands per meter strip along the Y-axis: Leber, 1982) during the peak season. An additional benefit from maintaining appropriate wave wash zonation is reduced feeding efficiency of predaceous fishes whose access is limited by the extreme water motion.

Enright (1978) models wave wash movements of *Synchelidium* as differential responses to hydrostatic pressure changes and turbulence. Amphipods located in the calm, upper wash zone burrow and feed in the top few millimeters of substrate. At this time in their endogenous tidal rhythm, they are negatively phototactic unless subjected to a sharp, but slight, change in hydrostatic pressure by wave uprush (5+ millibars or approximately 5-cm height of water), whereupon they become hyperactive and photopositive, swimming upward for 5–30 sec. This duration approximates that of the uprush period and so carries the animal upshore, where it again becomes photonegative and reburies. Repetition of the sequence with upshore progression of the wave wash actively maintains the animals in an optimal feeding area. At peak high tide, when stranding is imminent, the animals become both photopositive and spontaneously more active. Thus, they swim toward the surface and remain in the water column longer, effects which transport them seaward as the tide ebbs. Hence, their time in the water column is adjusted to the directional transport of the waves, and their locomotion per se is not oriented in the Y-axis. In fact, active horizontal orientation seems impossible for such small (2-mm length) animals in rapid and turbulent water flow. Mole crabs and surf isopods show endogenous tidal rhythms and responses to small hydrostatic pressure changes analogous to those of *Synchelidium*.

D. Subtidal Region

Offshore, tidal flux becomes less influential to crustaceans, and the Y-axis is no longer the major movement pathway. The numerous macrocrustaceans of this zone exhibit movement patterns in apparent accord with the spatial distribution of local shelter, food, competitors, and predators. The lack of movement studies on one hand and the diversity of species on the other make generalizations difficult. Therefore, several movement patterns appearing in a variety of animals, including crustaceans, will serve as reference points from which to view the zone's spatial attributes.

There exists a range of lifestyles from strict nomadism to home-ranging— actually a continuum from hermit crabs, ranging broadly with little devotion to a specific site, to stomatopods and alpheid shrimps, basing their operations from a self-dug burrow vigorously defended for long periods. Intermediate levels are represented by xanthids, majids, and the palinurid and nephropid lobsters, all of which are variously dependent on a shelter during inactive periods.

Excepting disasters (innundation by sediments, encroachment of red tide, lenses of low salinity water), day-to-day stresses of the subtidal zone are largely biotic, procuring food and mates while avoiding predation. Physical stresses are either seasonal or come about as a consequence of organismic changes with ontogeny (e.g., larval development, metamorphosis, sexual maturation). In either case, and often in a combination, the animals migrate as discussed elsewhere. Whereas the physiological stresses of the intertidal zone threaten immediate death, making the interpretation of adaptiveness straightforward, the impact of stresses in the subtidal zone is neither so obvious nor so easily measured.

Certain ecological models of foraging, home range, and spacing include risks of predation as a factor (Brown and Orians, 1970; Schoener, 1971; Covich, 1976). For example, Covich (1976) theorizes an optimal distance an animal should travel to maximize its resource (food, mate, refuge) yield, yet simultaneously minimize risk of predation. (Risk is assumed to be maximum if the animal cannot locate refuge at the terminus of its search period.) Predators of subtidal benthic crustaceans are collectively diverse, but include fishes, other crustaceans, and cephalopod mollusks. Among these are specialists which ambush, chase, flush, or forcefully extract. Some sense prey by sight, touch, distance chemosenses, contact chemosenses, near-field acoustics, or a combination (Ache and MacMillan, 1980). Certain biotopes house a few (sand plain) or many (coral reef) predatory species. One predicts that spatial patterns of daily movement will reflect adjustments to locally prevailing predation. Specific features include (1) whether or not an animal exhibits a home site to which it returns after feeding or when threatened; (2) the degree of restriction of movement within uniform, cryptic biotopes (e.g., seagrass beds: Virnstein, 1977; Heck and Thoman, 1981); and (3) the temporal regulation of emergence from refuge. Crustaceans seldom serve as models for general theory (Baker, 1978), although enough information exists to estimate their compliance or to make inferences (Hazlett and Rittschof, 1975).

Refuge type and local availability interact with predation in molding movement patterns. One can grossly classify refugia and attendant modes of movement as follows. Rare, infrequent, or otherwise limiting refuges, such as isolated crevices, empty shells, and colonial invertebrates (e.g., sponges, coral heads, bryozoans), serve as long-term residences, often defended, for gonodactylid stomatopods and lobsters (*Homarus* spp.). Occupants are typically center-oriented (Covich, 1976), moving about the vicinity to feed and mate, returning to the lair after feeding or when threatened. A similar situation exists for those species investing great energy in constructing burrows, as is true of the stone crab, *Menippe mercenaria* (Bert et al., 1978), some stomatopods (Dingle and Caldwell, 1978), and the lobster, *Nephrops norvegicus* (Chapman and Johnstone, 1975). Homing to one or several home lairs occurs in *Panulirus argus* and probably in other palinurids (Herrnkind *et al.*, 1975), *Nephrops norvegicus* (Chapman, 1980), the American lobster, *Homarus americanus* (Lund *et al.*, 1971), and the spider crab, *Mithrax spinosissimus* (Hazlett and Rittschof, 1975).

Homing is unlikely for crustaceans dwelling in uniform regions of seagrasses and benthic algae; rather, they move freely about within, but might be quite restricted to, those regions. For example, juvenile *Panulirus interruptus* remain for several years within the luxuriant surf grass (*Phyllospadix*), nearly impervious to large piscine and crustacean predators presumably unable to penetrate the thick blanketing leaves (Engle, 1979). Heck and Thoman (1981) demonstrated that predation rate increases dramatically for a variety of seagrass crustaceans exposed in adjacent nonvegetated areas. Occupants of crevices, such as *Carcinus maenas* and various species of porcellanid and xanthid crabs, rapidly orient to concealment upon translocation to a new area (Knudsen, 1960).

Dwellers of open sand and mud that do not occupy residences usually evade predators by burial and rapid flight (galatheids, portunids), or by being armored (hermit crabs), robust (majids, oxystomatids), or cryptic (dromiids, some majids). Movements there are not restricted by refuge location and might relate to distribution of food, mates, conspecifics, and other resources. Hermit crabs, *Paguristes oculatus,* occupying mud substrate off Trieste daily move over a distance averaging 20 m, but in an undirected way yielding a net positional change of approximately 2.5 m after 1 week (Stachowitsch, 1979). Nomadism, with vagility varying among species, probably represents the general situation. Furthermore, the small burying forms unable to defend themselves might be expected to space themselves out, i.e., to tend to over-disperse, thus reducing their risk of discovery by predators (Tinbergen *et al.*, 1967).

Competitors for refuge, including both conspecifics and species with overlapping needs, also potentially influence the frequency and type of movements shown by an animal. This is evidenced by the spacing of individuals resulting from aggressive encounters in stomatopods, alpheid (snapping) shrimps, and xanthids (Dingle and Caldwell, 1978). Local foraging is constrained by the location of aggressive, territorial neighbors. Further variation occurs as sexual differences in movement patterns. Adult males of some species, such as the xanthid, *Pilumnus sayi,* show higher vagility than adult females (Lindberg, 1980a,b); the opposite is true for the prawn *Macrobrachium rosenbergii* (Peebles, 1979). Restricted movements are noted for gravid females of several crustaceans, e.g., spiny lobsters (Kanciruk and Herrnkind, 1976). Male spider crabs, *Mithrax spinosissimus,* exhibit lengthier home range excursions than do females, perhaps as a consequence of seeking mates (Hazlett and Rittschof, 1975).

Feeding styles show incredible variety among crustaceans, matching the diverse material available as food. However, what matters with respect to movements is the pattern of distribution, abundance, and temporal–spatial stability of food material and its distance from refuge, i.e., accessibility. Adjustment of locomotory paths and orientational performance seems dependent on the relative distance an animal must travel to encounter food, as well as on dispersion or clumping, and on prey behavior, e.g., fixed, motile, securely attached, buried, and temporally variable in occurrence.

Relatively short foraging excursions may be sufficient for species whose refuge is amid a copious food supply, as occurs in both the herbivorous *Mithrax spinossimus* (Hazlett and Rittschof, 1975) on a reef and the carnivorous stone crab, *Menippe mercenaria*, dwelling among oysters (Sinclair, 1977). Movements of large *M. spinosissimus* average 4–6 m in radius about their den as they browse benthic algae and detritus. Longer movements and those showing a sexual differential appear in excess of that required merely for access to food. Various crabs feeding on epibiota, fouling assemblages, or other stationary prey may have similar patterns despite great dietary differences. Lobsters, both clawed and spiny, exhibit lengthy sinuous foraging paths, sometimes moving over 200 m from their dens, the path length much longer (Lund *et al.*, 1971; Herrnkind *et al.*, 1975). In *P. argus*, diets are diverse, including crustacean, molluscan, and echinoderm items in a single feeding period. Here, direct orientation occurs only in the transit to and from general regions (e.g., seagrass beds), but ranging movements serve to locate scattered prey during foraging. The blue crab, *Callinectes sapidus*, another opportunistic carnivore, shows this pattern in the submerged intertidal zone (Nishimoto, 1980).

Certain highly motile or well-armored species exhibit nomadic rather than center-oriented foraging. Nomadism is documented in blue crabs (Nishimoto, 1980), spiny lobsters (Herrnkind, 1980), and the hermit crabs, *Paguristes oculatus* (Stachowitsch, 1979) and *Clibanarius vittatus* (Hazlett and Herrnkind, 1980). In each case, feeding is implicated as, but not proven to be, a contributing factor. Tracking of *C. vittatus* over several days shows that they remain about oyster bars where feeding is frequent, but tend to move rapidly and continuously through areas of open sand or seagrass (Hazlett and Herrnkind, 1980). Spiny lobsters have a somewhat different situation in that refuge is sought for daylight periods of inactivity. Where refuge is sparse or food is scattered far from cover, the lobsters appear to adopt any available shelter that they happen across or to move on in some direction after foraging until chancing across shelter (Herrnkind, 1980). Such behavior could explain the observations of lobsters resting in otherwise open areas amid clumps of sea urchins (Davis, 1971), nestled in thick seagrass (W. Herrnkind, unpublished observation), or even in radial tail-to-tail rings about the base of sea whips (Herrnkind and Kanciruk, 1978). Each of the various movement patterns involves different orientational tasks and abilities, certain species exhibiting remarkable homing feats, some exhibiting efficient search patterns, others piloting accurately through new areas, and a few, such as lobsters, demonstrating the full range.

SPINY LOBSTERS

The palinurids exploit a variety of benthic habitats ranging from the shallow coral reef (*Panulirus* spp.) to moderate depths (*Palinurus* spp.; 100–400 m), from the tropics to high temperate latitudes (George and Main, 1967). They are characteristically large crustaceans (many species exceeding 1 kg in weight) capable of strong locomotion over virtually any topography, some actively entering heavily turbulent surge zones to feed (Berry, 1971; Engle, 1979). The movement patterns of the family collectively span the range from homing, with long term residency, to virtual nomadism (Herrnkind, 1980). Species of *Panulirus* such as *P. guttatus, P. versicolor,* and *P. longipus femoristriga* are reef dwellers, typically nonmigratory, remaining in the vicinity of their home reef perhaps for their entire benthic life. Den-dwelling *Panulirus argus* are known to inhabit one of several crevices within a few-hundred meters radius for continuous periods of weeks, leaving each night at or about sunset to forage up to 200 m away and returning to one of the "homes" before dawn (Clifton *et al.,* 1970; Herrnkind *et al.,* 1975). Homing from up to 1-km distance was documented by ultrasonic telemetry (Herrnkind *et al.,* 1975) and may occur over much longer distances (Creaser and Travis, 1950).

Preadult *P. argus* in the Florida Keys and Bahamas exhibit periods of extensive wandering over the sparsely sheltered shallow (2–10 m) banks. Nomadic movements averaging as much as 2–7 km per day for periods of weeks are recorded for individual lobsters not apparently taking part in a general population migration. Migrations in this taxon, however, are the most extensive of any benthic organism. For example, reproductive migrations of *Panulirus ornatus* may approach 500 km and those of *Jasus edwardsii* up to 200 km (Herrnkind, 1980). The migrations of *Panulirus longipus cygnus, P. argus,* and *Palinurus delagoae* occur as highly directional mass movements of many thousands of individuals, perhaps whole local populations.

Orientation mechanisms have been sought only for *P. argus*—a species, however, demonstrating the full range of palinurid movement patterns. Available information, while incomplete, establishes important mechanisms and guideposts as well as directions for future work with this and other benthic crustaceans.

The search for guideposts initially emphasized nonvisual cues because virtually all homing and most migration is nocturnal, excepting diurnal periods of mass movement. In addition, diurnal phases of mass movement

often occur in highly turbid waters obviating polarized light, light directivity, and sun compass under most conditions. Field experiments were
done initially to avoid artifacts and to provide adequate space for these
large, mobile animals.

Lobsters translocated from their home range to field arenas at 2–10 m
depths on uninterrupted sand substrate repeatedly walked on a given heading, the direction varying with environmental conditions and apparent indi-
viduality of each animal (Fig. 6a–d). Not infrequently, lobsters initially
walked in an essing path for several meters before straightening out and
walking rapidly away. Essing was reduced or not shown on repeated releases as the lobsters turned and walked in their preferred bearing. Many
lobsters, both eyecapped and normally sighted, exhibited upwave orientation (Fib. 6b). Wave tank studies, conducted in the facility described earlier
for blue crabs, confirmed the wave-surge orientation and suggested components of surge hydrodynamics necessary for discriminating direction (Walton and Herrnkind, 1977). As in the field, the mean direction of the eyecapped test group was upwave (Fig. 6f).

When both path straightness and degree of orientation for various wave
amplitudes and periods are separately plotted against velocity and displacement, an obvious trend appears only with velocity (Fig. 6e; Walton and
Herrnkind, 1977). For a range of displacements from 5 to 12 cm, path
straightness and orientational accuracy improve sharply above a 12 cm/sec
velocity (Fig. 6e; range 7–22 cm/sec). Apparently lobsters orient upwave
only in response to relatively strong surge, perhaps similar to that produced

(a)

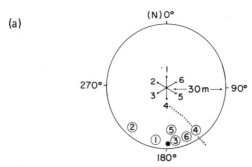

Fig. 6. (a) Field release arena for spiny lobster, *Panulirus argus*, showing hypothetical data
from six repetitive runs of one subject (encircled numbers) and the statistically significant mean
direction (filled circle). (b) Upwave mean direction (lower filled triangle) of group of *P. argus*
tested as in (a). S, wave approach. (c) *Panulirus argus*, tested as in (a), show individual directional responses in the presence of water currents. C, current direction. (d) In the absence of
current and surge, *P. argus* move only short distances, in scattered directions, indicating

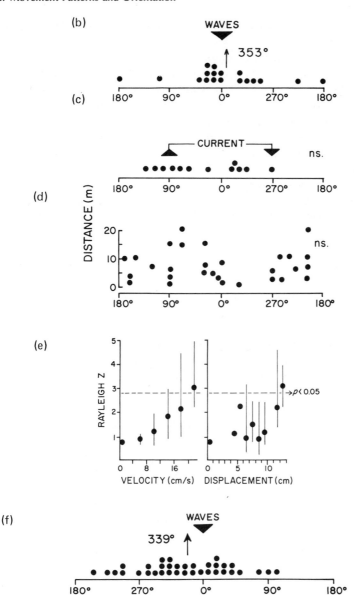

disorientation; each dot is a single run of an individual; eight lobsters were tested. (e) Surge orientation by *P. argus* in a wave tank (Fig. 4d). Directionality increases steadily with increasing velocity, becoming statistically significant above 12 cm/sec, but no trend appears with surge displacement. (f) *Panulirus argus,* grouped by individual directions after wave tank tests, show mean directionality (triangle) upwave. (All figures after Walton and Herrnkind, 1977.)

by oceanic swells passing over the coastal shallows (3–20 m depth) inhabited by this species.

The function of upwave orientation is speculated to lead lobsters offshore ultimately to a reef or other cover characteristically bounding the areas of open sand. Downwave movement would lead shoreward to ever shallower water. Hence, upwave movement conceivably is a shelter-seeking tactic for lobsters when they are outside the familiar home range.

Field tested lobsters also oriented in the presence of currents; i.e., unidirectional water flow at velocities over 5 cm/sec (Fig. 6c; Walton and Herrnkind, 1977). At slower rates, lobsters move circuitously, slowly, and for only short distances before coming to rest (Fig. 6d). The lobsters showed no collective propensity to move either directly upstream or downstream; rather they exhibited individual preferred directions, even when tested simultaneously. Again, the field sites were outside the home ranges of the tested animals.

Lobsters maintained in 2-m diameter circular tanks show a strong preference for movement about the periphery, either clockwise or counterclockwise, in the direction of current flow (flow rate = 20 cm/sec: J. Muller and W. Herrnkind, unpublished data). The movement direction reverses upon reversal of the current. The significance of the downcurrent direction to natural movements is difficult to interpret from studies in circular tanks, since the animal has little choice except upcurrent or downcurrent. Nevertheless, currents clearly influence orientation and probably guide certain natural movements. Moreover, the highly confused movements by eye-capped lobsters in the absence of both current and surge imply dependency on hydrodynamic information for movements in darkness and in turbid waters.

The repeated movements by numerous individual lobsters at some large angle from directly up/downwave or up/downcurrent suggest much more complex orientational ability than the bidirectional rheotaxis dogmatically assumed to be the basis of most crustacean (and other marine animal) migration (Creutzberg, 1975). In fact, such movements cannot explain the observed homing and migratory orientation of *P. argus*. For example, the nightly paths of lobsters moving several hundred meters from a den to forage, with return to the original vicinity, usually do not coincide with the predominant current and surge axes (Herrnkind and McLean, 1971; Herrnkind et al., 1975). Another case is the southerly bearing (approximately 190°) of mass migrants near Bimini, Bahamas, which brings lobsters through variable currents and surge consistently from the northwest (310°). A reasonable hypothesis is menotactic or compass orientation to hydrodynamic cues analogous to a light compass or, more appropriately, the anemomenotaxis reported for certain beetles and scorpions (Linsenmair, 1968). Such responses would likely depend on experience in a locality and be modifiable. Experimental testing of this hypothesis is essential, as its validation would represent an important advance in our understanding or orientation in the sea.

Amid the largely physiological research on chemical senses of lobsters is evidence of chemically mediated orientation (Ache et al., 1978; Reeder and Ache, 1980). The chemosensitive outer flagellae of the antennules direct accurate upcurrent movement by lobsters to the source of a variety of sapid substances (shrimp extract, various amino acids). Ache and co-workers examined the locomotory search pattern of P. argus over distances of approximately 1.5 m in a circular arena with maximum current flow of approximately 10 cm/sec across the center. Lobsters previously unfed for several days moved directly upstream with constant antennular "flicking" until touching the water inlet. Presumably, lobsters pursue the source of a chemostimulant in nature in the same manner. The dependence of chemoorientation on the water current per se, the concentration gradient, and the degree of turbulence is of interest since it relates to the functional significance and limitations of this mechanism in nature. As turbulence increases, the directional component of a chemostimulant decreases. The responses of nonfeeding lobsters to water currents suggest that water flow mediates long distance search for the source of chemicals present in low concentration or those mixed by minor turbulence. Field observations of chemically stimulated lobsters under selected conditions seem a reasonable next step in assessing the effectiveness of this orientational mode in nature.

The present knowledge of spiny lobster orientation suggests nonvisual guideposts potentially useful in their movements, but cannot satisfactorily explain how a lobster orients in its home range. Homing lobsters walk in straight paths across sand or algal areas where feeding occurs, but may follow reef contours leading circuitously to the home den (Clifton et al., 1970; Herrnkind et al., 1975). The former behavior suggests a menotactic response to prevailing surge or current serving steerage over long distances. The ability of deliberately displaced lobsters to proceed in a homeward direction and their change in course on contact with the reef suggest the action of releasing factors. Furthermore, when a lobster is displaced by the experimenter via an indirect course to the release point, it seems unlikely that it can integrate and backtrack. The high chemosensitivity to both waterborne substances (by the antennules) and substrate or contact substances (by the pereiopod aesthetasc hairs) implies that the choice of direction in a familiar area is set at a given internal state by the chemical milieu at that location. That is, the animal turns to the homebound course, say 90° cross current, when in seagrass, but turns downcurrent when in algae, in accord with the location of each relative to the den. Differing chemotactile features of the reef edge release directional changes or adjustments until the animal comes into the vicinity of the den area—perhaps recognizing it by waterborne chemicals emanating from conspecifics (Caldwell, 1979). This model and the known behavior conform to Baker's (1978) "familiar area map" hypothesis, consistent for local area orientation in many species.

Support for such an ability of spiny lobsters to integrate learning with the release and orientation of locomotion comes from Schöne's (1965) experiments with *P. argus* in a two-choice maze. While his test situation was artificial, requiring a dry lobster to discriminate correctly between two stimuli to reach water, the animals clearly improve both in latency of response in the situation (release) and in making the correct directional response (orientation). The guarded success of Schöne's technique invites experimental testing of the homing orientation hypothesis, just described, in a more natural situation. Again, other crustaceans represent potentially good models as well, e.g., nephropid lobsters, burrow dwelling xanthids, and stomatopods, all facing analogous homing problems.

IV. ORIENTATION OVER LONG DISTANCES

Bainbridge (1961) recognized crustacean migrations as persistent locomotory movements placing the animal under conditions suited to a particular life phase. Contemporary syntheses of migration literature affirm that an amazing variety of movement patterns are adaptive components of widely differing life-cycles (e.g., Baker, 1978; Dingle, 1980). This great variation yields a general definition of migration as *behavior especially evolved for the displacement of the animal in space* (Dingle, 1980). This definition is so broad as to include daily foraging-homing movements, tidal invasions, and various short-term intrazonal activities treated earlier in this chapter. The following discussion emphasizes the orientation of a subset of migratory phenomena in Crustacea: those which involve durations of locomotion ordinarily exceeding a daily cycle, often crossing several ecological zones, and those covering distances exceeding cyclic daily movements. This parallels Bainbridge's treatment of orienting factors in the "horizontal migrations" of crustaceans. Orientation of these events is interesting because in crossing physical zones, the organism must rely on guideposts either unavailable, or perhaps used differently, during intrazonal activities.

A. Terrestrial Migrations

Seaward reproductive migrations of several kilometers are documented in a number of ocypodids (Bliss, 1979; Bliss *et al.*, 1978) and coenobitids (de

Wilde, 1973). The movements include three phases: (1) coalescing of largely adult males and females at copulation sites in supralittoral dune or terrestrial habitats well inland of the surf zone; (2) nocturnal spawning movement by gravid females to the water's edge where larvae are released; and (3) inland return movement and dispersal. The path seaward may be quite indirect as a result of topographic obstacles (Fig. 7a), especially when gravid females detour low-salinity ponds (immersion triggers premature release of larvae: de Wilde, 1973; Bliss, 1979). Evidence suggests that spawning *C. clypeatus* in Curaçao follow the same pathways to the sea year after year, the paths more or less in the Y-axis, but not obviously demarcated (J. Langdon, personal communication). The immigrants are described anecdotally as thousands of animals in processions up to 100 m in length.

The orientational performance during migration is considerably more sophisticated than that shown during nightly feeding forays about the supralittoral zone (Vannini, 1975a), since the copulation sites and spawning beaches may not lie in a simple Y-axis direction from the various scattered inland locations of the crabs. It is assumed that some portion of the older adults have migrated previously but not those newly mature. Hence, it is unclear how individuals choose the migratory direction a *priori*. Secondly, the movements occur at night, sometimes under overcast and through foliage obscuring visual cues. Piloting is further complicated by the necessity to veer substantially from seaward during detours.

Migratory orientation has been subjected to little experimental analysis. de Wilde (1973) speculates that the tendency to congregate while moving results in the neophytes being essentially led by the experienced individuals. Observations on *Gecarcinus* by Klaassen (1975) and Bliss *et al.* (1978) suggest a role of visual cues during movements from the copulation sites to the water. Crabs moving seaward at this time veer toward bright sources of artificial illumination and even congregate beneath street lamps along coastal roads. Klaassen hypothesizes that crabs orient toward the most brightly illuminated part of the horizon lying in a broad band over the sea. This photic feature is also the guidepost for hatchling sea turtles, which face an analogous orientational problem (Mrosovsky, 1978).

The mechanism in hatchling green turtles involves both balancing illumination, so that orientation is approximately centered rather than widely scattered, and temporally compensating, such that brief periods of asymmetric intensities do not cause path deflections. This latter feature prevents briefly encountered shadows from nearby objects or passing clouds from causing disorientation. Another adaptive feature of this mechanism is the response to relative brightness. When a large object, such as a high dune or trees, blocks sufficient seaward light, the more landward region becomes comparatively brighter. The animal responds to this by detouring until the

(a)

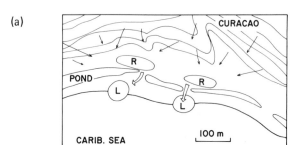

(b)

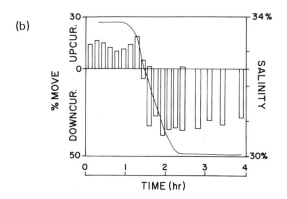

(c)

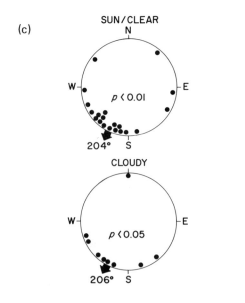

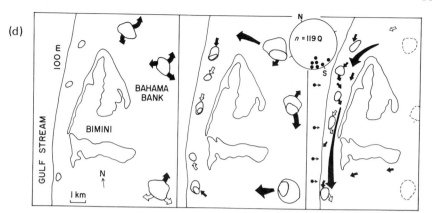

Fig. 7. (a) Terrestrial hermit crabs, *Coenobita clypeatus*, migrate at spawning time from inland (dots) to certain beach sites (large circles), detouring large ponds in the process (after de Wilde, 1973). (b) Juvenile pink shrimp, *Penaeus duorarum*, reverse movement direction (histograms) when salinity drops sharply (dashed lines) (after Hughes, 1969b). (c) Autumnal emigrant hermit crabs, *Pagurus longicarpus*, orient approximately seaward when the sky is clear or only partially clouded (after Rebach, 1978). (d) Mass migratory syndrome of spiny lobster, *Panulirus argus*. (Left), Nonmigratory lobsters move about nomadically over the shallows near Bimini, Bahamas. (Center), Migrants coalesce and move night and day in long queues oriented southward past Bimini. (Right), Scatter diagram shows bearing of queues west of Bimini at locations approximately as shown (after Herrnkind and Kanciruk, 1978).

view to the sea is once again unobscured. Whether an analogous mechanism operates in crustaceans is unknown. It is also not certain what role is played by the anisotropy of light radiance from the horizon and sky, as on a night of full moon (Verheijen, 1978). Once on the beach, substrate vibrations from the surf (Klaassen, 1975), downslope, and other cues (Vannini, 1975a) may become influential since, there, crabs veer only tangentially to a light source along shore, yet remain oriented generally seaward.

B. Movements to and from Estuaries

Estuaries serve as nurseries or seasonal feeding sites for numerous natant (penaeid and caridean shrimp, portunid crabs) and reptant (pagurid hermit crabs, *Carcinus*) crustaceans moving to and/or from the sea in a temporally organized manner (Allen, 1966; Antheunisse *et al.*, 1971; Klein-Breteler, 1976). For the migrant crustacean, the orientational task is discriminating unambiguous guideposts to the goal, whether it be seaward or into the estuary. The task is complicated by the complex physical topography of estuaries and the recurring tidal reversals in current direction. Availability of photic and optical cues is often greatly restricted by turbid waters and the nocturnal activity pattern of the migrants. Work on natant forms in the past two decades by Verwey (1958), Creutzberg (1961, 1963, 1975), Hughes (1969a,b, 1972), and others (Morgan, 1967) suggests that salinity, hydrostatic pressure, and chemical cues act in concert with endogenous rhythms to enhance up-estuary or down-estuary travel, either by active swimming or in combination with current transport (Fig. 7b). Ontogenetic factors, as well as seasonal environmental events such as sharply falling autumn temperatures, rapid salinity decreases at wet season, or an interplay of such factors, may trigger population-wide migration (Boddeke, 1975; Dorgelo, 1976).

The western Atlantic pink shrimp, *Penaeus duorarum*, spends the postlarval–juvenile period in estuaries (Hughes, 1972). The postlarvae move into the estuary largely at night by swimming in the water column or near the surface on flooding tides. Juveniles leave the estuary on ebbing tides. Hence, each stage times its swimming activity to take advantage of transport in the appropriate direction. The mechanism of orientation actually constitutes the response to factors correctly identifying the appropriate tidal flow.

In the case of juveniles, water current in a circular chamber elicits strong upstream swimming or, if especially strong, burrowing in the substrate. This response acts adaptively by preventing up-estuary displacement during flood phase. A drop in salinity as small as 2 ‰ over 20 min, as occurs during tidal outflow, triggers passive drifting or actual downstream swimming. The duration of the downstream phase is regulated by endogenous factors such that it lasts longest when stimulation occurs at the time corresponding to ebb tide.

The postlarvae are relatively weak swimmers (compared to the juveniles) and potentially subject to seaward (i.e., wrong way) displacement at ebbing tides. However, by swimming in the water column at high salinity, corresponding to up-estuary flow of tidal water, they are transported in the correct direction. Decreasing the salinity induces cessation of swimming and settlement, or burial in the substrate, preventing down-estuary displacement. Furthermore, the postlarvae remain in the high salinity bottom water by actively avoiding the less saline water layers (Hughes, 1969a,b). An upward swimming postlarva turns and swims downward upon contact with a halocline interface; i.e., where less saline water is layered atop higher salinity water.

Other estuarine crustaceans, especially other penaeids and carideans (Allen, 1966), as well as brachyuran larvae, exhibit behavior suggesting that modified models of the pink shrimp mechanism can be reasonably hypothesized for them as well (Cronin and Forward, 1979; Forward and Cronin, 1980). Modifications include either responses to differing combinations of salinity and rhythmic phases, or addition and subsitution of alternate releasing cues. Evidence exists that the portunid, *Macropipus holsatus*, when migrating (Venema and Creutzberg, 1973), actively swims in response to hydrostatic pressure changes corresponding to 0.5 m of water (Morgan, 1967). The blue crab, *Callinectes sapidus,* under some conditions actively avoids low pH outflow from certain estuaries (Laughlin et al., 1978). It is reasonable to postulate that crustaceans discriminate chemicals signifying different water masses, as known for fishes (Creutzberg, 1961; Hasler and Scholz, 1978). One attribute of mechanisms using passive transport is that the migrant does not require constant course correction or active piloting. The advantages conferred seem especially adapted for turbid waters and allow the migrants to avoid visual predators by confining activity to night or by adjusting depth (Forward and Cronin, 1978; Cronin and Forward, 1980).

Reptant crustaceans migrating into and out of estuaries not only must adjust to current reversals as do natant forms, but also, being restricted to the substrate, they cannot avail themselves of guidance information inherent in the water column, e.g., visual cues and vertical physical–chemical gradients. Furthermore, they cannot as effectively utilize passive transport. The hermit crab *Pagurus longicarpus,* like most other temperate zone nearshore crabs (Fotheringham 1975; Dorgelo, 1976) seasonally migrates between estuaries and deeper bays, as well as from the littoral to the subtidal (Rebach, 1978). Crabs invade the estuaries in spring, correlated with increased temperature, feed and breed there over summer, then move out to the sea in autumn as temperature declines. There, winter temperatures induce burying in the substrate. Hermit crabs tested in autumn in large circular chambers confine their movement at the tank edge corresponding to seaward at the summer habitat. This orientation is visually mediated through a sun-compass mechanism, although it is either variable or subject to modification by as yet unspecified factors (Fig. 7c; Rebach, 1978). Other cues, especially the bottom slope and chemical signature of the tidal flow, are hypothesized to play complementary roles, since the celestial cues are often obscured and the irregular shoreline obviates using a highly focused compass bearing (Rebach, 1981).

Movements of riverene forms, particularly crayfish, suggest up- and downcurrent orientation serving several functions, depending on the species, habitat type, and environmental conditions. Localized home range movements (*Orconectes juvenalis:* Merkle, 1969; *O. virilis:* Hazlett et al., 1974) apparently involve nocturnal journeys more or less equally distributed over time between up- and downstream. Upstream movement in *Orconectes nais* occurs after lengthy downstream displacement caused by heavy flooding that seasonally characterizes certain areas (Momot, 1966). Sex-related and reproductive differences in migration are known for *O. virilis.* In particular, ovigerous females in spring tend to remain in an area until the young are released, at which time they move downstream for the summer, returning upstream in fall prior to burrowing-in over winter (Hazlett et al., 1979). Factors possibly regulating or triggering the direction of movement are derived from field correlates, due to the paucity of experimental studies. For example, *O. nais* moves upstream from silty substrate until reaching rock rubble (Momot, 1966). The lake dwelling *O. virilis* (Momot and Gowing, 1972) show differential reproductive inshore–offshore movements, suggesting internal gonadal-endocrine influences on migratory behavior (Aiken, 1969). Drops in water level of 5–26 cm caused increased activity, and levels raised 6–29 cm caused decreased activity of stream-dwelling *O. virilis*, although migration was not induced (Hazlett et al., 1979).

C. Benthic Subtidal Migrations: Spiny Lobsters

The autumnal and storm-associated "marches" of *Panulirus argus* are synchronous mass migrations wherein the lobsters walk single-file, in so-called queues, both day and night for periods lasting several days. Estimates of these localized movements range up to 200,000 individuals moving unidirectionally over 2–5 days through areas less than 2 km in width (Herrnkind, 1969; Herrnkind and Kanciruk, 1978). The distance traversed is estimated to be 30–50 km during the period of continuous marching.

Migrants in an area such as Bimini, Bahamas, show high directionality; e.g., mean headings of lobsters in 1969 ranged from 159 to 210° with strong modes at approximately 180°–190° over a distance exceeding 10 km (Fig. 7d). This directionality occurred during four witnessed migrations (1961, 1963, 1969, and 1971). The migratory direction differs in other localities, being roughly northward at Boca Raton, Florida, in 1965 and southward off Isla Contoi, Yucatan, in 1975 (Herrnkind, 1969 and unpublished data). The localized directions suggest orientation to local features rather than geographically persistent guideposts such as celestial cues or geomagnetism.

The guideposts for palinurid migrations are not yet known, although hydrodynamic cues, particularly water currents, are suspected for *Jasus edwardsii* (Street, 1971), *Palinurus delagoae* (Berry, 1973), and *Panulirus argus* (Herrnkind, 1980), based on hydrographic features of the respective migratory regions. Further inference derives from the high orientational responsiveness to hydrodynamic cues by nonmigrating *P. argus* (Walton and Herrnkind, 1977). The tendency of some individuals to orient consistently up- or downcurrent in the field, and especially of those stimulated to hyperactivity in the laboratory, roughly match the migratory headings in certain areas. In addition, a menotactic response to wave surge would provide a redundant mechanism, since swells are typically very consistent during migratory conditions and superimpose their effect in areas where tides and topography modify current direction.

At present the most parsimonious model hypothesizes hydrodynamic guideposts with bicomponent steerage. The lobsters, as they come into migratory state, orient themselves in some relation to prevailing current or surge, perhaps downcurrent. They maintain directionality by substituting orientation to the alternate cue when either becomes obscured, variable, or ceases. Ongoing research aims to induce the migratory state under controlled conditions in order to test for changes in orientational preferences for current, surge, and other potential guideposts (J. Muller and W. Herrnkind, unpublished data). Consistent results can then serve to predict directionality of migrants in the field under natural conditions, thus allowing tests of hypotheses.

Certain cues including vision, chemical factors, and the geomagnetic field, are not yet eliminated. Migration occurs unabated at night and in turbid waters. Orientation other than upcurrent over long distances obviates primary guidance by chemical factors, although they may well act as releasing factors, as is also suggested for thermal cues (Kanciruk and Herrnkind, 1978). Field tests of nonmigratory lobsters indicate no orientational influence from disruption of the earth's magnetic field (Walton and Herrnkind, 1977), although this cue might be effective only when lobsters are in the migratory state, or it may act as a synergistic component with other cues (Able, 1980).

Long-distance migration is recorded for other reptant crustaceans, including the Alaskan king crab, *Paralithodes camtschatica* (Allen, 1966), and the lobster *Homarus americanus* (Cooper and Uzmann, 1971, 1980). The orientational bases have not been examined, although in the latter species, offshore–onshore movement correlates with seasonal thermal conditions. Recent models of fish migration suggest that thermal (and other) features influence orientation, the animal moving to maintain position within its thermal preferendum (Leggett and Trump, 1978). An analogous model seems applicable to open-water movements of natant crustaceans.

Breeding horseshoe crabs (*Limulus polyphemus*) invade intertidal sand beaches on the Atlantic and Gulf of Mexico (Rudloe and Herrnkind, 1976). Females move shoreward on flooding spring tides where they are intercepted by males, which attach themselves to the females. The pair emerges onto the beach, burrowing into the sand at the upper reach of the wave wash. There, eggs are released and fertilized, after which the pair emerges from the sand, moves downshore and seaward into the subtidal zone. Handling a horseshoe crab induces rapid locomotion, consistently into the wave surge, independent of bottom slope and water current flow. Occasionally, horseshoe crabs released offshore turn and proceed downwave onto shore, mimicking the normal breeding behavior. Experiments in both field arenas (Rudloe and Herrnkind, 1976) and a wave tank (Rudloe and Herrnkind, 1980) show that wave surge is a fundamental guidepost to the Y-axis orientation described, and that visual cues, slope, and water flow are unnecessary. However, visual information probably plays some orienting or releasing role (Ireland and Barlow, 1978); other cues, including steep slopes and currents near the substrate, may affect offshore movements (Rudloe and Herrnkind, 1976).

V. CONCLUDING COMMENTS

From the ecoevolutionary point of view, the central questions of orientation are those asking by what means an animal performs its movements under natural conditions. The value of elegant sensorial studies not withstanding, I was here concerned with studies demonstrating the extent to which certain guideposts, sensors, and neural integrating processes actually serve behavior. Characterization and analysis of oriented movement in any crustacean seem incomplete as yet, although there have been major advances in the past two decades and active work continues. Good examples include the talitrid amphipods, wherein the early work of Pardi and Papi and of Williamson has been critiqued and focused by Craig, Hartwick, and, especially, Enright, Ercolini, and Scapini. Ontogenetic approaches proved valuable to understanding adaptive changes in orientation of decapods, as shown by Herrnkind, Forward, and Cronin. Herrnkind, Nishimoto, Rudloe, and Walton collectively provide a useful perspective, not just by examining different species, but by analyzing the orientational repertoire and mechanisms of each in differing milieus of guidance information. Work on fiddler crabs, blue crabs, *Limulus,* and spiny lobster showed distinct abilities reflecting the adjustments to available guideposts and greater sophistication in orienting ability than realized previously, and uncovered several heretofore unknown mechanisms; e.g., wave surge and, hypothetically, menotaxis to both waves and currents. In addition, the approaches of Enright, Hughes, and Vannini to crustacean orientation in differing biotopes are especially fruitful ones.

I attempted neither to compare systematically nor to contrast the orientation of crustaceans with that of other taxa, although several points deserve mention here. First, reliance upon multiple guideposts to accomplish an oriented activity, such as home ranging or a Y-axis circuit, is commonplace in crustaceans just as in other groups (Creutzberg, 1975; Able, 1980). Different guideposts serve not only to provide unique directionality within a single movement pattern (e.g., fiddler crab orientation to a visual object for shelter, following an upshore track guided by celestial cues), but also may serve redundantly (e.g., upshore orientation by either polarized skylight or sun disc). Second, crustaceans collectively exhibit a range of oriented movements representative of most of the animal kingdom, with the exception of true long-distance navigation. The range of sensory mechanisms, too, seems generally comparable to those of insects, motile mollusks, and vertebrates. Exceptions to this generalization, which may reflect lack of research rather than phyletic capability, include low reliance on water or airborne acoustics, geomagnetism and the absence of any emitted energy mechanism (e.g., echolocating, electrolocating). Finally, the diversity in the physical milieus of crustacean species makes it unlikely that particular mechanisms serve the group generally. For example, spiny lobsters and other deep-water forms simply cannot avail themselves of celestial (visual) cues, whereas land crabs do not have access to waves and currents. Widespread or sophisticated celestial orienting and navigating mechanisms, such as appear in insects, birds, and, probably, cetaceans and pelagic fish migrants, seem unlikely to have evolved in crustaceans, which lack flight or rapid swimming over long distances.

Despite substantial gains in understanding of both the ecological contexts and mechanisms of crustacean orientation, there remain obvious gaps and enigmas. Foremost, we do not know the features of path during homing, migration and daily movement by most natant and reptant species. Without such spatial and physical information, one is poorly equipped to hypothesize the guideposts and neurosensory mechanisms. The ever more commonplace use of scuba and tagging-telemetry promises to provide needed information. Another apparent shortcoming is the insufficiency of presently known mechanisms to explain certain observed feats of directionality, e.g., homing by spiny lobsters, migrating performances of land decapods, various "lobsters" and other benthic forms. More attention might be paid to diagnosing the releasing conditions, whereby the animal avails itself of appropriate guidance cues or passive transport, such as current flow, to achieve directionality. Comparatively more sophisticated mechanisms should be further tested, including menotactic orientation to water currents and balancing of optical stimulation. Also, we must better examine the roles of recently demonstrated guideposts such as wave surge, wind, and anisotropic radiance, and continue to search for still others. The origin of directional choice remains mysterious, in particular the innate escape directions in talitrid amphipods and the highly labile escape directions in blue crabs. Simply put, the crustaceans provide much challenging research for the orientational biologist at both the ecological and physiological level.

ACKNOWLEDGMENT

I am grateful to R. T.Nishimoto for extensive assistance in preparing this chapter and for his superior research in crustacean orientation.

REFERENCES

Abele, L., and Means, B. (1977). *Sesarma jarvisi* and *Sesarma cookei:* Montane, terrestrial crabs in Jamaica (Decapoda). *Crustaceana* **32,** 91–93.

Able, K. (1980). Mechanisms of orientation, navigation and homing. *In* "Animal Migration, Orientation, and Navigation" (S. Gauthreaux, Jr., ed.), pp. 284–373. Academic Press, New York.

Ache, B., and Case, J. (1969). An analysis of antennular chemoreception in two commensal shrimps of the genus *Betaeus. Physiol. Zool.* **42,** 361–371.

Ache, B., and MacMillan, D. (1980). Neurobiology. *In* "The Biology and Management of Lobster" (S. Cobb and B. Phillips, eds.), Vol. I, pp. 165–213. Academic Press, New York.

Ache, B., Johnson, B., and Clark, E. (1978). Chemical attractants of the Florida spiny lobster, *Panulirus argus. Sea Grant Program, Tech. Pap.* **10,** 1–28.

Adler, H. (1971). Orientation: Sensory basis. *Ann. N.Y. Acad. Sci.* **188,** 1–408.

Aiken, D. (1969). Photoperiod, endocrinology, and the crustacean molt cycle. *Science* **164,** 149–155.

Allen, J. (1966). The rhythms and population dynamics of decapod crustacea. *Oceanogr. Mar. Biol.* **4,** 247–265.

Altevogt, R. (1965). Lichtkompass- und Landmarken-dressuren bei *Uca tangeri* in Andalusien. *Z. Morphol. Ökol. Tiere* **55,** 641–655.

Altevogt, R., and von Hagen, H. (1964). Zur Orientierung von *Uca tangeri* in Freiland. *Z. Morphol. Ökol. Tiere* **53,** 636–656.

Antheunisse, L., Lammeas, J., and van den Hoven, N. (1971). Diurnal activities and tidal migrations of the brackish water prawn *Palaemonetes various* (Leach) (Decapoda, Caridea). *Crustaceana* **21,** 203–217.

Atema, J., and Engstrom, D. (1971). Sex pheromone in the lobster, *Homarus americanus. Nature (London)* **232,** 261–263.

Bainbridge, R. (1961). Migration. *In* "Physiology of Crustacea" (T. Waterman, ed.), Vol. 2, pp. 431–436. Academic Press, New York.

Baker, R. (1978). "The Evolution and Ecology of Animal Migration." Holmes and Meier, New York.

Barnes, T. (1932). Salt requirements and space orientation of the littoral isopod *Ligia* in Bermuda. *Biol. Bull. (Woods Hole, Mass.)* **63,** 496–504.

Barnes, T. (1935). Salt requirements and orientation of *Ligia* in Bermuda. III. *Biol. Bull. (Woods Hole, Mass.)* **69,** 259–268.

Berry, P. (1971). The biology of the spiny lobster *Panulirus homarus* (Linnaeus) off the coast of southern Africa. *Oceanogr. Res. Inst. (Durban), Invest. Rep.* **27,** 1–75.

Berry, P. (1973). The biology of the spiny lobster *Palinurus delegoae* Barnard, off the coast of Natal, South Africa. *Oceanogr. Res. Inst. (Durban), Invest. Rep.* **31,** 1–27.

Bert, T., Warner, R., and Kessler, L. (1978). The biology and Florida fishery of the stone crab, *Menippe mercenaria* (Say), with emphasis on southwest Florida. *Fla. Sea Grant Tech. Pap.* **9,** 1–82.

Bliss, D. (1979). From sea to tree: Saga of a land crab. *Am. Zool.* **19,** 385–410.

Bliss, D., van Montfrans, J., van Montfrans, M., and Boyer, J. (1978). Behavior and growth of the land crab *Gecarcinus lateralis* (Freminville) in southern Florida. *Bull. Am. Mus. Natl. Hist.* **160,** 114–151.

Boddeke, R. (1975). Autumn migration and vertical distribution of the brown shrimp *Crangon crangon* L. in relation to environmental conditions. *Proc. 9th Eur. Mar. Biol. Symp.* pp. 483–494.

Bowers, D. (1964). Natural history of two beach hoppers of the genus *Orchestoidea* (Crustacea: Amphipoda) with reference to their complemental distribution. *Ecology* **45**, 677–696.

Bright, D. (1966). The land crabs of the world. *Rev. Biol. Trop.* **14**, 183–203.

Brown, J., and Orians, G. (1970). Spacing patterns in mobile animals. *Annu. Rev. Ecol. Syst.* **1**, 239–262.

Caldwell, R. (1979). Cavity occupation and defensive behavior in the stomatopod *Gonodactylus festai*: Evidence for chemical mediated individual recognition. *Anim. Behav.* **27**, 194–201.

Carthy, J. (1958). "An Introduction to the behavior of Invertebrates." Allen and Unwin, London.

Chapman, C. (1980). Ecology of juveniles and adult *Nephrops*. In "Biology and Management of Lobsters" (J. Cobb and B. Phillips, eds.), pp. 143–178. Academic Press, New York.

Chapman, C., and Johnstone, A. (1975). The behavior and ecology of the Norway lobster, *Nephrops norvegicus* (L.). *Proc. 9th Eur. Mar. Biol. Symp.* pp. 58–74.

Chelazzi, G., and Ferrara, F. (1978). Researches on the coast of Somalia—the Shore and dune of Sar Uanle.19. Zonation and activity of terrestrial isopods (Oniscoidea). *Monit. Zool. Ital. NS Suppl.* **11**, 189–219.

Clifton, H., Mahnken, C., Van Derwalker, J., and Waller, R. (1970). Tektite I., Man-in-the-sea project: Marine science program. *Science* **168**, 659–663.

Cloudsley-Thompson, J. (1956)). Studies on diurnal rhythms. VII. Humidity response and diurnal activity in wood-lice (Isopoda). *J. Exp. Biol.* **33**, 576–582.

Cooper, R., and Uzmann, J. (1971). Migrations and growth of deep-sea lobsters, *Homarus americanus*. *Science* **171**, 288–290.

Cooper, R., and Uzmann, J. (1980). Ecology of juvenile and adult *Homarus*. In "The Biology and Management of Lobsters" (S. Cobb and B. Phillips, ed.), Vol. II, pp. 97–142. Academic Press, New York.

Covich, A. (1976). Analyzing shapes of foraging areas: Some ecological and economic theories. *Annu. Rev. Ecol. Syst.* **7**, 235–257.

Cowles, R. (1908). Habits, reactions, and associations in *Ocypoda arenaria*. *Pap. Tortugas Lab. Carnegie Inst. Wash.* **2**, 1–41.

Craig, P. (1971). An analysis of the concept of lunar orientation in *Orchestoidea corniculata* (Amphipoda). *Anim. Behav.* **19**, 368–374.

Craig, P. (1973a). Orientation of the sand beach amphipod, *Orchestoidea corniculata*. *Anim. Behav.* **21**, 699–706.

Craig, P. (1973b). Behaviour and distribution of the sand beach amphipod *Orchestoidea corniculata*. *Mar. Biol.* **23**, 101–109.

Crane, J. (1975). "The Fiddler Crabs of the World." Princeton Univ. Press, New Jersey.

Creaser, E., and Travis, D. (1950). Evidence of a homing instinct in the Bermuda spiny lobster. *Science* **112**, 169–170.

Creutzberg, F. (1961). On the orientation of migrating elvers (*Anguilla vulgaris* Turt.) in a tidal area. *Neth. J. Sea Res.* **1**, 257–338.

Creutzberg, F. (1963). The role of tidal streams in the navigation of migrating elvers (*Anguila vulgaris* Turt.). In "Animal Orientation" (H. Autrum, ed.), pp. 118–127. Springer-Verlag, Berlin and New York.

Creutzberg, F. (1975). Orientation in space: Animals. Invertebrates. In "Marine Ecology" (O. Kinne, ed.), Vol. II, pp. 555–655. Wiley, New York.

Cronin, T., and Forward, R. Jr. (1979). Tidal vertical migration: an endogenous rhythm in estuarine crab larvae. *Science* **205**, 1020–1022.

Cronin, T., and Forward, R. Jr. (1980). The effects of starvation on phototaxis and swimming of larvae of the crab *Rhithropanopeus harrisii*. *Biol. Bull. (Woods Hole, Mass.)* **128**, 283–294.

Daumer, K., Jander, R., and Waterman, T. (1963). Orientation of the ghost crab *Ocypode* in polarized light. *Z. Vgl. Physiol.* **47**, 56–76.

Davis, G. (1971). Aggregations of spiny sea urchins, *Diadema antillarum*, as shelter for young spiny lobsters, *Panulirus argus. Trans. Am. Fish. Soc.* **100**, 586–587.

den Boer, P. (1961). The ecological significance of activity patterns in the wood-louse *Porcellio scaber. Arch. Neerl. Zool.* **14**, 283–409.

de Wilde, P. (1973). On the ecology of *Coenobita clypeatus* in Curacao with reference to reproduction, water economy and osmoregulation in terrestrial hermit crabs. *Stud. Fauna Curacao* **44**, 1–138.

Dingle, H. (1980). Ecology and evolution of migration. *In* "Animal Migration, Orientation, and Navigation" (S. Gauthreaux, Jr., ed.), pp. 1–101. Academic Press, New York.

Dingle, H., and Caldwell, R. (1978). Ecology and morphology of feeding and agonistic behavior in mudflat stomatopods (Squillidae). *Biol. Bull. (Woods Hole, Mass.)* **155**, 134–149.

Dorgelo, J. (1976). Salt tolerance in Crustacea and the influence of temperature upon it. *Biol. Rev.* **51**, 255–290.

Edney, E. (1960). Terrestrial adaptations. *Physiol. Crustacea* **1**, 367–393.

Edney, E. (1968). Transition from water to land in isopod crustaceans. *Am. Zool.* **8**, 309–326.

Engle, J. (1979). Ecology and growth of juvenile California spiny lobster, *Panulirus interruptus* (Randall). Sea Grant Dissertation Series, USCSG-TD-03-79, pp. 1–298.

Enright, J. (1961). Lunar orientation of *Orchestoidea corniculata* Stout (Amphipoda). *Biol. Bull. (Woods Hole, Mass.)* **120**, 148–156.

Enright, J. (1972). When the beach-hopper looks at the moon: The moon-compass hypothesis revisited. *In* "Animal Orientation and Navigation" (S. Galler, K. Schmidt-Koenig, G. Jacobs, and R. Belleville, eds.), pp. 523–555. U.S. Govt. Printing Office, Washington, D.C.

Enright, J. (1978). Migration and homing of marine invertebrates: A potpourri of strategies. *In* "Animal Migration, Navigation, and Homing" (K. Schmidt-Koenig and W. Keeton, eds.), pp. 440–446. Springer-Verlag, Berlin and New York.

Ercolini, A., and Scapini, F. (1972). On the non-visual orientation of littoral amphipods. *Monit. Zool. Ital. NS* **6**, 75–84.

Ercolini, A., and Scapini, F. (1974). Sun compass and shore slope in the orientation of littoral amphipods (*Talitrus saltator* Montagu). *Monit. Zool. Ital. NS* **8**, 85–115.

Ercolini, A., and Scapini, F. (1976). Sensitivity and response to light in the laboratory of the littoral amphipod *Talitrus saltator* Montagu. *Monit. Zool. Ital. NS* **10**, 293–309.

Farr, J. (1978). Orientation and social behavior in the supralittoral isopod *Ligia exotica* (Crustacea: Oniscoidea). *Bull. Mar. Sci.* **28**, 659–666.

Ferguson, D. (1967). Sun-compass orientation in anurans. *In* "Animal Orientation and Navigation" (R. Storm, ed.), pp. 21–31. Oregon State Univ. Press, Corvallis, Oregon.

Forward, R., Jr. (1976). Occurrence of a shadow response among brachyuran larvae. *Mar. Biol. (Berlin)* **39**, 331–34.

Forward, R., Jr., and Costlow, J., Jr. (1974). The ontogeny of phototaxis by larvae of the crab *Rithropanopeus harisii. Mar. Biol. (Berlin)* **25**, 27–33.

Forward, R. Jr., and Cronin, T. (1978). Crustacean larval phototaxis: possible functional significance. *In* "Physiology and Behavior of Marine Organisms" (D. McLusky and J. Berry, eds.), pp. 253–261. Pergamon, New York.

Foward, R., Jr., and Cronin, T. (1979). Spectral sensitivity of larvae from intertidal crustaceans. *J. Comp. Physiol.* **133**, 311–315.

Forward, R., Jr., and Cronin, T. (1980). Tidal rhythms of activity and phototaxis of an estuarine crab larva. *Biol. Bull. (Woods Hole, Mass.)* **158**, 295–303.

Fotheringham, N. (1975). Structure of seasonal migrations of the littoral hermit crab *Clibanarius vittatus* (Bosc.). *J. Exp. Mar. Biol. Ecol.* **18,** 47–53.

Fraenkel, G., and Gunn, D. (1961). "The Orientation of Animals." Dover, New York.

Friedlander, C. (1965). Aggregation in *Oniscus asellus* L. *Anim. Behav.* **13,** 342–346.

Galler, S., Schmidt-Koenig, K., Jacobs, G., and Belleville, R. (1972). "Animal Orientation and Navigation." U.S. Govt. Printing Office, Washington, D.C.

Gauthreaux, S. (1980). "Animal Migration, Orientation, and Navigation." Academic Press, New York.

Gendron, R. (1977). Habitat selection and migratory behaviour of the intertidal gastropod *Littorina littorea* (L.). *J. Anim. Ecol.* **46,** 79–92.

George, R., and Main, A. (1967). The evolution of spiny lobsters (Palinuridae): A study of evolution in the marine environment. *Evolution* **21,** 803–820.

Gepetti, L., and Tongiorgi, P. (1967a). Nocturnal migrations of *Talitrus saltator* (Montagu) (Crustacea Amphipoda). *Monit. Zool. Ital. NS* **1,** 37–40.

Gepetti, L., and Tongiorgi, P. (1967b). Ricerche ecologiche sugli artropodi di una spiaggia sabbiosa del littorele tierrenico. II. le migrazioni di *Talitrus saltator* (Montagu) (Crustacea-Amphipoda). *Redia* **50,** 309–336.

Gifford, C. (1962). Some observations on the general biology of the land crab, *Cardisoma guanhumi* (Latreille), in south Florida. *Biol. Bull. (Woods Hole, Mass.)* **123,** 207–223.

Gorner, P. (1958). Die optische und kinasthetische Orientierung der Trichterspinne *Agelena labryinthica* (Cl.). *Z. Vgl. Physiol.* **41,** 111–153.

Hamilton, P. (1976). Predation on *Littorina irrorata* (Mollusca: Gastropoda) by *Callinectes sapidus* (Crustacea: Portunidae). *Bull. Mar. Sci.* **26,** 403–409.

Hamner, W., Smyth, M., and Mulford, E., Jr. (1968). Orientation of the sand-beach isopod *Tylos punctatus. Anim. Behav.* **16,** 405–409.

Hartwick, R. (1976a). Beach orientation in talitrid ampipods: Capacities and strategies. *Behav. Ecol. Sociobiol.* **1,** 447–458.

Hartwick, R. (1976b). Aspects of celestial orientation behavior in talitrid amphipods. *In* "Biological Rhythms in the Marine Environment" (P. DeCoursey, ed.), pp. 189–197. Univ. of South Carolina Press, Columbia, South Carolina.

Hasler, A. (1966). "Underwater Guideposts." Univ. of Wisconsin Press, Madison, Wisconsin.

Hasler, A., and Scholz, A. (1978). Olfactory imprinting in coho salmon *(Onchorhynchus kisutch). In* "Animal Migration, Navigation, and Homing" (K. Schmidt-Koenig and W. Keeton, eds.), pp. 356–369. Springer-Verlag, Berlin and New York.

Hazlett, B., and Herrnkind, W. (1980). Orientation to shell events by the hermit crab *Clibanarius vittatus* (Bosc.) (Decapoda, Paguridae). *Crustaceana* **39,** 311–314.

Hazlett, B., and Ritschoff, D. (1975). Daily movements and home range in *Mithrax spinosissimus* (Majidae, Decapoda). *Mar. Behav. Physiol.* **3,** 101–118.

Hazlett, B., Rittschof, D., and Rubenstein, D. (1974). Behavioral biology of the crayfish *Orconectes virilis.* I. Home range. *Am. Midl. Nat.* **92,** 301–320.

Hazlett, B., Rittschof, D., and Ameyaw-Akumfi, C. (1979). Factors affecting the daily movements of the crayfish *Orconectes virilis* (Hagen, 1870) (Decapoda, Cambaridae). *Crustaceana* (Suppl. 5), 121–130.

Heck, K., and Thoman, T. (1981). Experiments on predator-prey interactions in vegetated aquatic habitats. *J. Exp. Mar. Biol. Ecol.* **53,** 125–134.

Herrnkind, W. (1968a). Adaptive visually-directed orientation in *Uca pugilator. Am. Zool.* **8,** 585–598.

Herrnkind, W. (1968b). Breeding of adult *Uca pugilator* and mass rearing of the larvae with comments on the behavior of the larval and early crab stages. *Crustaceana* (Suppl. 2), pp. 214–224.

Herrnkind, W. (1969). Queueing behavior of spiny lobsters. *Science* **164**, 1425–1427.

Herrnkind, W. (1972). Orientation in shore-living arthropods, especially the sand fiddler crab. *In* "Behavior of Marine Animals, Invertebrates" (H. Winn and B. Olla, eds.), Vol. 2, pp. 1–59. Plenum, New York.

Herrnkind, W. (1980). Spiny lobsters: Patterns of movement. *In* "Biology and Management of Lobsters" (B. Phillips and J. Cobb, eds.), Vol. I, pp. 349–407. Academic Press, New York.

Herrnkind, W., and Kanciruk, P. (1978). Mass migration of spiny lobster, *Panulirus argus* (Crustacea: Palinuridae); synopsis and orientation. *In* "Animal Migration, Navigation, and Homing" (K. Schmidt-Koenig and W. Keeton, eds.), pp. 430–439. Springer-Verlag, Berlin and New York.

Herrnkind, W., and McLean, R. (1971). Field studies of homing, mass emigration and orientation in the spiny lobster, *Panulirus argus. Ann. N.Y. Acad. Sci.* **188**, 359–377.

Herrnkind, W., Van Der walker, J., and Barr, L. (1975). Population dynamics, ecology and behavior of the spiny lobster, *Panulirus argus*, of St. John, U.S. Virgin Islands: Habitation and pattern of movements. Results of the Tektite Program, Vol. 2. *Bull. Nat. Hist. Mus. L.A. County* **20**, 31–45.

Hill, B. (1978). Activity, track, and speed movement of the crab *Scylla serrata* in an estuary. *Mar. Biol. (Berlin)* **47**, 135–141.

Horch, R., and Salmon, M. (1972). Responses of the ghost crab, *Ocypode*, to acoustic stimuli. *Z. Tierpsychol.* **30**, 1–13.

Horridge, G. (1967). Perception of polarization plane, colour, and movement by the crab, *Carcinus. Z. Vgl. Physiol.* **55**, 207–224.

Hueftle, K. (1977). Near-orientation in the homing of the fiddler crab, *Uca crenulata* (Lockington). M.S. Thesis, pp. 1–40. San Diego State Univ., San Diego, California.

Hughes, D. (1966). Behavioural and ecological investigations of the crab *Ocypode ceratopthalmus* (Crustacea: Ocypodidae). *J. Zool.* **150**, 129–143.

Hughes, D. (1969a). On the mechanism underlying tide-associated movements of *Penaeus duorarum* Burkenroad. *FAO Fish. Rep.* **57**, 867–874.

Hughes, D. (1969b). Responses to salinity change as a tidal transport mechanism of pink shrimp, *Penaeus duorarum. Biol. Bull. (Woods Hole, Mass.)* **136**, 43–53.

Hughes, D. (1972). On the endogenous control of tide-associated displacements of pink shrimp, *Penaeus duorarum* Burkenroad. *Biol. Bull. (Woods Hole, Mass.)* **142**, 271–280.

Hunt, L. (1961). "Wave Generated Oscillatory Currents Along the Bottom in the Eulittoral and Sublittoral Zones." Mine Advisory Committee, Natl. Res. Counc. Natl. Acad. Sci. Washington, D.C.

Hurley, D. (1959). Notes on the ecology and environmental adaptations of the terrestrial Amphipoda. *Pac. Sci.* **13**, 107–129.

Hurley, D. (1968). Transition from water to land in amphipod crustaceans. *Am. Zool.* **8**, 327–353.

Hyatt, G. (1974). Behavioural evidence for light intensity discrimination by the fiddler crab, *Uca pugilator* (Brachyura, Ocypodidae). *Anim. Behav.* **22**, 796–801.

Hyatt, G. (1975). Physiological and behavioral evidence for color discrimination by fiddler crabs (Brachyura, Ocypodidae, genus *Uca*). *In* "Physiological Ecology of Estuarine Organisms" (V. Vernberg, ed.), pp. 333–365. Univ. of South Carolina Press, Columbia, South Carolina.

Ireland, L., and Barlow, R. (1978). Tracking normal and blind-folded *Limulus* in the ocean by means of acoustic telemetry. *Biol. Bull. (Woods Hole, Mass.)* **155**, 445–446.

Jander, R. (1975). Ecological aspects of spatial orientation. *Annu. Rev. Ecol. Syst.* **6**, 171–188.

Kanciruk, P., and Herrnkind, W. (1976). Autumnal reproduction in *Panulirus argus* at Bimini, Bahamas. *Bull. Mar. Sci.* **26,** 417–432.

Kanciruk, P., and Herrnkind, W. (1978). Mass migration of spiny lobster, *Panulirus argus* (Crustacea: Palinuridae): Behavior and environmental correlates. *Bull. Mar. Sci.* **28,** 601–623.

Kinne, O. (1975). "Marine Ecology," Vol. II. Wiley, New York.

Klaassen, F. (1975). Ökologische und ethologische Untersuchungen zur Fortpflanzungsbiologie von *Gecarcinus lateralis* (Decapoda, Brachyura). *Forma Functio* **8,** 101–174.

Klein-Breteler, W. (1976). Migrations of the shore crab *Carcinus maenas* in the Dutch Wadden Sea. *Neth. J. Sea Res.* **10,** 338–353.

Knudsen, J. (1960). Aspects of the ecology of the California pebble crabs (Crustacea: Xanthidae). *Ecol. Monogr.* **30,** 165–185.

Korte, R. (1966). Untersuchungen zum Sehervermogen einiger Dekapoden inbesondere von *Uca tangeri. Z. Morphol. Ökol. Tiere* **58,** 1–37.

Kuenen, D., and Nooteboom, H. (1963). Olfactory orientation in some land isopods (Onisc.-Crust.). *Entomol. Exp. Appl.* **6,** 133–142.

Langdon, J. (1971). Shape discrimination and learning in the fiddler crab *Uca pugilator.* Ph.D. Diss., pp. 1–102. Florida State Univ., Tallahassee, Florida.

Laughlin, R. (1979). Trophic ecological and population distribution of the blue crab (*Callinectes sapidus* Rathbun) in the Apalachicola estuary (north Florida, U.S.A.). Ph.D. Diss., pp. 1–143. Florida State Univ., Tallahassee, Florida.

Laughlin, R., Cripe, R., and Livingston, R. (1978). Field and laboratory avoidance reactions by blue crabs *(Callinectes sapidus)* to storm water runoff. *Trans. Am. Fish. Soc.* **107,** 78–86.

Leber, K. (1982). Seasonality of macroinvertebrates on a temperate, high wave energy sandy beach. *Bull. Mar. Sci.* **32,** 86–98.

Leggett, W., and Trump, C. (1978). Energetics of migration in American shad. *In* "Animal Migration, Navigation, and Homing" (K. Schmidt-Koenig and W. Keeton, eds.), pp. 370–377. Springer-Verlag, Berlin and New York.

Lindberg, W. (1980a). Patterns of resource use within a population of xanthid crabs occupying broyozoan colonies. *Oecologia* **46,** 338–342.

Lindberg, W. (1980b). Behavior of a xanthid crab occupying bryozoan colonies, and patterns of resource used with reference to mating systems. Ph.D. Diss., Florida State Univ. Tallahassee, Florida.

Linsenmair, K. (1967). Konstrucktion und Signalfunktion der Sandpyramide der Reiterkrabbe *Ocypode saratan* Forsk (Decapoda Brachyura Ocypodidae). *Z. Tierpsychol.* **24,** 403–456.

Linsenmair, K. (1968). Anemomenotakische Orientierung bei Skorpionen (Chelicerata, Scorpiones). *Z. Vgl. Physiol.* **60,** 445–449.

Linsenmair, K. (1972). Die Bedeutung familienspezifischer "Abseichen" für den Familienzusammenhalt bei der sozialen Wüstenassel *Hemilepistus reaumuri* Audouin. u. Savigny (Crustacea: Isopoda, Oniscoidea). *Z. Tierpsychol.* **31,** 134–155.

Lund, W., Stewart, L., and Weiss, H. (1971). Investigation on the lobster. Final Report for Commercial Fish. Res. Development Act. 3-44-12.

McLean, R. (1974). Direct shell acquisition by hermit crabs from gastropods. *Experientia* **30,** 206–208.

Magnus, D. (1960). Zur Ökologie des Landeinsiedlers *Coenobita joussaumei* Bouvier unde der Krabbe *Ocypode aegyptica* Gerstaecker am Roten Mer. *Verh. Zool. Ges. Bonn,* 316–329.

Marchionni, V. (1962). Modificazione sperimentale della direzione innata di fuga in *Tal-*

orchestia deshayesei Aud. (Crustacea Amphipoda). *Boll. Inst. Mus. Zool. Univ. Torino* **6,** 29–39.

Marler, P., and Hamilton, W. (1966). "Mechanisms of Animal Behavior." Wiley, New York.

Menzel, R., and Nichy, F. (1958). Studies of the distribution and feeding habits of some oyster predators in Alligator Harbor, Florida. *Bull. Mar. Sci. Gulf Caribb.* **8,** 125–145.

Merkle, L. (1969). Home range of crayfish *Orconectes juvenalis. Am. Midl. Nat.* **81,** 228–235.

Momot, W. (1966). Upstream movement of crayfish in an intermittent Oklahoma stream. *Am. Midl. Nat.* **75,** 150–159.

Momot, W., and Gowing, H. (1972). Differential seasonal migration of the crayfish *Orconectes virilis* (Hagen), in marl lakes. *Ecology* **53,** 479–483.

Morgan, E. (1967). The pressure sense of the swimming crab *Macropipus holsatus* (Fabricius), and its possible role in the migration of the species. *Crustaceana* **13,** 275–280.

Mrosovsky, N. (1978). Orientation mechanisms of marine turtles. *In* "Animal Migration, Navigation, and Homing" (K. Schmidt-Koenig and W. Keeton, eds.), pp. 413–419. Springer-Verlag, Berlin and New York.

Newell, R. (1970). "Biology of Intertidal Animals." Paul Elek, London.

Nishimoto, R. (1980). Orientation, movement patterns and behavior of *Callinectes sapidus* Rathbun (Crustacea: Portunidae) in the intertidal. Ph.D. Diss. Florida State Univ., Tallahassee, Florida.

Nishimoto, R., and Herrnkind, W. (1978). Directional orientation in blue crabs, *Callinectes sapidus* Rathbun: Escape responses and influence of wave direction. *J. Exp. Mar. Biol. Ecol.* **33,** 93–112.

Nishimoto, R., and Herrnkind, W. (1982) Orientation of the blue crab, *Callinectes sapidus* Rathbun: Role of celestial cues. *Mar. Behav. Physiol.* **9,** 1–11.

Norse, E. (1978). An experimental gradient analysis: hyposalinity as an "upstress" distributional determinant for Caribbean portunid crabs. *Biol. Bull. (Woods Hole, Mass.)* **155,** 586–598.

Oesterling, M. (1976). Reproduction, growth, and migration of blue crabs along Florida Gulf coast. *Fla. Sea Grant Publ. SUSF-SG-76-003, Fla. Coop. Ext. Service,* pp. 1–19.

Palmer, J. (1964). A persistent light-preference rhythm in the fiddler crab, *Uca pugnax* and its possible adaptive significance. *Am. Nat.* **98,** 431–434.

Papi, F. (1960). Orientation by night: The moon. *Cold Spring Harbor Symp. Quant. Biol.* **25,** 475–480.

Papi, F., and Pardi, L. (1953). Richerche sull'orientamento di *Talitrus saltator* (Montagu) (Crustacea Amphipoda). *Z. Vgl. Physiol.* **35,** 490–518.

Papi, F., and Pardi, L. (1959). Nuovi reperti sull'orientamento lunare di *Talitrus saltator* Montagu (Crustacea Amphipoda). *Z. Vgl. Physiol.* **41,** 583–596.

Papi, F., and Pardi, L. (1963). On the lunar orientation of sandhoppers (amphipoda Talitridae). *Biol. Bull. (Woods Hole, Mass.)* **124,** 97–105.

Pardi, L. (1954). Uber die Orientierung von *Tylos laterillii* (Isopoda terrestria). *Z. Tierpsychol.* **11,** 175–181.

Pardi, L. (1960). Innate components in the solar orientation of littoral amphipods. *Cold Spring Harbor Symp. Quant. Biol.* **25,** 395–401.

Pardi, L., and Grassi, M. (1955). Experimental modification of direction finding in *Talitrus saltator* (Montagu) and *Talorchestia deshayesei* (Aud.) (Crustacea-Amphipodia). *Experientia* **11,** 202–205.

Pardi, L., and Papi, F. (1952). Die Sonne als Kompass bei *Talitrus saltator* (Montagu) (Amphipoda, Talitridae). *Naturwissenschaften* **34,** 262–263.

Pardi, L., and Papi, F. (1953). Richerche sull'orientamento di *Talitrus saltator* Montagu (Crustacea, Amphipoda). I. L'orientamento durante il giorno in una popolazione del litorale Terrenico. *Z. Vgl. Physiol.* **35,** 459–489.

Pardi, L., and Papi, F. (1961). Kinetic and tactic responses. *In* "The Physiology of Crustacea" (T. Waterman, ed.), pp. 365–399. Academic Press, New York.

Paris, D. (1963). The ecology of *Armadillidium vulgare* (Isopoda: Oniscoidea) in California grassland: Food, enemies and weather. *Ecol. Monogr.* **33,** 1–22.

Pearse, A., Humm, H., and Wharton, G. (1942). Ecology of sand beaches of Beaufort, N.C. *Ecol. Monogr.* **12,** 135–190.

Pearson, W., and Olla, B. (1977). Chemoreception in the blue crab, *Callinectes sapidus*. *Biol. Bull. (Woods Hole, Mass.)* **153,** 346–354.

Peebles, J. (1979). Molting, movement and dispersion in the freshwater prawn *Macrobrachium rosenbergii*. *J. Fish. Res. Board Can.* **36,** 1080–1088.

Perttunen, V. (1963). Effect of desiccation on the light reaction of some terrestrial arthropods. *In* "Animal Orientation" (H. Autrum, ed.), pp. 90–97. Springer-Verlag, Berlin and New York.

Pyke, G., Pulliam, H., and Charnov, E. (1977). Optimal foraging: A selective review of theory and tests. *Q. Rev. Biol.* **52,** 137–154.

Rebach, S. (1978). The role of celestial cues in short range migrations of the hermit crab, *Pagurus longicarpus*. *Anim. Behav.* **26,** 835–842.

Rebach, S. (1981). Use of multiple cues in short-range migrations of Crustacea. *Am. Midl. Nat.* **105,** 168–180.

Reeder, P., and Ache, B. (1980). Chemotaxis in the Florida spiny lobster, *Panulirus argus*. *Anim. Behav.* **28,** 831–839.

Rittschof, D. (1980). Chemical attraction of hermit crabs and other attendants to simulated gastropod predation sites. *J. Chem. Ecol.* **6,** 103–118.

Roe, J. (1980). Zonation and movements of the ghost crab, *Ocypode quadrata*. M.S. Thesis, pp. 1–84. University of Florida, Gainesville, Florida.

Rudloe, A., and Herrnkind, W. (1976). Orientation of *Limulus polyphemus* in the vicinity of breeding beaches. *Mar. Behav. Physiol.* **4,** 75–89.

Rudloe, A., and Herrnkind, W. (1980). Orientation by horseshoe crabs, *Limulus polyphemus*, in a wave tank. *Mar. Behav. Physiol.* **7,** 199–211.

Ryan, E. (1966). Pheromone: Evidence in a marine crustacean. *Science* **151,** 340–341.

Salmon, M., and Atsaides, S. (1968). Visual and acoustic signalling during courtship by fiddler crabs (Genus *Uca*). *Am. Zool.* **8,** 623–639.

Salmon, M., and Horch, K. (1972). Acoustic signalling and detection by semiterrestrial crabs of the family Ocypodidae. *In* "Behavior of Marine Animals" (H. Winn and B. Olla, eds.), Vol. 1, pp. 60–96. Plenum, New York.

Scapini, F. (1979). Orientation of *Talitrus saltator* Montagu (Crustacea Amphipoda) in fresh, sea and diluted sea water. *Monit. Zool. Ital. NS* **13,** 71–76.

Scapini, F., and Ercolini, A. (1973). Research on the non-visual orientation of littoral amphipods: Experiments with young born in captivity and adults from a Somalian population of *Talorchestia martensii* Weber (Crustacea Amphipoda). *Monit. Zool. Ital. NS Suppl.* **5,** 23–30.

Scapini, F., and Pardi, L. (1979). Nuovi dati sulla tendenza direzionale innata nell'orientamento solare degli Anfipodi litorali. *Lincei* **46,** 592–597.

Schmidt-Koenig, K., and Keeton, W. (1978). "Animal Migration, Navigation, and Homing." Springer-Verlag, Berlin and New York.

Schoener, T. (1971). Theory of feeding strategies. *Annu. Rev. Ecol. Syst.* **2,** 269–404.

Schöne, H. (1963). Menotaktische Orientierung nach polarisiertem und unpolarisiertem Licht bei der Mangrovekrabbe *Goniopsis*. *Z. Vgl. Physiol.* **46,** 496–514.

Schöne, H. (1965). Release and orientation of behaviour and the role of learning as demonstrated in crustacea. *Anim. Behav. Suppl.* **1,** 135–144.

Schöne, H. (1975). Orientation in space: Animals. General Introduction. *In* "Marine Ecology" (O. Kinne, ed.), Vol. II, pp. 499–553. Wiley, New York.

Shepard, F. (1963). "Submarine Geology." Harper and Row, New York.

Sinclair, M. (1977). Agonistic behaviour of the stone crab, *Menippe mercenaria* (Say). *Anim. Behav.* **25,** 193–207.

Siniff, D., and Jessen, C. (1969). A simulation model of animal movement patterns. *Adv. Ecol. Res.* **6,** 185–219.

Snyder, N., and Snyder, H. (1970). Alarm response of *Diadema antillarum. Science* **168,** 276–278.

Southwood, T. (1962). Migration of terrestrial arthropods in relation to habitat. *Biol. Rev.* **37,** 171–214.

Stachowitsch, M. (1979). Movement, activity pattern, and role of hermit crab population in a sublittoral epifaunal community. *J. Exp. Mar. Biol. Ecol.* **39,** 135–150.

Street, R. (1971). Rock lobster migration off Otago. *N. Z. Comm. Fish. June 1971,* pp. 16–17.

Sulkin, S. (1975). Factors influencing blue crab population size: nutrition of larvae and migration of juveniles. Center for Environmental and Estuarine Studies: Chesapeake Biological Laboratory Annual Report, pp. 32–56.

Tesch, F. (1975). Orientation in space: Fishes. *In* "Marine Ecology" (O. Kinne, ed.), Vol. II, pp. 657–707. Wiley, New York.

Tinbergen, N., Impekoven, M., and Franck, D. (1967). An experiment on spacing out as a defence against predation. *Behaviour* **28,** 307–321.

van den Bercken, J. Broekheuzen, S., Ringelberg, J., and Velthuis, H. (1967). Non-visual orientation in *Talitrus saltator. Experientia* **23,** 44–45.

van Tets, G. (1956). A study of solar and spatial orientation of *Hemigrapsus oregonensis* (Dana) and *Hemigrapsus nudus* (Dana). B.A. Thesis, Univ. of British Columbia, Vancouver.

Vannini, M. (1975a). Researches on the coast of Somalia. The shore and the dune of Sar Uanle. 4. Orientation and anemotaxis in the land hermit crab, *Coenobita rugosus* Milne Edwards. *Monit. Zool. Ital. NS Suppl.* **6,** 57–90.

Vannini, M. (1975b). Researches on the coast of Somalia. The shore and the dune of Sar Uanle. 5. Description and rhythmicity of digging behaviour in *Coenobita rugosus* Milne Edwards. *Monit. Zool. Ital. NS Suppl.* **6,** 233–242.

Vannini, M. (1976a). Researches on the coast of Somalia. The shore and the dune of Sar Uanle. 7. Field observations on the periodical transdunal migrations of the hermit crab *Coenobita rugosus* Milne Edwards. *Monit. Zool. Ital. NS Suppl.* **7,** 145–185.

Vannini, M. (1976b). Researches on the coast of Somalia. The shore and the dune of Sar Uanle. 10. Sandy beach decapods. *Monit. Zool. Ital. NS Suppl.* **8,** 255–286.

Vannini, M., and Chelazzi, G. (1981). Orientation of *Coenobita rugosus* (Crustacea: Anomura): A field study in Aldabra. *Mar. Biol.* **64,** 135–140.

Venema, S., and Creutzberg, F. (1973). Seasonal migration of the swimming crab, *Macropipus holsatus,* in an estuarine area controlled by tidal streams. *Neth. J. Sea Res.* **7,** 94–102.

Verheijen, F. (1978). Orientation based on directivity, a directional parameter of the animals radiant environment. *In* "Animal Migration Navigation, and Homing" (K. Schmidt-Koenig and W. T. Keeton, eds.), pp. 447–458. Springer-Verlag, Berlin and New York.

Verwey, J. (1958). Orientation in migrating marine animals and a comparison with that of other migrants. *Arch. Neerl. Zool.* **13** (Suppl.), 418–445.

Virnstein, R. (1977). The importance of predation by crabs and fishes on benthic infauna in Chesapeake Bay. *Ecology* **58,** 1199–1217.

von Frisch, K. (1967). "The Dance Language and Orientation of Bees." Harvard Univ. Press, Cambridge, Massachusetts.

von Hagen, H. (1967). Nachweis einer kinasthetischen Orientierung bei *Uca rapax. Z. Morphol. Ökol. Tiere* **58**, 301–320.

Walton, A., and Herrnkind, W. (1977). Hydrodynamic orientation of spiny lobster, *Panulirus argus* (Crustacea: Palinuridae): Wave surge and unidirectional currents. *Memorial Univ. of Newfoundland Mar. Sci. Res. Lab. Tech. Rept. No. 20,* pp. 184–211.

Warburg, M. (1964). The response of isopods towards temperature, humidity and light. *Anim. Behav.* **12**, 175–186.

Warburg, M. (1965a). The microclimate in the habitats of two isopod species in southern Arizona. *Am. Midl. Nat.* **73**, 363–375.

Warburg, M. (1965b). Water relations and internal body temperature of isopods from mesic and xeric habitats. *Physiol. Zool.* **37**, 99–109.

Warburg, M. (1968). Behavioral adaptations of terrestrial isopods. *Am. Zool.* **8**, 545–559.

Waterman, T. (1966). Polarotaxis and primary photoreceptor events in Crustacea. *In* "The Functional Organization of the Compound Eye" (C. Bernhard, ed.), pp. 493–511. Pergamon, Oxford.

Waterman, T. (1972). Visual direction finding by fishes. *In* "Animal Orientation and Navigation" (S. Galler, K. Schmidt-Koenig, G. Jacobs, and R. Belleville, eds.), pp. 437–456. NASA, Washington, D.C.

Waterman, T. (1975). The optics of polarization sensitivity. *In* "Photoreceptor Optics" (A. Snyder and R. Menzel, eds.), pp. 339–371. Springer-Verlag, Berlin.

Waterman, T., and Horch, K. (1966). Mechanism of polarized light perception. *Science* **154**, 467–475.

Williams, A. (1974). The swimming crabs of the genus *Callinectes* (Decapoda: Portunidae). *Fish. Bull.* **72**, 685–798.

Williamson, D. (1951). Studies in the biology of Talitridae (Crustacea, Amphipoda): Visual orientation in *Talitrus saltator. J. Mar. Biol. Assoc. U.K.* **30**, 91–99.

Williamson, D. (1954). Land and seaward movements of the sandhopper. *Adv. Sci.* **11**, 71–73.

Wolcott, T. (1978). Ecological role of ghost crabs, *Ocypode quadrata* (Fabricius) on an ocean beach: Scavengers or predators? *J. Exp. Mar. Biol. Ecol.* **31**, 67–82.

Young, D., and Ambrose, H. (1978). Underwater orientation in the sand fiddler crab, *Uca pugilator. Biol. Bull (Woods Hole, Mass.)* **155**, 246–258.

3

Biological Timing

PATRICIA J. DeCOURSEY

I. INTRODUCTION

The importance of rhythmic time structuring in organisms from protists to man has been demonstrated repeatedly at all levels of functional organization, as an adaptation to cyclic parameters of an animal's environment. Detailed studies over the past 25 years have demonstrated that most rhythmic functions in animals are not mere passive reflections of environmental cycles such as the day–night cycle, but depend instead on an underlying

THE BIOLOGY OF CRUSTACEA, VOL. 7

physiological pacemaker, kept in precise phase with the prevailing environment by slight corrections of specific environmental agents (reviews in Menaker, 1971; Enright, 1975; Aschoff, 1981). Biological rhythms are composites, therefore, consisting of an endogenous physiological pacemaker which generates the basic frequency, coupled to an environmental cycle which adjusts the phase of the internal clock to local time. In spite of the importance of biological timing for animals, major gaps occur in knowledge of its mechanism and ecological significance in crustaceans.

The Crustacea hold a place in the Earth's biota as one of the largest groups in numbers of species and individuals, as well as one of the most diverse in behavior, adapted superbly for an array of highly specialized habitats. Along creeks of the vast intertidal marshes of temperate estuaries, adult fiddler crabs of the genus *Uca* may form a living carpet, reaching numbers of thousands per hectare, while *Uca* larvae in the water column may reach peak densities over 100,000/m^3 (DeCoursey, 1979). In coastal and open ocean waters, copepods are present in immense numbers, and myriads of krill swarm in arctic waters. Crabs are found even in the hot water vent community of the Galápagos abyss (Corliss et al., 1979; Enright et al., 1981). Thus, from arctic to tropical waters, in ocean depths or in estuarine habitats, on the supralittoral strand, in freshwater streams, and even in terrestrial habitats, crustaceans abound. An awareness of the role of biological clock adaptations for survival in these varied habitats is rapidly growing, but no comprehensive review of the physiology and ecology of rhythms for crustaceans exists at present.

This chapter will summarize selected aspects of current knowledge about biological timing in Crustacea, concentrating on the theme of clocks as a means of matching animal functions with the time structure of their environment. A single chapter cannot cover all aspects of crustacean rhythmicity. A mere catalog of reported rhythms would not serve a useful purpose. The theme of crustaceans as models for understanding biological clocks is equally unfeasible, since a majority of crustacean rhythm studies have been field studies lacking any concrete information either on the underlying mechanism or about the presumed adaptive and survival value of rhythmicity. The few comprehensive crustacean clock studies have used a particularly favorable organism to study a specific feature, such as a tidal phase response curve in *Excirolana chiltoni*[1] (Enright, 1976b), and the adaptive importance of the rhythm has usually been assumed or ignored.

Faced with these limitations, the most useful approach will be to look at

[1]A recent Biosis computer search located over 55,000 rhythm papers published in the period 1970–1982. Repeatedly cited were a small number of species which were useful research models due to their availability and to the high precision of their behavioral rhythms. For ease in reading, therefore, the first mention of a species in the text will include the complete scientific name; subsequently, where no ambiguity exists, only the generic name will be used.

crustacean clocks to understand the forms and functions of biological rhythms in a very large and diverse taxonomic group: form in terms of the periodicities expressed, and function in terms of the correspondence of biological rhythm with cyclic environmental constraints. In this review, therefore, the matching of animal rhythms with physical habitat cycles will first be illustrated briefly, followed by a resumé of basic concepts of mechanisms in biological timing. A survey of crustacean rhythms will then be presented, emphasizing endogenous pacemakers and environmental entraining agents for each major type of rhythmic behavior. Speculations on the ecological importance of temporal adaptation in crustaceans will be followed by a look forward to probable future directions of research in this field.

II. MATCHING OF BIOLOGICAL RHYTHMS AND ENVIRONMENTAL CYCLES

The dominant cyclic parameters of the earth arise from rotation of the celestial bodies of the solar system. The rotation of the earth on its axis relative to the sun imposes profound daily light intensity changes in terrestrial or freshwater habitats, and on all photic zones of the oceans. Over 70% of the earth's surface is covered by oceans, whose fluid masses are not rigid, but greatly influenced by the tug of the sun's and moon's gravitational fields. One result of the moon's pull is tidal water movement, involving the ebb and flood of tides in estuarine and littoral environments, usually at 12.4-hr intervals. The alignment of sun and moon every 14.9 days is the basis of the semilunar and lunar cycles of coastal waters. Alternating with the extreme spring tides at the new and full moon, are the neap tides in which vertical change in tidal elevation is minimal. The actual magnitude of water movement and its temporal pattern varies systematically around the world, providing three basic patterns in variation of the form of the tide. Semidiurnal equal tides, in which the two high tides per day are nearly equal, are typical along the Atlantic coast of the United States; semidiurnal unequal tides, with great differences in the height and spacing of the two daily tides occur in such areas as the Pacific Coast of the United States. Diurnal tides with one tidal excursion per day are typified by those along the Gulf Coast of the United States. For an excellent discussion of the patterns of submersion of the intertidal zone that result from the various differences in spatial and timing relationships of the tides see Barnwell (1976). Finally, the annual cycle of temperature and daylength exerts an influence on crustaceans in many ways. Other physical periodicities are known but tidal, daily, semilunar, lunar, and annual cycles of the environment have had a major impact on animal periodicity, shaping rhythmicity of species which have survived strong temporal pressures.

These periodicities are mirrored directly by living animals in their natural habitats. Five environmentally related behavioral periodicities may be found, with elements of time of day, tide height, spring–neap phase, lunar phase, and time of year: tidal, diurnal, semilunar, lunar, and annual components (Table I). A few selected examples of these five chief frequencies of rhythms in the animal kingdom will illustrate the rhythmic complexity of animals, and will provide a backdrop for comparing crustacean timing with other major phylogenetic groups. In this initial overview, examples have been chosen to illustrate precisely timed behavior throughout the animal kingdom, not necessarily because the underlying biological timing system has been studied in detail. For this reason, the label *rhythm* has been employed rather than biological timer. Tidal, semilunar, and lunar rhythms frequently occur as multiple components in a single organism and will therefore be considered together for each phylum in order of increasing phylogenetic complexity. The planktonic foraminiferan protozoan, *Hastigerina pelagica,* spawns on a lunar schedule, with gametogenesis taking place 3–7 days before full moon (Spindler et al., 1979). In platyhelminths, the flatworm *Convoluta roscoffensis* is noted for its movements to the surface at low tide (Palmer, 1974). Many annelids are famous for their spectacularly rhythmic breeding habits. The Bermuda fireworm swarms for breeding on the dark nights of the new moon (Huntsman, 1948); *Platynereis dumerilii* breeds under similar conditions (Hauenschild, 1960); but best known is the Pacific Palolo worm, which spawns at dawn 1 week after the full moon (Korringa, 1957). Among mollusks, species of Hawaiian *Littorina* spawn daily at the time of highest tide (Struhsaker, 1966), while in the snail *Melampus bidentatus* egg laying, hatching, and settlement occur with semilunar synchrony (Russell-Hunter et al., 1972). Among crustaceans, tidal locomotor rhythms are well-developed in isopods (Enright, 1965a, 1972), in amphipods (Enright, 1963), and in decapods (extensive coverage in DeCoursey, 1976a). A semilunar rhythm of locomotor activity and molt synchrony is also found in the amphipod *Talitrus saltator* (Williams, 1979).

TABLE I

Endogenous Pacemakers in Relation to the Environment

Type animal rhythm	Type of clock	Period of expressed rhythm (τ)	Environmental period (T)
Tidal	Circatidal	≈12.4 hr	12.4 hr
Daily	Circadian	≈24 hr	24 hr
Semilunar	Circasemilunar	≈15 days	15 days
Lunar	Circalunar	≈29 days	29 days
Annual	Circannual	≈365 days	365 days

Even more clearly developed are the semilunar reproductive rhythms of courtship and spawning in crustaceans such as *Uca* spp. (Christy, 1978; DeCoursey, 1979), *Sesarma* spp. (Saigusa and Hidaka, 1978), or *Excirolana* (Klapow, 1976). In the intertidal insect *Clunio marinus,* semilunar reproduction is also well-developed (Neumann, 1976a,b). Among vertebrates, tidal locomotor rhythms have been described for fish (Gibson, 1970). The spawning of grunion is another well-known example of lunar-related rhythmicity; females deposit their eggs in moist sand just after the turning of the highest spring tides, to await hatching on the next spring high tide 2 weeks later (Thompson and Muench, 1976). Similarly, semilunar egg laying and hatching occurs in the Atlantic silversides, *Menidia menidia* (Middaugh, 1981), and the whitebait *Galaxius attenuatus* (Benzie, 1968).

Diurnal rhythms are widespread in every terrestrial and freshwater group studied, as well as in some marine organisms, at all taxonomic levels (major reviews in Chauvnick, 1960; Aschoff, 1965a, 1981; Menaker, 1971). While repeated reference will be made in later sections of this paper to the large body of information on terrestrial vertebrates, the emphasis in this volume must necessarily be on crustaceans. Daily rhythms occur among crustaceans in several situations. Freshwater crayfish have daily locomotor rhythms (Page and Larimer, 1972), and semiterrestrial crabs such as *Ocypode quadrata,* living well above the high tide benchmark, are strictly nocturnal in locomotor activity (Palmer, 1971). Daily rhythms are evident in vertical migration of copepods in estuaries (Bosch and Taylor, 1973) and open ocean waters (Enright and Hamner, 1967; Kampa, 1976). Egg laying by deep-water species such as the lobster *Nephrops norvegicus* (Moller and Branford, 1979) or *Homarus gammarus* (Branford, 1978) is diurnal, and orienting beach amphipods (Hartwick, 1976) use a diurnal clock in connection with a celestial compass.

Annual rhythms have been extensively documented in mammals where yearly hibernation, migration, or breeding cycles are pronounced (Pengelley, 1974); further cases of annual breeding patterns can be found in practically every group, including Crustacea.

These examples illustrate the widespread nature of the five types of environmentally related rhythms for many varied functions in crustaceans, as well as in the remainder of the animal kingdom, stressing the repetition of behavioral events of free-living animals on a regular schedule without any statement about the role of biological clocks and entraining agents in such phenomena. The next section will summarize current ideas about the clock mechanism, using the clearest examples, regardless of the taxon involved. The importance of this section is to establish the endogeneity of timers for "biological clocks" as a reasonable basis for subsequent discussions of temporal adaptiveness.

III. REGULATION OF BIOLOGICAL RHYTHMS: BASIC CONCEPTS

A. The "Clock Question": An Overview

Rhythmic behavior has been observed and documented for centuries, but a real knowledge of control mechanisms and adaptiveness has only been discerned in the past three decades. Four key workers in the early 1950s gave impetus to the new field of "biological clocks." Gustav Kramer (1951), in studying the orientation of starlings, coined the term "biologische Uhr" and thereby sharpened awareness of clocks as precise mechanisms for biological time measurement; for a bird to use a moving sun as a compass, it must have an internal sense of time. Elsewhere in Germany, Jürgen Aschoff (1951) isolated mice in constant environmental conditions to show that rhythmic activity persisted in a nonfluctuating environment and must have an endogenous, physiological source. Colin Pittendrigh (1954, 1960) demonstrated that a free-running rhythm of fruitfly eclosion from pupal cases persisted in constant temperature conditions with near-temperature independence. In his matchless prose, he presented a compelling picture of temporal organization: a time framework for the factory that constitutes a living organism. Also working in Germany was Erwin Bünning (1936, 1959), whose studies of plant rhythms were of great importance in understanding photoperiodic time measurement.

The contribution of these four scientists was to point out that two main elements operate in practically all rhythmic biological systems. One endogenous, physiological element acts very much like an alarm clock which runs a little too fast or too slow each day. The second component is an external synchronizer or entraining agent, acting to correct the "free-running" endogenous element each day and keep the clock in phase with prevailing environmental conditions. Coordinating these two elements is the hormonal and nervous system machinery to detect the environmental signals, transmit them to the pacemaker, and eventually carry a message from pacemaker to an output system. The relationship of the exogenous and endogenous elements is expressed in elegant simplicity by a brief equation (Pittendrigh, 1973); for stable synchronization or entrainment to occur, both phase and frequency of the biological timer must occur in such a way that τ, the period of the free-running clock under constant conditions, must be capable of a correction, $\Delta\phi$, in order to match the period of the driving environmental oscillation, T (Table I). In an entrained organism, $\tau + \Delta\phi = T$.

In direct counterpoint to these endogenous clock proponents stood a group supporting exogenous control of rhythmicity. Frank Brown, Jr., and his students, working at Northwestern University, were first attracted to the color change rhythm based on chromatophore contraction and expansion in the integument of the fiddler crab, Uca pugnax. Using a semiquantitative color scale, they examined crabs in a dim light at hourly intervals and rated them in color intensity. The data were first analyzed by a five-point moving average method; then selected frequencies were arbitrarily processed to detect energy of rhythmicity (Brown et al., 1953). The resultant "exact 24-hour" or "exact 12.4-hour rhythms" were cited as conclusive proof of a direct response of the chromatophores to some unknown exogenous environmental frequency which had not been excluded from the so-called constant environment. Numerous authors have developed more accurate techniques of recording (DeCoursey, 1961) and of analysis (Enright, 1965b,c; Mercer, 1965; Rawson and DeCoursey, 1976) that have demolished the credibility of the theory of exclusively exogenous control. A schematic diagram of frequency analysis suggests the dangers of selecting particular components out of the context of a frequency spectrum of noisy data (Fig. 1A); in contrast is a continuous frequency spectrum by periodogram analysis for sharply rhythmic data (Fig. 1B). Both sides of the controversy have marshalled new evidence and arguments (Brown, 1972; Pittendrigh and Daan, 1976a; Brady, 1979). In the long run, the issue has served in a beneficial way to stimulate expansion and development of the field of biological timing.

The study of biological clocks has grown from the early 1950s to include many major laboratories around the world. A vast literature on diurnal

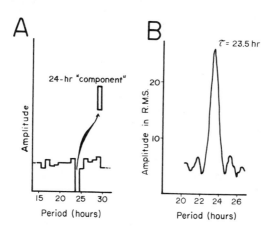

Fig. 1. Methods of frequency analysis in times series data. (A) Dangers inherent in selection of isolated "components" in a schematic periodogram of an arrhythmic time series. Arrow depicts lifting out of the 24-hr component from background noise to measure its amplitude (Enright, 1965b). (B) Continuously plotted periodogram of the free-running rhythm of *Peromyscus maniculatus austerus* in constant dark conditions, showing true peak with period of 23.5 hr (after Rawson and DeCoursey, 1976).

rhythms and their circadian basis has arisen, punctuated with many symposium volumes (Withrow, 1959; Chauvnick, 1960; Aschoff, 1965a; Menaker, 1971; Bierhuizen, 1972; Thorpe, 1978), many monographs or extensive reviews (Bünning, 1959; Harker, 1964; Pavlidis, 1973; Bennett, 1974; Rusak and Zucker, 1975; Palmer, 1976; Saunders, 1977; Menaker et al., 1978; Brady, 1979), and several thousand research publications. Some of the work is complex and highly technical; a specialized terminology has arisen, and controversies abound that confuse the nonspecialist.

Information on biological timing in crustaceans is more limited. The reasons are evident: small size of many crustaceans, relatively inaccessible habitats, extreme complexity of the cyclic environmental input, resultant noisy biological output, and the fragility of most marine crustaceans under long-term laboratory conditions. Several reviews, monographs, and symposia include certain aspects of crustacean rhythmicity (Korringa, 1957; Enright, 1970, 1975, 1977a,b; Palmer, 1973, 1974; DeCoursey, 1976a; McDowall, 1976; Aréchiga, 1977; Naylor and Hartnoll, 1979). The results of all these studies have confirmed the thesis of biological pacemakers operating in rhythmic functions of animals.

B. Problems in Demonstrating Endogenous Timers

Until recently it has been difficult to directly demonstrate a living clock; physiological examples are presented and discussed in Section III,G. By analogy with a wrist watch, the hands or rhythmic output may be readily observed, but rough attempts to open the case and look inside may break off the hands and damage the pacemaker. As a result, biological timing has been studied primarily through indirect means, using rhythmic output to study formal clock properties. Rate of oxygen consumption, color change, susceptibility to toxins or to X-ray damage, and blood levels of various chemical constituents are some of the biochemical and physiological indicators of a biological clock. The majority of indicators, however, have been behavioral ones, since these can be most easily examined over extended periods of time either by direct observation or by telemetry with minimal disturbance to the animal. Locomotor activity and color change, as well as egg laying or hatching, have been favorite functions for studying biological clocks in crustaceans.

C. Specific Criteria for Demonstrating Endogenous Timers Indirectly

Several indirect methods have been used to distinguish purely exogenous rhythms from endogenous clocks. The most commonly used standard is a free-running rhythm of an indicator process under constant environmental conditions differing in frequency from any known cyclic environmental parameter and dependent in most species on light intensity (Aschoff, 1960; Pittendrigh, 1960). The usefulness of certain processes such as locomotor activity in rodents (DeCoursey, 1961; Suter and Rawson, 1968) or calling in crickets (Loher, 1972) lies in the clarity and lack of noise in these systems; consequently, frequency is readily apparent by simple inspection (Fig. 2A). A drawback in applying this criterion to crustacean rhythmicity has been the noisy nature of locomotor activity in most aquatic animals, compounded by the short-term duration of most published crustacean locomotor activity records and the common practice of pooling data for large groups of individuals. Clear examples, however, include the swimming rhythm in the isopod *Excirolana* (Enright, 1972), the swimming rhythm in the amphipod *Talitrus* (Williams, 1979) or *Talorchestia quoyana* (Benson and Lewis, 1976), locomotor activity rhythm in the purple crab *Gecarcinus lateralis* (Fig. 2E; Palmer, 1971), and locomotor activity and ERG amplitude rhythms of the crayfish *Procambarus clarkii* (Page and Larimer, 1972, 1975a,b). For noisy systems, computer techniques such as periodogram analysis (Fig. 1B) or power spectral analysis have been developed to help detect the exact period of the rhythm (Enright, 1965a; Suter and Rawson, 1968; Binkley, 1976; Lehmann, 1976; Rawson and DeCoursey, 1976).

A second criterion for endogeneity is an innate basis. Experiments with chicks (Aschoff and Meyer-Lohmann, 1954) and with lizards (Hoffmann, 1955) removed parental and sibling learning influences; the resultant free-running rhythms of newly hatched organisms, even after several generations of rearing in isolation, verified the concept of an internal timer. Unfortunately, no comparable data have been published for crustaceans.

A third approach to the demonstration of an internal timer has been phase shifting of an environmental cycle relative to the pacemaker, either by a transoceanic flight or by changing the phase of a light schedule. The clearest examples are for mammals (DeCoursey, 1961), but a few records are available for decapod crustaceans. Experiments have used shifts of a light cycle (Palmer, 1971; Page and Larimer, 1972), of tidal flooding cycles (Honegger, 1973a,b, 1976; Moller and Jones, 1975; Lehmann, 1976; Hammond and Naylor, 1977), of tidal immersion and temperature changes (Williams and Naylor, 1969), of tidal simulated turbulance (Enright, 1965a; Jones and Naylor, 1970; Klapow, 1972) or of cycles of salinity change (Taylor and Naylor, 1977). The gradual entrainment to a cyclic schedule out of phase with the animal's activity period is strongly suggestive of internal timing, since the animal does not immediately respond to the exogenous clues, but appears to reset its timer in stepwise increments.

The endogenous nature of clocks can also be demonstrated in a fourth method by celestial navigation of animals. If a bird is able to fly north in the spring using a moving compass such as the sun (Kramer, 1951; Hoffmann,

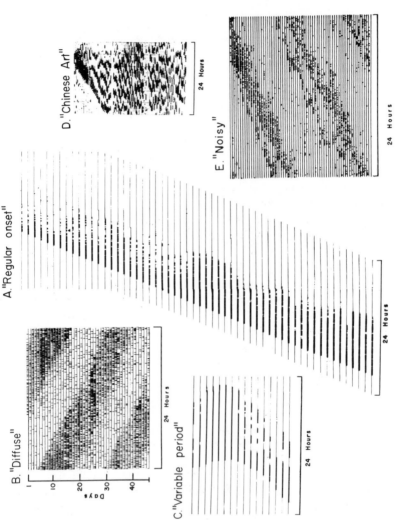

Fig. 2. "Portrait Gallery" of free-running rhythms of locomotor activity showing the range of patterns in constant dark conditions. (A) Regular onset of locomotor activity in a running wheel in the flying squirrel, *Glaucomys volans* (P. J. DeCoursey, unpublished). (B) Diffuse activity pattern in the house sparrow, *Passer domesticus* (J. S. Takahasi, unpublished). (C) Variable period in the hazel mouse, *Muscardinus avellanarius* (P. J. DeCoursey, unpublished). (D) Polyphasic in the red-backed mouse *Clethrionomys rutilus* (from Pittendrigh, 1960). (E) Noisy onset in the purple land crab, *Gecarcinus lateralis*, with days 1–15 at 30°C, days 16–26 at 25°C and days 27–65 at 20°C (from Palmer, 1971).

1965b) or an amphipod to use celestial orientation and navigation (Pardi, 1960), then it must be able to calculate the rate of angular movement of the sun with an internal clock. For details of time-compensated navigation in crustaceans, see Chapter 2.

Fifth, a compelling line of evidence concerns temperature compensation of rhythmic output. Characteristically, the Q_{10} for behavioral rhythms lies close to but not exactly 1.0 for poikilotherms (Pittendrigh, 1960, 1973), for heterothermic mammals, or for mammals with body temperatures lowered under Nembutal anesthesia (Rawson, 1960). The Q_{10} for circadian locomotor activity of two semiterrestrial crabs at three different temperatures ranged from 0.996 to 1.08 (Palmer, 1971). Temperature-independent functions controlled by exogenous environmental factors yield a Q_{10} of 1.0, while the temperature-dependent metabolic processes show a Q_{10} of 2–3. Clearly, the Q_{10}'s of biological clocks differ from 1.0, reflecting temperature compensation of a physiological clock in order to buffer it from thermal disruptions (Pittendrigh, 1973).

Finally, a sixth type of evidence for an endogenous basis of biological rhythmicity has been the dose-dependent, reversible slowing of locomotor cycles in *Excirolana* by D_2O (Enright, 1971a), suggesting a physiological, metabolic pacemaker rather than a direct exogenous response to an environmental cycle. While a number of chemicals have been tried, only D_2O has given consistent, uniform results in a variety of species, ranging from algae through crustaceans to rodents (Suter and Rawson, 1968; Enright, 1971a; Daan and Pittendrigh, 1976a).

The demonstration of endogenous timing has been carried out primarily in terrestrial vertebrates with LD-entrained circadian timers, and throughout this group many similarities in clock features have been found. For this reason, circadian clocks will serve as a model for detailed consideration of basic properties of endogenous timers, and reference will be made to non-crustacean examples where these serve better to illustrate major points. Where available, comparisons will be made with circatidal and circasemilunar timers.

D. Basic Properties of Free-Running Pacemakers

A first set of pacemaker properties concerns pattern and frequency variations. In terms of locomotor activity patterns, a range is seen from clear-cut envelopes of activity with sharp onsets and cutoffs to nearly continuous activity or erratic envelopes where pattern and frequency are unclear (Fig. 2). These patterns are mere variations on a common theme of widely recurring features which include (1) a specific temporal pattern, (2) considerable precision of the free-running rhythm, (3) a narrow range of free-running τ

values, with (4) slight malleability due to age, prior conditions, or reproductive state, (5) a temperature-compensated frequency, and, finally, (6) a frequency dependent on the intensity level of constant illumination.

The patterning of circadian functions has been analyzed in detail for several vertebrate species (Aschoff and Homa, 1950; Aschoff, 1960; De-Coursey, 1961; Aschoff et al., 1971b; Davis and Menaker, 1980), but to a much lesser extent for crustaceans (Enright, 1963, 1972, 1975; Palmer, 1971; Honegger, 1976; Rawson and DeCoursey, 1976). A large fraction of these studies have assayed locomotor activity. In many terrestrial vertebrate species, a highly precise pattern of activity and rest occurs, as in the rodent *Glaucomys volans* in a wheel cage (Fig. 2A). In other species, dissociation into polyphasic patterns takes place (Fig. 2D). "Splitting" may occur in several species after long-term maintenance in constant conditions (Pittendrigh and Daan, 1976a). The noisy nature of crustacean locomotor activity patterns (Fig. 2E) has already been emphasized, but some circadian exceptions are found; in the supralittoral amphipod *Talitrus*, clear free-running rhythms comparable to rodent patterns are seen (Williams, 1979). Other good tidal-semilunar examples include *Excirolana, Emerita analoga,* and *Synchelidium* spp. (Fig. 3; Enright, 1975); for these species, modulation in the activity pattern of a group of animals closely approximates the amplitude of the semidiurnal, unequal tides after isolation of animals under constant laboratory conditions for many days.

The interspecific range of free-running circadian rhythms in constant darkness for the animal kingdom ranges only from approximately 22–26 hr (Hoffmann, 1965a), and for all constant light conditions, the range rarely exceeds 21–28 hr. The intraspecific range of most species is more restricted. The most precise, free-running data come from terrestrial rodents in DD: for *Glaucomys* (DeCoursey, 1961) a range of 22 hr 58 min to 24 hr 21 min; for the hamster *Mesocricetus auratus,* from approximately 23.5 to 25 hr (De-Coursey, 1964; Pittendrigh and Daan, 1976a); and for 2 species of *Peromyscus,* from approximately 22 to 25 hr (Pittendrigh and Daan, 1976a). The strictly freshwater crustaceans, exemplified by stream crayfish, exhibit fairly sharp circadian rhythms (Page and Larimer, 1972; Hammond and Fingerman, 1975) with an intraspecific range for τ_{DD} from 23.3 to 23.6 hr. In deep-water species such as the lobster *Homarus,* egg hatching is also circadian, with frequencies close to 24 hr in constant conditions (Branford, 1978). For locomotor activity of terrestrial purple crabs, a value of 23 hr, 12 min was found for one individual at 25°C, and 11 other crabs were very close in value (Palmer, 1971). In contrast, intertidal species of crustaceans exposed to the multiple cycles of tides (12.4 or 24.8 hr), day–night cycle (24 hr), and semilunar cycle (14.9 days), often showed both circadian and tidal frequencies (Barnwell, 1966; Webb, 1971, 1972; Enright, 1972; Williams,

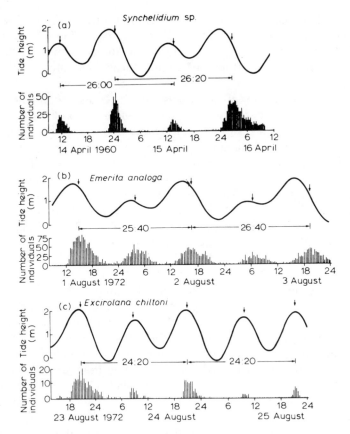

Fig. 3. Tidal rhythms of swimming activity in groups of freshly collected individuals for three species of sandy beach crustaceans. Activity is based on photographs of the number of animals swimming. (From Enright, 1975.)

1979). Values for bitidal free-running swimming rhythms of six *Excirolana* in dim light ranged from 24.25 to 25.05 hr (Enright, 1971b) and in another series of experiments with 52 *Excirolana* from 23 hr 50 min to 25 hr 10 min (Enright, 1972). For *Uca crenulata* in dim light, the values ranged from 24.3 to 25.2 hr (Honegger, 1973a), for *Orchestia* spp. in dim light from 24.68 to 24.88 hr (Wildish, 1970), and for 15 *Talorchestia* in DD from 24.1 to 26.3 hr (Benson and Lewis, 1976).

The frequency of a free-running rhythm for most circadian rhythms is dependent on the level of constant illumination in a predictable way, summarized by Aschoff's Rule (Hoffmann, 1965a; Aschoff et al., 1971b). Nocturnal terrestrial species increase the period length with increased intensity of constant illumination and reduce the amount of activity per cycle at higher intensities of constant light, while the reverse is true for day-active species. Moller and Branford (1979) state that nocturnal hatching rhythms in *Nephrops* follow Aschoff's Rule, but no numerical values are available for crustacean rhythms.

The free-running rhythms of most terrestrial species under constant conditions are extremely precise, particularly with reference to onset of activity (Fig. 2A). The greatest precision is on the order of 0.1% error in predicting the realized free-running circadian frequency for the squirrel *Glaucomys* (DeCoursey, 1961), and comparable values are available for a wide variety of species, including *Mesocricetus* (DeCoursey, 1964), *Peromyscus* spp. (Suter and Rawson, 1968; Pittendrigh and Daan, 1976a), *Rhinolophus ferroequinum* (DeCoursey and DeCoursey, 1964), *Drosophila pseudoobscura* (Pittendrigh, 1960), *Teleogryllus commodus* (Loher, 1972), and cockroaches (Roberts, 1960). Day-active rodents (DeCoursey, 1973), microtine rodents (Swade and Pittendrigh, 1967), and birds (Eskin, 1971) tend to have much less precision in free-running circadian locomotor rhythms. For almost all crustaceans, the noisy nature of the rhythm plus short-term recording make precision measurements difficult. However, a value of the standard deviation for six *Excirolana*, recorded individually in continuous dim light, is quoted as ±33 to ±53 min (Enright, 1971b). The relative precision of rodent and crustacean locomotor rhythms are compared in Rawson and DeCoursey (1976) for a series of habitats ranging from strictly terrestrial through intertidal sites. All mice were precise, while all crab species examined were too noisy to analyze for frequency even with a periodogram analysis.

The value of the free-run under any given conditions tends to be relatively constant, but for a number of species a lability or malleability has been demonstrated (Pittendrigh and Daan, 1976a). History-dependent changes related to prior entrainment, lengthening of free-running endogenous rhythms with increasing age of an individual (Pittendrigh and Daan, 1974), or progressive short-term changes in the "profile of the free-running rhythms" (Eskin, 1971) have been noted. These history dependent features are relatively small, and normally require very careful analysis to detect. The more diffuse, short-term crustacean rhythms do not lend themselves to such rigorous analysis, and consequently information is lacking.

In summary, most free-running circadian rhythms throughout the animal kingdom fall within the range of 22–26 hr in period length. Fairly large changes in period are induced by changing the level of constant illumination, while minor deviations in period are attributable to age, prior conditions, and unpredictable spontaneous changes. Much less can be generalized, as yet, about patterning and frequency modulations of free-running tidal, bitidal, semilunar, lunar, or annual rhythms in crustaceans. However, the innate basis has now been firmly established for five frequencies which approximate five major environmental cycles.

E. Entrainment

A second set of fundamental properties of biological timers concerns entrainment features. It is the ability of a few specific environmental agents to entrain a biological rhythm in a predictable way, through a species-specific phase response system, which further supports the idea of adaptiveness. The very nature of free-running clocks necessitates some correction by agents called entraining agents (Pittendrigh, 1960), synchronizers (De-Coursey, 1961), or Zeitgeber (Aschoff, 1960), in order for the animal's frequency to match the environmental regime. The prerequisites for successful entrainment are frequency control and phase control. As described below, an entraining agent corrects the period of the endogenous rhythm to match the period of the entraining cycle, and locks the biological process in a specific phase to the environmental cycle (Pittendrigh and Daan, 1976b).

Only a small number of cyclic environmental stimuli are actually effective in entrainment. Examples of a terrestrial mammal entrained by a light–dark cycle and of an intertidal crustacean entrained by tides are contrasted in Fig. 4. In the light entrainment example (Fig. 4A), the pattern of phase and frequency control depend on the initial placement of the light schedule with reference to the active period (DeCoursey and DeCoursey, 1964). The clock, indicated by the start of flight activity in the nocturnal bat, readjusts in small daily increments until it stabilizes with the start of activity at the lights out or "dusk" transition, with its corrected period equal to the 24 hr of the entraining light cycle. In this manner, entrainment has been achieved through phase and frequency control. In the tidal entrainment example (Lehmann et al., 1974), the locomotor activity of the fiddler crab, Uca urvillei, follows the 12.4-hr simulated tidal cycle in the laboratory, but is unaffected by the superimposed 24-hr light–dark cycle during the first 25 days of the experiment (Fig. 4B).

For circadian timers, the chief entraining agent is a light–dark cycle (Hoffmann, 1969). In semiterrestrial crabs, a circadian rhythm entrained by light–dark cycles has been demonstrated (Palmer, 1971), and for supratidal amphipods, the circadian rhythm of nocturnal foraging was entrained by light–dark cycles (Benson and Lewis, 1976; Williams, 1979). Similarly, for freshwater crayfish (Page and Larimer, 1972), for the subtidal lobsters Homarus and Nephrops (Hammond and Naylor, 1977; Branford, 1978; Moller and Branford, 1979), for prawns (Moller and Jones, 1975), or for vertical migration of many deep-water copepods (Enright and Hamner,

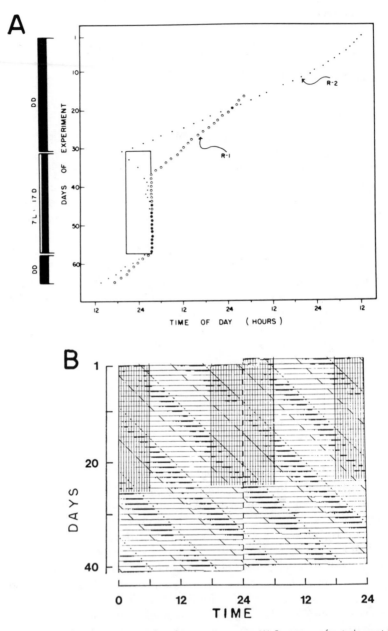

Fig. 4. Entrainment in a mammal and in a crustacean. (A) Summary of entrainment by a light–dark regime in two isolated bats, *Rhinolophus*, under DD → LD → DD light conditions as indicated on left, with ○ = onset of flight activity for bat R-1; • = onset for bat R-2; ● = simultaneous onset for bat R-1 and R-2 (adapted from DeCoursey and DeCoursey, 1964). (B) Entrainment by a tidal regime, but failure to entrain to a light–dark cycle, for the fiddler crab, *Uca urvillei*, in laboratory conditions. Symbols: grid on days 1–25 indicates light schedule, – – – – for high tide and . . . . for low tide; record is doubled with a 48-hr time scale, for ease in reading (from Lehmann *et al.*, 1974).

1967), the dominant rhythmicity is circadian and the synchronizer is an LD cycle. Neither the intensity of light, nor the duration of the photoperiod, nor the wave form of the LD entraining cycle is important over large ranges; as a result, most terrestrial species of animals will utilize a variety of intensities, waveforms, and photofractions of LD synchronizers. More significant is the fact that the range of entrainment for most circadian species spans 23–26 hr (Enright, 1965d). Just beyond the limits, the phenomenon of "relative coordination" is seen, in which the animal free-runs through the light cycle at a variable rate (Enright, 1965d). Of the remaining circadian entraining agents, temperature cycles are effective only in several insects (Pittendrigh, 1960; Roberts, 1962) and poikilothermic vertebrates such as lizards (Hoffmann, 1968), with effectiveness directly related to amplitude of the temperature cycle. Even fewer cases of sound synchronization have been unequivocally substantiated, chiefly in social finches (Gwinner, 1966) and in house sparrows (Menaker and Eskin, 1966). Entrainment of human circadian rhythms in continuous darkness by social clues has been noted (Aschoff et al., 1971a).

Not surprisingly, the predominently circatidal rhythms of intertidal crustaceans respond little to light–dark cycles (Barnwell, 1966; Webb, 1971; Honegger, 1973a,b, 1976; Lehmann, 1976). Circatidal rhythms entrain to artificial tides in Uca spp. (Honegger, 1973b, 1976; Lehmann, 1976), to simulated wave action in Excirolana (Enright, 1965a) or in the isopod Eurydice pulchra (Jones and Naylor, 1970), and to temperature changes associated with tides in the amphipod, Bathyporeia pelagica (Fincham, 1970a) or in Carcinus maenas (Williams and Naylor, 1969; Naylor et al., 1971). Cycles of salinity change were effective for Carcinus (Taylor and Naylor, 1977). Much less is known about the entrainment of circasemilunar and circalunar rhythms. Stimuli arising from turbulent waves moving across the beach or from shaking in the laboratory act as entraining agents for Excirolana (Enright, 1965a, 1976b). Moonlight cycles acted as entraining agents for 2 species of Sesarma (Saigusa, 1980). For the intertidal midge Clunio, artificial moonlight cycles also acted as entraining agents, and in some cases the phase relationship between 24-hr LD cycles and 24.8-hr vibrational cycles acted as synchronizers for a fortnightly eclosion cycle (Neumann, 1976a,b). For the polychaete Platynereis (Hauenschild, 1960), lunar cycles of artificial moonlight were effective entraining agents in the lunar reproductive cycle. In Talitrus, neither substrate vibration cycles nor artificial moonlight cycles, nor water temperature cycles were capable of entraining the circasemilunar rhythms of locomotor activity, molting, or reproduction (Williams, 1979). Table I illustrates the interaction between the biological timing element τ and the period of the environmental entraining agent T to produce an environmentally synchronized organism.

F. Phase Response Systems

In order to entrain to a particular environmental variable, an organism must exhibit a differential responsiveness to phase shifting by that stimulus. Lacking such rhythmically changing responsiveness, an animal could never lock on to an appropriate phase of the synchronizer. Thus, an integral part of phase and frequency control is a phase response system: the complex reaction in terms of phase shifting to a specific type of stimulus, related in a temporal sense to the animal's internal sense of time. A phase response curve is one very limited definition of a part of the phase response system, for each curve is intimately related to the stimulus assay method as well as to the modality, intensity, and duration of the stimulus tested.

The most frequently used method in constructing a phase response curve (PRC) has involved pulse-type test signals (Aschoff, 1965b). Typical methods are shown for high precision hamsters (Fig. 5A–D) and for isopods (Fig. 6). PRCs have now been standardized (Pittendrigh, 1973; Daan and Pittendrigh, 1976b) to portray magnitude of response to a test signal plotted against subjective animal time. A comparison of Fig. 2A with Fig. 2E points out the problems of obtaining a detailed PRC for all but the most precise individuals by this method, and explains the paucity of complete published phase response curves. A PRC atlas has summarized the approximately 60 known curves or families of curves for species ranging from protists to vertebrates (DeCoursey, 1977): 12 curve for protists, 9 for higher plants, 25 for invertebrates, including one for the crustacean *Excirolana* (Enright, 1976a,b), one for a bird, and 13 for rodents. The response systems have been measured for visible light pulses in 34 cases, for ultraviolet light in 2 cases, for vibration in the case of *Excirolana*, for temperature in 4 cases, and for chemical pulses in 14 cases. The general feature of all PRC curves is a differential response dependent on the phase at which the stimulus is administered.

Of greatest interest for this review is the sole crustacean phase response curve for the isopod *Excirolana* (Fig. 6; Enright, 1976a,b). The test stimuli of 2-hr "simulated waves" were administered by an oscillatory shaker bench; plotting of the resultant phase shifts yielded a unique symmetrically biphasic response curve (Fig. 6). As Enright (1976a) pointed out, such a bimodal curve would normally be maladaptive, but ". . . only in tidal synchronization of a circadian system can an ecological appropriateness for such a pattern of responses be recognized."

A comparison of the remaining 27 PRCs having at least 12 test points indicates that most are curves for responses to light and that the response in all is very large in relation to the stimulus. PRC amplitude is generally greatest in invertebrates, with much smaller amplitude in mammals. Data

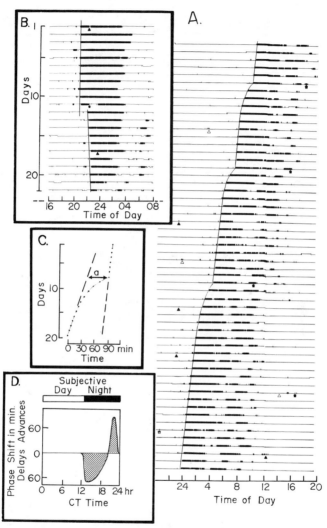

Fig. 5. Phase response curve for a hamster, *Mesocricetus*, based on single isolated pulses of 10-min duration and 0.5 foot-candle intensity. (A) Advance shifts. (B) Delay phase shifts. (C) Schematic of calculations for final steady state advance phase shifts, a, after disappearance of transients. (D) PRC plotted with circadian time (CT) 0 = subjective dawn, and CT 12 = subjective dusk; ↑ = time of light pulse. (Adapted from DeCoursey, 1964.)

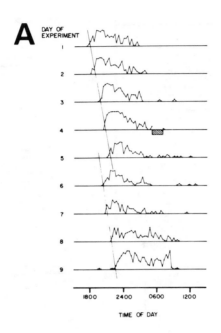

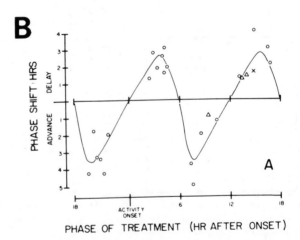

Fig. 6. Circatidal phase response curve for the isopod *Excirolana*. (A) Swimming activity of a single isopod, with shifting stimulus of 2 hr shaking, ▨, on day 4, and phase shift measured on day 5. (B) Phase response curve. (From Enright, 1976b.)

for the hamster, *Mesocricetus,* demonstrate the typical form of light PRCs (Fig. 5D), with a delaying segment and advancing segment during subjective night, and an unresponsive period during subjective day. The ecological and behavioral importance of light PRCs lies in their effectiveness in adjusting endogenous, free-running pacemakers to the pervasive day–night schedule throughout the year. A model by DeCoursey (1961, 1964) or Daan and Pittendrigh (1976b) proposes dusk entrainment by delay resetting when the free-running rhythm is shorter than the environmental driving period, and dawn entrainment for free-runs longer than the environmental period.

For other phase response modalities, much less is known. Temperature PRCs have been reported in cockroaches (Roberts, 1962), and in a lizard by Hoffmann (1968). Although temperature synchronization has been demonstrated in both species, the relationship of the PRCs to the entrainment process is not clear. Partial PRCs for various chemicals have also been reported, but the chemical PRCs are even more poorly understood than the temperature ones. Some appear to mimic light PRCs.

G. Physiology of Clocks

Until recently, knowledge of the physiological nature of clocks and their mode of responding to environmental entraining agents was very limited. In summarizing research progress on physiology of biological clocks, three issues should be addressed (Fig. 7):

1. What and where is the oscillatory center(s), and if multiple, how are they coupled?
2. What sensory systems operate in entrainment and what is the neurohumoral pathway to the pacemaker system?
3. What are the neural and humoral pathways from oscillator to effector organs?

Insight into these questions has been gained by rapid advances in 5 noncrustacean systems: the sea hare *Aplysia californica,* the giant silkworms *Antheraea* spp., the cockroach *Periplaneta americana,* the bird *Passer domesticus,* and such mammals as the hamster *Mesocricetus* (reviews in Menaker et al., 1978; Eskin, 1979; Takahashi and Menaker, 1979). The *Aplysia* compound eye continues to oscillate under constant conditions *in vitro* (Jacklet, 1969; Eskin, 1979); entrainment is by way of light signals to the eye. In silkworm moth pupae, the brain contains the photoreceptor as well as a circadian oscillator regulating rhythmic pupal hatching (Truman, 1972, 1974). In cockroaches, circadian pacemakers are located in the optic lobes, and entrainment utilizes the compound eyes (Brady, 1967; Roberts, 1974).

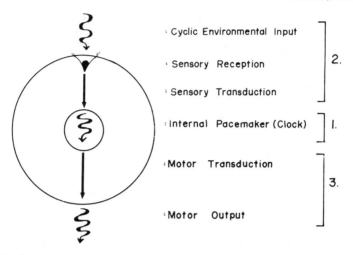

Fig. 7. Schematic representation of components in the physiology of a biological clock. Numbers refer to text items.

Species differences occur in birds, but the pineal appears to be a dominant pacemaker. Photoreception in entrainment utilizes both retinal and extraretinal pathways, and a hormonal output of the pineal probably mediates expression of circadian rhythmicity. In mammals, research centers on the suprachiasmatic nucleus (SCN) of the hypothalamus in its role in maintenance of many circadian functions (reviews in Moore, 1978; Rusak and Zucker, 1979); photic input for entrainment is primarily by way of a special retinohypothalamic tract direct from eye to SCN.

The crustacean system promises to be another useful model for examining physiology of timers (Fig. 7). Pronounced rhythms of retinal distal pigment migration (Aréchiga and Mena, 1975; Aréchiga, 1977), of evoked electroretinogram (ERG) response to light (Aréchiga and Wiersma, 1969; Page and Larimer, 1975b) and a locomotor activity rhythm (Page and Larimer, 1972, 1975a; Hammond and Fingerman, 1975), all provide detailed output measures of a circadian system. The major organs postulated as pacemakers, sensory receptors, and neurohumoral transducers are large and easily assessible in the anterior brain, the eyestalks, and the eyes. Originally, one group of workers favored the supraesophageal ganglion of the brain as pacemaker site (Page and Larimer, 1975a,b) with extraretinal photoreception for entrainment (Page and Larimer, 1976). In contrast, another group supported the sinus glands of the eyestalks as the primary oscillator sites, since the ERG rhythm persisted in organ culture of excised eyestalks (Aréchiga, 1977). A recent review of technical details partially resolves the issue by proposing that multiple centers of circadian oscillation exist in crayfish nervous systems; this complex model is "comprised of at least three pairs of intercoupled oscillators and two pairs of entraining photoreceptors" (Larimer and Smith, 1980). Further details are discussed in Section IV,A. The crayfish model supports the hypothesis (Menaker et al., 1978) that the path from photoreceptor to pacemaker in *Aplysia, Antheraea, Periplaneta, Passer,* and *Mesocricetus,* is very short, and that an intimate connection between photoreceptors and circadian systems exists. No physiological studies of tidal and lunar timing have been carried out. However, the evidence that now exists for circadian pacemakers is important in establishing an actual physiological and anatomical entity for timing. The demonstration of endogenous clocks no longer depends entirely on indirect demonstrations (Section III,B and C).

IV. SURVEY OF BIOLOGICAL RHYTHMS IN CRUSTACEA

The research summarized in Section III, D–G illustrates animal clocks as endogenous, self-sustaining oscillators, whose origin lies in the living protoplasm, with sensory "windows" on the rhythmic environments in which each organism lives. This section will focus on the degree of matching of biological rhythms in crustaceans with environmental parameters of their niches, emphasizing the physiological basis of the clock and the method of entrainment. It is well to remember that data on many facets of these issues are still lacking. Nevertheless, rhythms of animals in natural habitat have been included where the information helps clarify the theme or points to potential future experiments. Published papers encompass a great diversity of functions, from cellular to organismic.

A. Cellular and Physiological Rhythms

Among the cellular rhythms described are cyclic pigment movements in compound eyes. In crayfish, the distal pigments disperse to shield the rhabdome of the omatidium during daytime, and concentrate during the dark-adapted state at night (Welsh, 1941; Fingerman and Lowe, 1957). The rhythm persists in constant dark conditions with circadian frequency in the crayfish (Aréchiga, 1977) and also in 2 species of hermit crabs (Ball, 1968). Closely affiliated with retinal pigment movement is the responsiveness of compound eyes to light stimulation, as reflected by the amplitude of the electroretinogram (ERG); amplitude is greatest when pigments are in a concentrated, dark-adapted state, and lowest when they are dispersed (Aréchiga and Wiersma, 1969; Page and Larimer, 1975b). The rhythm persists under constant laboratory conditions in intact animals. In the extensive studies of

Page and Larimer (1975a,b) free-run values ranged from approximately 23 to 25 hr. Physiological aspects of the crayfish compound eyes are discussed in Section III,G. These examples suggest a circadian adaptation for greater visual efficiency at night under attenuated light conditions.

The question of integumentary chromatophore cycles in crustaceans is a strife-ridden topic (review in Palmer, 1973). A persistent daily rhythm of color change in *Uca* exoskeleton (Brown et al., 1953) with a period of exactly 24 hr, but modulated by a tidal periodicity, was the basis of the exogenous–endogenous controversy (Section III,A). Later work, which eliminated hourly light clues, resulted in a free-running rhythm of melanophore expansion (Stephens, 1962). Furthermore, translocation of fiddler crabs across three time zones (3.5-hr difference from Massachusetts to California) resulted in only a 22-min difference on the following day (Brown et al., 1955), suggesting that crab chromatophores were under endogenous control, not reacting exogenously to clues of the local site. A circadian rhythm of color change was also reported for *Carcinus* (Powell, 1962a). In *Uca* the color change rhythm is said to persist in eyestalkless crabs under constant conditions (Fingerman and Yamamoto, 1964), whereas in *Carcinus* the rhythm was abolished in eyestalkless crabs but was reinstated when eyestalks from rhythmic crabs were implanted (Powell, 1966). Adaptiveness of color change may relate to antipredation strategies, and also to social signaling.

Cycles of several cell or system constituents have been reported for several crustaceans. Cyclic output of a neurosecretory cell unique to Hanström's organ occurred in the eyestalk of the crab *Carcinus* (Williams et al., 1979) and persisted during *in vitro* culture of isolated eyestalks under constant conditions. Similarly, circadian variations in concentration of the light-adapting hormone for the retinal distal pigment occurred in crayfish optic ganglia (Aréchiga and Mena, 1975). A rhythmic tidal cycle of blood sugar concentration occurs in *Carcinus* (Rajan et al., 1979), with a fall in glucose at the time of local high tide for crabs maintained in atidal conditions. A similar picture is known for *Crangon crangon* (Poolsanguan and Uglow, 1974), and an example has been published for an endogenous circadian rhythm of blood sugar level for the crayfish *Orconectes limosus* (Hamann, 1974). Respiration cycles are also known for crustaceans. In two species of amphipods, respiration is highest during the nighttime active period (Cederwall, 1979). In *Uca*, peaks of respiration occur in constant conditions at the time of low tide, but with daily components as well (Brown et al., 1954). For *Carcinus*, a persistent rhythm of gill ventilation rate and oxygen consumption persists in LL with tidal periodicity (Arudpragasam and Naylor, 1964). In the mole crab *Emerita asiatica*, persistent free-running rhythms of O_2-consumption occur, directly related to the locomotor activity rhythm (Chandrashekaran, 1965). In all these cases, data were limited to short-term documentation of the cycle under controlled laboratory conditions, with no information on possible adaptiveness. However, such data fit the picture of temporal organization of internal cycles of animals for greatest biochemical efficiency (Pittendrigh, 1973).

B. Organismic Cycles

Much more is known for organismic levels of function in crustaceans, especially for the conspicuous locomotor activity rhythms (Table II). In spite of widely varying results, several generalizations are apparent. Most of the studies have been laboratory experiments, presumably because of difficulties in long-term observation in aquatic habitats. While a large number of taxonomic groups have been involved, most have been macromembers of a few groups: barnacles, stomatopods, mysids, isopods, amphipods, euphausids, and decapods (Table II). A large proportion have been conspicuous, littoral, day-active animals such as *Uca*. The data were often collected for only 3 or 4 days, since damping out of the rhythm or early death of the animals commonly occurred. Data were often very noisy.

The characteristics of the rhythms center around habitat. For each main group, at least one crustacean member has been extensively studied (Table III). In lightless environments such as caves, rhythmic environmental input may be lacking, but the blind cave crayfish *Orconectes pellucidus* retained a circadian clock regulation of locomotor activity and oxygen consumption in spite of isolation from day–night cycles for many generations; values of τ for three individuals in DD were 26, 27, and 34 hr. Only one of three individuals showed clear entrainment in an LD cycle (Jegla and Poulson, 1968). These data suggest that the rhythm is retained, but is drifting from 24 hr in the absence of environmental selection for daily period.

A second habitat type consists of freshwater streams or lakes. The locomotor activity of stream crayfish in continuous darkness free-runs with circadian frequency (Page and Larimer, 1972, 1975a,b; Hammond and Fingerman, 1975). Although minor species differences occur, locomotor rhythms are usually bimodal, with an endogenous dusk peak entrained to the lights-off signal (Page and Larimer, 1972). Pacemaker and entrainment mechanisms are discussed in Section III,G.

The littoral *Talitrus* (Williams, 1979) and the semiterrestrial *Gecarcinus* and *Cardisoma guanhumi* (Palmer, 1971) maintained free-running circadian locomotor activity rhythms for many weeks; LD cycles were the entraining agent. The subtidal *Nephrops* was circadian in its activity, with the light–dark regime as the entraining agent (Atkinson and Naylor, 1976; Hammond and Naylor, 1977). The spiny lobster, *Panulirus argus*, was also

TABLE II

Locomotor Activity Rhythms in Crustacea

Order	Species	Location Field	Location Lab	Type	Basis	Entraining agent	Reference
Cirripedia	Semibalanus balanoides	X		Tidal	—	—	Arnold (1970)
			X	Tidal and semilunar	—	—	Fish and Fish (1972)
Isopoda	Eurydice pulchra	X	X	Tidal	Endogenous	Wave action	Jones and Naylor (1970)
			X	Tidal and semilunar	Endogenous	Wave action	Alheit and Naylor (1976)
	Excirolana chiltoni[a]		X	Tidal	Endogenous	Wave action	Enright (1965a, 1971a, 1971b, 1972, 1976b)
Amphipoda	Talorchestia quoyana	X	X	Tidal	Endogenous	Wave action	Klapow (1972, 1976)
			X	Daily	Endogenous	LD	Benson and Lewis (1976)
	Corophium volutator		X	Tidal	Endogenous	Tide variables	Morgan (1965)
	Talitrus saltator[a]		X	Daily, semilunar	Endogenous	LD	Williams (1979)
			X	Daily	Endogenous	—	Bregazzi and Naylor (1972)
	Synchelidium spp.[a]		X	Daily	Endogenous	—	Bregazzi (1972)
	Bathyporeia pelagica		X	Tidal	Endogenous	Wave action	Enright (1963)
	Ligia oceanica		X	Tidal	Endogenous	?	Fincham (1970a)
	Orchestia cavimana and O. mediterranea		X	Daily	Endogenous?	—	Powell (1962b)
			X	Tidal and/or circadian	Endogenous	?	Wildish (1970)

	Species						Reference
	O. gamarella		X	Circadian	Endogenous	?	Wildish (1970)
	Gammarus spp.	X	X	Tidal	Endogenous	Tide variables	Dieleman (1977)
Decapoda	Nephrops norvegicus	X		Daily	Endogenous	LD	Atkinson and Naylor (1976)
			X	Daily	Endogenous	—	Hammond and Naylor (1977)
	Dichelopandalus bonnieri	X		Daily	—	—	Al-Adhub and Naylor (1976)
	Procambarus clarkii[a]		X	Daily	Endogenous	LD	Page and Larimer (1972, 1975a, 1976)
	Faxonella clypeata		X	Daily	Endogenous	LD	Hammond and Fingerman (1975)
	Orconectes pellucidus		X	Daily	Endogenous	LD?	Jegla and Poulson (1968)
	Penaeus duorarum		X	Tidal	Endogenous	—	Hughes (1972)
	Penaeus semisulcatus		X	Daily	Endogenous	LD cycle	Moller and Jones (1975)
	Palaemon elegans and P. serratus	X		Tidal	Endogenous	Tidal variation	Rodriguez and Naylor (1972)
	Crangon crangon		X	Daily	Exogenous	—	Hagerman (1970)
		X		Tidal	Endogenous	LD	Al-Adhub and Naylor (1975)
	Uca[a] pugnax		X	Tidal, daily	Endogenous	LD, tides	Webb (1971, 1972)
			X	Tidal	Endogenous	—	Bennett et al. (1957)
			X	Daily, tidal	Endogenous	LD, tides	Barnwell (1966, 1968)
	crenulata		X	Tidal	Endogenous	Tide, LD	Honegger (1973a,b)
	urvillei		X	Tidal, daily	Endogenous	Tides	Lehmann et al. (1974)
			X	Tidal, daily	Endogenous	LD, tides	Lehmann (1976)
	pugilator		X	Tidal?	Too noisy for analysis	—	Rawson and De-Coursey (1976)

(continued)

TABLE II Continued

Order	Species	Location		Type	Basis	Entraining agent	Reference
		Field	Lab				
	Sesarma reticulatum	X		Tidal, daily Semilunar	Endogenous	—	Palmer (1967)
	Rhithropanopeus harrisii larvae		X	Tidal	Endogenous	—	Forward and Cronin (1980)
	Carcinus maenas[a]		X	Tidal, daily	Endogenous	Tidal variables	Naylor (1958)
			X	Tidal	Endogenous?	—	Powell (1962b)
			X	Tidal, daily	Endogenous	Tide variables	Williams and Naylor (1969)
			X	Tidal	Endogenous	Salinity cycle	Taylor and Naylor (1977)
	Hemigrapsus edwardsi		X	Tidal	Exogenous	Tide variables	Naylor et al. (1971)
			X	Tidal	Endogenous	—	Williams (1969)
	Emerita asiatica		X	Tidal	Endogenous	—	Chandrashekaran (1965)
	Gecarcinus lateralis	X	X		Endogenous	LD	Palmer (1971)
	Ocypode quadrata[a]	X	X	Daily	Endogenous	LD	Palmer (1971)
	Coenobita clypeatus	X	X	Daily	Endogenous	LD	Palmer (1971)
	Cardisoma guanhumi	X	X	Daily	Endogenous	LD	Palmer (1971)

[a]Extensively studied.

subtidal, with circadian free-running locomotor rhythms in constant dark or continuous illumination, and with LD entrainment (Lipcius et al., 1979). In all of these, no tidal element was detectable, but in Talitrus a free-running semilunar modulation of the amount of activity was seen (Williams, 1979) for which neither wave action nor moonlight was an effective entraining agent.

Markedly different were intertidal Synchelidium, Excirolana, Carcinus, Emerita, and Uca, exposed in natural habitat to the rhythmic ebb and flow of the tides. In Uca, rhythmicity was barely detectable under laboratory conditions (Honegger, 1976; Rawson and DeCoursey, 1976), while in Enright's (1972) "virtuoso isopods" both timing of swimming activity and amplitude matched the semidiurnal unequal peaks of water height, even after isolation from all tidal influences for several weeks (Fig. 3). Most Excirolana, when analyzed in the laboratory, were strongly tidal, sometimes with circadian elements as well as longer semilunar components. In almost all cases, the day–night cycles were relatively ineffective as a synchronizer, and variables such as water turbulence or artificial tides were effective in entrainment (Enright, 1963, 1965a, 1972, 1975, 1976b). In these locomotor rhythm

TABLE III

Patterns of Locomotor Rhythms Relative to Habitat

Habitat	Animal	Locomotor rhythm frequency (τ)	Dominant synchronizer
A. Circadian rhythms			
Cave	Orconectes	Circadian	Light–dark cycles (partial)
Freshwater streams	Procambarus	Circadian	Light–dark cycles
Supralittoral or terrestrial	Talitrus	Circadian, circasemilunar	Light–dark cycles
	Gecarcinus	Circadian	Light–dark cycles
Subtidal	Nephrops	Circadian	Light–dark cycles
	Homarus	Circadian	Light–dark cycles
	Panulirus	Circadian	Light–dark cycles
B. Circatidal, circasemilunar rhythms			
Intertidal	Uca	Circatidal (circadian?), circasemilunar	Tides (light–dark cycles)
	Carcinus	Circatidal	Tide-related variables
	Excirolana	Circabitidal, circasemilunar	Wave action
	Synchelidium	Circatidal	Wave action

studies discussed above, the correlation between dominant cyclic environmental parameter and frequency of locomotor activity was very strong. In terms of fitness, such synchronized locomotor activity could hypothetically be important in spatiotemporal orientation for greater efficiency or enhanced survival.

Closely related to locomotor rhythms are the cyclic phototactic responses of several crustacean species. The fiddler crab, *Uca pugnax*, showed a 24-hr fluctuation in phototactic reaction under LL conditions, with a positive response during subjective daytime and a negative phototaxis during subjective night (Palmer, 1964). In studies on phototaxis of the amphipod *Synchelidium*, freshly collected animals were positively phototactic on falling tides, and photonegative on rising tides (Forward, 1980). Forward and Cronin (1980) also detected endogenous tidal rhythms of phototaxis of larval *Rhithropanopeus harrisii*. These studies have confirmed short-term persistence of phototactic rhythms under constant conditions in the laboratory, but almost nothing about entrainment is known.

Still another rhythmic function of crustaceans related to locomotor activity is vertical migration, for which a very large literature now exists. Vertical migration of crustaceans in coastal and open ocean waters was early described in the work of Russell (1927), as well as Hardy and Paton (1947). The work of Cushing (1951, 1955), however, first drew large-scale scientific attention to the magnitude and extent of the phenomenon and its ecological significance. Several overviews are available (Baylor and Smith, 1957; Bainbridge, 1961; Raymont, 1963; Allen, 1966, 1972; Kerfoot, 1970; Rudjakov, 1970; Forward, 1976; Kampa, 1976; Longhurst, 1976; Palmer, 1976; Enright, 1977a,b; Enright and Honegger, 1977). Many different approaches have been used to study diverse examples of vertical migration as a rhythmic function. The examples are widely distributed in various aquatic habitats and in crustacean taxa.

Only recently has it been shown that endogenous timers in part control vertical migration, with physical cycles acting as entraining agents (Harris 1963; Rudjakov, 1970). Enright and Hamner (1967) monitored vertical migration of 13 species of crustacean zooplankton in a large seawater holding tank under constant conditions and by surface net sampling detected many species-specific patterns of appearance and disappearance from the surface layer on a daily basis. Endogenous vertical migration rhythms in a simulated natural environment have been demonstrated for larval *Rhithropanopeus* by Cronin and Forward (1979); an endogenous rhythm of phototactic response in vertical migration has been examined by Forward (1980).

The majority of vertical migration studies have been much less useful in demonstrating an endogenous pacemaker entrained to environmental cycles in an organism's niche. Initially, the development of sonar during

World War II facilitated detection of the deep scattering layer (DSL) of massed planktonic crustaceans and fish as well as its cyclic movement up and down from day to day (Dietz, 1962). The construction of deep-water submersibles spurred direct observation of the DSL to identify its composition and clarify its behavior (Barham, 1963, 1966; Donaldson and Pearcy, 1972). The extensive sampling of the 1965 SOND Cruise helped characterize the DSL, particularly for calanoid copepods (Roe, 1972) and for pelagic decapods (Foxton, 1970). These early studies point out one conspicuous trend in vertical migration research: extensive periodic field sampling to identify and enumerate the submicroscopic planktonic crustacean constituents. Recent studies have verified vertical migration in open ocean waters for copepod species (Tsalkina, 1970, 1972; Guerderat and Friess, 1971; Rudjakov and Veronina, 1973; McLaren, 1974; Donaldson, 1975; Marlow and Miller, 1975; Enright, 1977b), for cumacean crustaceans (Corey, 1970), and for shrimp (Pearcy, 1970). Relatively few studies of vertical migration in estuarine waters or surf zones have been carried out, possibly due to greater complexity of environmental factors, and sampling problems in shallow waters. A few instances, however, have been documented for the cladoceran *Podon polyphemoides* (Bosch and Taylor, 1973), for mysids (Hulbert, 1957; Herman, 1963), for copepods (Stickney and Knowles, 1975), for a surf amphipod (Fincham, 1970a,b), for *Uca* zoeae (DeCoursey, 1976b), and for an assemblage of mysids, isopods, and amphipods (Wooldridge, 1976). Freshwater vertical migration studies are available for *Daphnia magna* (Ringelberg, 1964; Haney and Hall, 1975) and for the amphipod *Pontoporeia affinis* (Marzolf, 1965). These studies serve primarily to establish a cyclic 24-hr movement in natural habitat.

Another dominant theme in vertical migration research has been correlation of position of organisms with specific field conditions, in order to deduce adaptive significance. Among the factors considered have been presence of predators (Zaret and Suffern, 1976; Robertson and Howard, 1978), dispersal and retention (DeCoursey, 1976b; Bosch and Taylor, 1973), or bioenergetic considerations of food availability, metabolic rate, and growth rate (Kerfoot, 1970; Rudjakov, 1970; Dumont, 1972; McLaren, 1974; Donaldson, 1975; Foulds and Roff, 1976; Enright, 1977a; Enright and Honegger, 1977; Hu, 1978).

Related to the correlation studies in vertical migration research have been field or laboratory experiments on environmental factors causing or influencing vertical position and migration of crustaceans. Most of these studies have proceeded from the *a priori* assumption that vertical migration was a direct exogenous response to physical variables in the environment. The primary factor influencing vertical migration, particularly in open ocean waters, has long been considered to be light. Light was clearly influential in studies of DSL responses in solar eclipses (Caruthers et al., 1970; Bright et al., 1972; Kampa, 1975). The response of the DSL to natural day–night changes has been documented by several workers (Kampa and Boden, 1954; Boden and Kampa, 1967; Currie et al., 1969; Kampa, 1976). Other workers have lowered artificial lights into the DSL (Blaxter and Currie, 1967). Laboratory studies on light influences have centered on phototaxis (review by Forward, 1976; other papers by Thorson, 1964; Sulkin, 1973; Forward, 1974; Forward and Costlow, 1974; Forward and Cronin, 1980). Additional factors considered with respect to their role in vertical migration include salinity (Lance, 1962; Grindley, 1964) or hydrostatic pressure (Lincoln, 1970, 1971; Ennis, 1973a).

The composite picture from these data is that vertical migration rhythms are widely distributed in copepods, mysids, amphipods, euphausids, and decapods from both freshwater and marine environments. Vertical migration is more pronounced from deep ocean waters where the DSL is well-developed. In contrast to the earlier work postulating direct effects of light or temperature or salinity, recent research emphasizes the importance in some species of endogenously controlled locomotor rhythms or photoactic response rhythms in regulating vertical migration. The information available implies that vertical migration increases the potential for survival by reduction of predation, increased feeding and metabolic efficiency, and avoidance of lethal habitat zones.

Another aspect of biological timing in spatial distribution involves horizontal movements: migration, homing, navigation, and orientation. These terms have often been loosely used to cover a variety of processes in crustaceans. In an excellent review, Enright (1978) has aptly emphasized the diversity and complexity of the broad spectrum of crustacean movements by referring to them as "a potpourri of strategies." A detailed analysis of the spatial aspects of migration is found in Chapter 2. Most appropriate for this review are the crustacean examples of time-compensated movement using a continuously consulted chronometer, which permits extensive trips or orientation of the individual in a fixed direction by means of a moving compass. Migration in this context will be restricted to the regular cyclic movement of a species on a predictable spatial and temporal round-trip path, using an endogenous clock, while homing will be limited to an animal finding its way home when displaced, by means of time-compensated celestial navigation. Orientation of crustaceans will refer to the return to a suitable zone of beach, rather than to a specific home site. These narrow definitions of time-compensated movements do not include a number of cases of extensive crustacean journeys. For example, the impressive "queuing" of spiny lobsters, Panulirus (Herrnkind, 1969, 1970), following the first severe autumn squall, entails synchronously moving lobsters in a mass movement day and night, using water currents and wave surge patterns as orientation guideposts; no evidence exists, either for the initial queuing or for the return phase, that any time-compensated migration occurred (Herrnkind and Kanciruk, 1978; Kanciruk and Herrnkind, 1978).

The most intensively studied of crustacean migrants, the amphipod beachhoppers, live on sandy beaches but are unable to withstand either the dessication of the dry upper beach or immersion in seawater. Consequently, they are active at night, browsing on debris, then taking shelter by day in the moist zone just above the high tide mark. When displaced, they orient landward by several means. *Talitrus* (Pardi and Papi, 1952; Papi, 1960; Pardi, 1960) and *Orchestoidea* spp. (Hartwick, 1976; Enright, 1978) may use time-compensated celestial orientation, but in some cases sun-compass orientation may be overridden by other cues, such as local landmarks, moisture gradient, slope of substrate, or wind direction (Enright, 1978). A functional clock has also been shown for several of these beachhoppers by means of free-running locomotor rhythms in *Talitrus* (Bregazzi, 1972; Williams, 1979), in *Talorchestia* (Benson and Lewis, 1976), and in *Orchestia* (Wildish, 1970). Very different is the tidal migration of the surf zone isopod, *Excirolana*, the surf crab, *Emerita*, and a surf zone amphipod, *Synchelidium*. By means of endogenous swimming rhythms, they maintain their position in the uprush zone (Enright, 1972, 1978).

Among the most pronounced of all crustacean rhythms are reproductive ones: courtship, mating, egg-laying, egg-hatching, as well as final metamorphosis or settlement. A number of ecological studies have defined seasonal changes in population levels of crustaceans, including copepods (Harris, 1972; Coull and Vernberg, 1975; Lonsdale and Coull, 1977); decapod larvae (Williams, 1971; Sandifer, 1973), blue crabs (Dudley and Judy, 1973), pebble crabs (Knudsen, 1960), or *Cancer magister* (Lough, 1976). More relevant to the theme of biological timing are studies of reproductive parameters relative to cyclic habitat features. Little is known for copepods, but in *Acartia clausi* (Landry, 1975), darkness suppressed hatching, and a short pulse of light exposure for these dark-inhibited eggs brought about a synchronous hatching 15 min after exposure, regardless of time of day. Reaka (1976) found semilunar oviposition and related molting rhythms in 16 species of stomatopods. The isopod *Excirolana* molted and reproduced synchronously with peaks of hatching every 2 weeks just prior to the new or full moon under field conditions (Klapow, 1972, 1976). Among the amphipods, *Talitrus* is known to exhibit synchronous embryonic development and hatching in field-collected specimens, but little is known of control (Williams, 1979). In decapods, the shrimp *Anchistioides* swarmed for spawning on the nights following the new moon (Wheeler, 1937), whereas the land crab *Cardisoma* migrated to the ocean to release its larvae at the time of high spring tides (Gifford, 1962).

In contrast to such exogenously controlled hatching are numerous cases of endogenous regulation. The lobsters *Homarus* and *Nephrops* have many cyclic hatching features in common; in both species the ovigerous female carries the developing embryos for many months, and over a period of several weeks, hatches a new group each night. In *Nephrops*, the free-running rhythm of hatching persists in LL and DD (Moller and Branford, 1979). In *Homarus*, the rhythm is also free-running and depends on the female for hatching. Attached eggs hatch over a period of several minutes each night, whereas only a small percent of isolated eggs hatch, spread over a period of many hours. The rhythm is entrained by the light–dark cycle, but temperature, daylength, and other possible factors may influence the phase of the hatch interval with respect to dusk (Ennis, 1973b; Branford, 1978.)

A concise study of hatching rhythms of two species of the crab *Sesarma* was carried out in the field over a period of 2 years (Saigusa and Hidaka, 1978). The semiterrestrial females walked to the riverbank, particularly at the spring tides of the full and new moon, to release their zoeal larvae at the time of dusk high tides. The data illustrate a strongly tidal as well as semi-lunar element, modified slightly by the inhibiting effect of daylight. Thus, when the spring high tides of full or new moon fell several hours before dusk, no hatching occurred until darkness; entrainment in controlled laboratory conditions was accomplished by a moonlight cycle (Saigusa, 1980).

The most complete data with respect to biological timing of reproduction in crustaceans concerns the fiddler crab, *Uca*. All elements of reproduction appear to be highly rhythmic, from courtship (Barnwell, 1968; Zucker, 1976; Christy, 1978) to egg hatching (Bergin, 1978, 1981; Wheeler, 1978; DeCoursey, 1979) to larval metamorphosis (J. H. Christy, personal communication). Hatching peaks of *Uca* in estuaries were concentrated at the time of the nighttime high spring tides (Fig. 8; also DeCoursey, 1979). In the laboratory, complete hatching of an egg sponge took place over several minutes with an extremely sharp "hatch profile" (Fig. 9; also DeCoursey, 1979). Using this hatch peak as a marker of reproductive timing, the rhythm was shown to persist in the laboratory in LL or LD conditions after isolation from tidal conditions for as long as several weeks, with peak hatching related to time of nighttime high tide on the beach of origin (Fig. 10). In a detailed analysis of reproductive timing in *Uca*, P. J. DeCoursey and D. P. Middaugh (unpublished) conducted a lunar scan of multiple reproductive parameters over a 6-week period in the laboratory and field (Fig. 11). Population samples of ovigerous females at the spring tides of new or full moon yielded peaks of light-gray egg sponges, indicating imminent hatching; im-

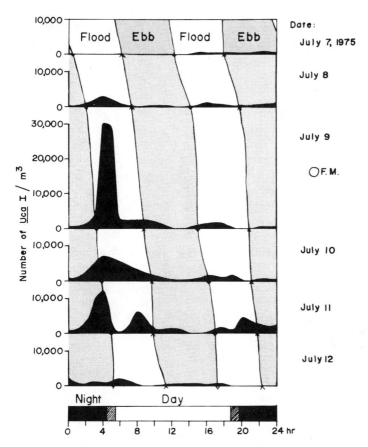

Fig. 8. Peak hatch in a semilunar rhythm of egg hatching for fiddler crabs, *Uca pugilator*, based on number of newly hatched (Stage I) larvae/m³ from plankton samples pumped in a vertical column at 2-hr intervals from the old Man Creek Station, North Inlet Estuary, South Carolina. Symbols: flood □ and ebb ▨ based on on-site current velocity measurements; day–night schedule shown below; ○ F.M. = full moon. (P. J. DeCoursey, unpublished.)

mediately afterwards a rise in freshly laid, dark purple sponges occurred. Maximum courtship and mating paralleled the egg-laying activity with a peak immediately prior to egg laying. Ovigerous, gray stage females were observed at the full and new moon to walk from their burrows as peak water levels reached the zone of greatest *Uca* burrow concentration shortly after dusk, and deposit hatching zoeae in the shallow creek waters. The dramatic rise of newly hatched *Uca* zoeae in tidal creeks took place concurrently with the movement of the females to the water's edge. Numbers rose at these periods of nighttime high spring tide as high as 100,000/m³, while on a

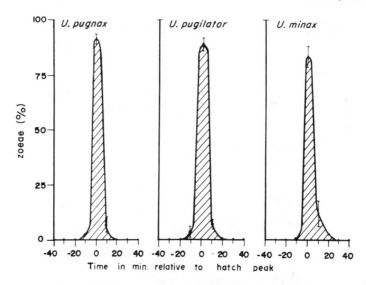

Fig. 9. "Hatch profiles" for three species of fiddler crab, *Uca*, under nontidal, natural daylight conditions in the laboratory at 30 ‰ salinity and 25° C, based on collections of hatching zoeae for eight individuals for each species, at 10-min intervals. Curves for each individual are normalized to percent total/10 min relative to peak hatch time. (From De-Coursey, 1979.)

comparable neap tide the numbers reached only several hundred or less. The endogenous component in timing of hatching was illustrated by holding ovigerous females under nontidal conditions of either DD, LD, or LL at least 10 days until hatching occurred. In all cases, peak hatching coincided with field hatching peaks (Fig. 11). Light was shown to be ineffective as an entraining agent, but the true synchronizer has not yet been demonstrated for *Uca*.

These reproductive examples are still another demonstration of activities driven by an internal timer and synchronized by a critical cyclic environmental factor, in a way that culminates in releasing the activity at what appears to be the time when offspring are most likely to survive.

V. A PERSPECTIVE OF THE ADAPTIVE VALUE OF ENVIRONMENTALLY ENTRAINED CLOCKS

The examples of crustacean rhythmic processes in the preceeding section have touched on a large portion of the research on crustaceans, from the cellular to the organismic level, and from field observations to the complexities of neurophysiological recording. The starting point of this chapter was the suggestion that biological timing is widespread as a mechanism in these rhythmic events. In subsequent sections of this chapter, a pervading theme has been the hypothesis of fitness or adaptiveness of biological clocks in crustaceans as well as other animals, through their molding of multiple animal functions for greater efficiency and survival. It has often been presumed that genetic fitness, the ability to survive and place offspring in future generations, is enhanced by appropriate temporal and spatial programming of functions.

A major difficulty in evaluating this hypothesis is the almost total lack of rigorous testing of the adaptiveness of specific physiological and behavioral functions in any species. As pointed out by Enright (1970), the field of

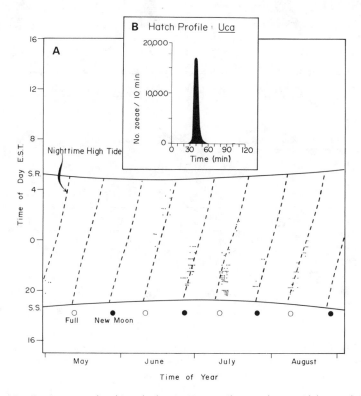

Fig. 10. Persistent egg-hatching rhythm in *Uca pugilator* under nontidal natural daylight conditions. (A) "Hatching profile" for a single female *Uca pugilator* in the laboratory at 25°C using manual transfer of the female at 10-min intervals, with subsequent preservation and counting of zoeae (P. J. DeCoursey, unpublished). (B) Peak hatch times for 175 *Uca pugilator* females in LD 12:12 in the laboratory at 25°C under tide-free conditions, using manual transfer at 30-min intervals (adapted from Bergin, 1981). Symbols: SS, sunset; SR, sunrise; ○, full moon; ●, new moon, - - -, time of predicted nighttime high tide (adapted from Bergin, 1978).

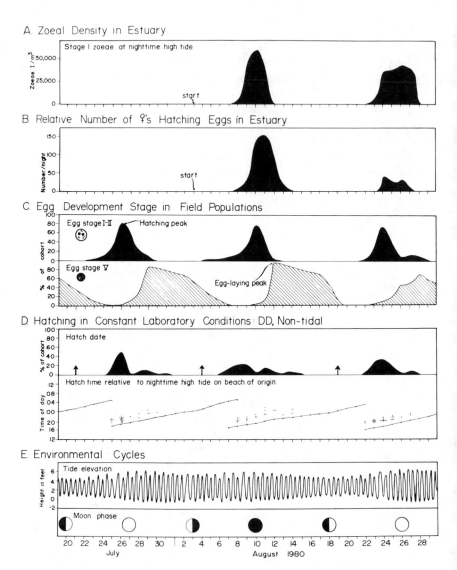

A. Zoeal Density in Estuary

B. Relative Number of ♀'s Hatching Eggs in Estuary

C. Egg Development Stage in Field Populations

D. Hatching in Constant Laboratory Conditions: DD, Non-tidal

E. Environmental Cycles

Fig. 11. Rhythmic reproductive parameters in the fiddler crab *Uca pugilator* from a site in North Inlet, South Carolina. (A) Number of newly hatched zoeae/m³ pumped over the 90 min following nighttime high tide at a shallow creek station. (B) Relative number of ovigerous females walking down to water's edge to release larvae, based on counts per evening along 50 m of typical habitat. (C) Egg development stage based on standard trenched samples of 50 to 100 individuals staged immediately after collection under a dissecting scope using morphological characters; above for late egg stage (I and II) within 24 hr of hatching and below for early egg stage V. (D) Persistent hatching rhythm of 3 cohorts of 32 ovigerous crabs collected 7 days before full or new moon as indicated by ↑, then maintained and monitored individually for peak hatch time in nontidal conditions of constant darkness at 25°C, with semilunar rhythm plotted on the basis of daily number of hatches (above), and tidal rhythm relative to nighttime

biological timing is no exception. Most authors have merely ignored or speculated about fitness. It would seem logical, for example, that hatching of larval fiddler crabs predominantly at the turn of the nighttime high spring tides twice each month, based on a genetic timer, would benefit the species. The effort to rigorously test the role of biological timing in fitness for most of the examples in Section IV is very great and usually not feasible.

Evolutionary change operates through an ultimate cause and a proximate mechanism. The proximate mechanism may be experimentally dissected by a variety of methods. Thus, a genetic basis for timing in several periodicities, particularly circadian ones, has been demonstrated along with the ability of a specific key factor to adjust frequency and phase to match local time (Section III). Frequencies not advantageous or no longer functional may be deleted. The cave crayfish is a possible example; after long isolation in a lightless environment, free-running rhythm precision and ability to entrain to a light–dark cycle were reduced in comparison with stream crayfish (Jegla and Poulson, 1968).

More difficult to assess are the ultimate questions: Why use an endogenous clock? What selective forces operate in deleting nonadaptive changes or in favoring favorable genetic changes? The problem could best be approached with experiments comparing individual and offspring survival of a group with normal timing and a clock-impaired group or clockless group when exposed to a wide variety of environmental selective forces for a number of generations. Selective forces, however, may be complex, multiple, and intertwined. A simple example from crabs will illustrate this point. Rhythmic color change in beach fiddler crabs may be selected for as an antipredatory feature relative to the day–night cycle, but must also account for superimposed tide-related variations in water level that scan across the day–night frequency. Furthermore, color may play other roles than protection, such as in territorial defense and courtship. Many Panamanian crabs on exposed sandy beaches undergo "brightening" of the courting male about 1 hr before low tide, when display and courtship become most intense. The female remains cryptically colored, while the male assumes the conspicuous courtship colors characteristic of each species (P. J. DeCoursey, unpublished). Interaction between courtship and protection needs is seen.

In the absence of extensive data about selective advantages of clocks, a few comments may be helpful. With respect to why an endogenous clock is an asset, advantages are seen in several categories of functions. For many

high tide based on continuous monitoring (below). Symbols: ●, time of each individual hatch; ————, time of nighttime high tide on beach of origin. (E) Environmental cycles shown as predicted tidal elevations on beach of origin (above), and moon phase (below); ○ = full moon, ● = new, ◑ and ◐ = quarters. (P. J. DeCoursey and Middaugh, unpublished.)

animals, an endogenous timer acts as a wake-up timer, to alert it in advance of a specific event in its niche, thereby allowing for preparation in anticipation of the event. This element may be particularly important for animals lacking a direct environmental clue to the start of favorable conditions; some animals in light-poor environments of caves and the aphotic zones of the oceans may have activities which are tied to the day–night changes. Good examples for crustaceans are burrowing lobsters, or vertical migrants who move from the aphotic zones into upper water levels at night to avoid predation. The warm-up timer of bats (DeCoursey and DeCoursey, 1964), serving to wake up a torpid bat and bring about raising of temperature and metabolic rate prior to dusk flights, may have a counterpart in cold water invertebrates. Still another advantage of endogenous timing concerns animals in niches with very noisy environmental signals; a probability approach to the start of activity could be successful in intertidal organisms. Another major class of functions depending upon internal timing is navigation, using a moving celestial reference point in migration; such continuously consulted clocks need a reliable internal time base. A further need for internal timing is the entrainment of a large variety of body functions with different phase relationships to each other, for which appropriate individual triggering stimuli are not available in the environment; the "temporal organization" of animals has been graphically described by Pittendrigh (1973).

The question of why a self-sustained pacemaker is necessary is still not clear. Most clocks are capable of generating precise rhythms for the lifetime of the organism, but most organisms do not live for more than a few days in constant environmental conditions. Enright (1970) has commented that an hourglass-type timer, tipped each day and capable of generating one or at most a few cycles might work as well; he has suggested that the self-sustained nature may be an intrinsic feature of the mechanism which evolved, or perhaps it may be needed to retain flexible learned behavior such as timed feeding patterns.

The final issue of what selective forces are important lacks rigorous experimental evidence. Insight may be gained by relating endogenous timing examples with critical life history features. Because the direct information on selective advantages is so limited, the following paragraphs will focus on ecology of clocks in crustaceans within the framework of "uses" of clocks in the animal kingdom. Rhythmic color changes of several types have been presented earlier. Integumentary color as an antipredator device in Uca is presumably effective, with maximal pigment dispersal at times of daytime low tide when fiddlers are most active and very exposed to predators (Powell, 1962a). The endogenous nature of the color rhythm has been controversial, and little is known of entrainment. Far more certain is the rhythm of pigment dispersal in crayfish eyes (Aréchiga, 1977); the rhythm may function in increased light sensitivity at night and greater acuity by daylight.

Locomotor activity is clearly under clock control in several crustacean species. Locomotor rhythms in species such as *Uca* are rather noisy in contrast to the precise locomotor rhythms of terrestrial mammals, birds, and lizards, or of terrestrial invertebrates such as cockroaches (Roberts, 1960). However, in some crustaceans, activity is exact, not only in frequency, but also in pattern. Nocturnal activity in the amphipod *Talitrus* (Williams, 1979) or the semiterrestrial crabs *Ocypode* and *Gecarcinus* (Palmer, 1971) may provide safer foraging on exposed beaches. Swimming patterns matching the pattern of tide height of the semidiurnal unequal tides in the beachhopper, *Excirolana,* trigger movement up and down the beach with the wave fronts and may prevent stranding.

Vertical migration is widespread among most of the crustacean subclasses, as well as in myctopid lantern fish. While vertical migration is found in freshwater species and estuarine species, it is best developed in openocean species. Here it probably serves many functions from metabolic efficiency to predator avoidance. Relatively few instances of clock-controlled vertical migrations have been shown (Enright and Hamner, 1967). Whether an inadequate technology for handling the difficult monitoring problems lies at fault is still unclear. However, many physiological response systems of phototaxis, pressure, temperature, and salinity responses have been shown to directly affect vertical migration, and it may well be that an internal clock is not operating in the responses of many species.

Migration and homing have developed among many organisms, both marine and terrestrial. Against the background of the long-distance exploits of salmon, eels, sea turtles and birds, the migratory feats of crustaceans are weak by comparison. Giant land crabs, *Cardisoma,* may wander a few miles to the ocean for spawning (Gifford, 1962), or spiny lobsters may queue for several miles (Herrnkind, 1970), but the fact is clear that crustaceans are basically minute to small in size with relatively weak locomotory apparatus. The two prime functions of long-distance migration in vertebrates and insects are escape from inclement weather and procurement of a food supply for breeding. In crustaceans, complex dispersal mechanisms of planktonic larvae or vertical migration may accomplish this purpose rather than extensive migration of adults. Perhaps the need for celestial navigation has therefore not arisen; closest to this need is spatial orientation in the escape response of beachhoppers (Papi, 1960; Pardi, 1960; Hartwick, 1976) using solar and lunar navigation.

Rhythmic reproduction is highly developed in many crustaceans. The precision of the rhythmic egg hatching discussed earlier for *Uca* has parallels in other crabs, including *Cardisoma* and *Sesarma.* Placing offspring in a suitable environment at a time when they can best survive may be vitally important, especially for intertidal crustaceans which hatch a large mass of planktonic larvae. A timing error of several hours could have very deleterious effects. All rhythmic aspects of reproduction from courtship to egg laying, egg hatching, larval dispersal, and larval settlement may be adaptive. Nighttime hatching at high tide provides protection to the egg-laden female Uca or Sesarma with minimal exposure to visual predators as she walks to the water's edge (Fig. 11). The pulse-type hatching mode is a well-known device to "saturate" the environment periodically and thereby confound predators. Release of larvae on a turning high tide prevents stranding on inhospitable mud or in stagnant pools, and ensures transport to areas where conditions are optimal for development. In addition, megalopa may possibly ride the flooding spring tides back up into the estuary to reach suitable mudflats (J. H. Christy, personal communication). Rhythmic courtship may have its own selective advantages, or may be a consequence of hatching, dispersal, and recruitment needs. At least the egg-hatching function has been shown to be accurately timed, even after maintenance in nontidal laboratory conditions for several weeks. Similar reproductive rhythms occur in other marine phyla. Lunar spawning is found in the Palolo worm and the Atlantic fireworm (Hauenschild, 1960); semilunar rhythms of reproduction are seen in the snails Melampus (Russel-Hunter et al., 1972; Berry and Chew, 1973) or Littorina (Struhsaker, 1966); whether endogenous timing regulates these species remains to be studied. The reproductive feat of one insect, Clunio, entails highly precise endogenous timing; ecolsion of adults, mating, and egg deposition occur during the few hours of each semilunar cycle when the algal substrate of the larvae is exposed to air. The rhythm is entrained by interaction of semilunar moonlight cycles and the day–night cycle (Neumann, 1976a,b). Lunar reproductive rhythms also occur in echinoderms (Pearse, 1975). The spawning of the grunion (Walker, 1952; Thompson and Muench, 1976) or spawning of the lesser known Atlantic silversides (Middaugh, 1981) is also semilunar, timed to the high spring tides to ensure deposition on moist sand for development during the 2-week period before the next spring tide washes out the hatching eggs. In contrast, terrestrial organisms rarely exhibit mass population synchrony of reproduction to tidal or lunar signals. The majority, instead, are confined by the seasonal restrictions of food supply and avoidance of winter rigors or tropical dry seasons. The element of clock control of reproduction in terrestrial environments is often a photoperiodic time measurement to predict appropriate time of year for breeding. Highly precise circadian time-measurement in photoperiodic control of reproduction is now known for birds (Follett, 1973; Follett et al., 1981) and mammals (Elliott, 1976, 1981) and probably will be shown to operate in many more species in the future. Only a few instances of photoperiodic control of reproduction have been suggested for crustaceans (Barnes, 1963; Little, 1968), and no evidence is available for circadian time measurement in these instances.

VI. SUMMARY

The earth's main habitats are strongly cyclic in their physical factors. Dominating the freshwater and terrestrial cycles are the day–night solar cycle, reinforced by much more irregular temperature, humidity, and other solar-related cycles. Close to these major habitat divisions, in a rhythmic sense, is the photic zone of the open ocean, where the solar cycle still predominates in spite of the marine qualities of the habitat. In contrast, the littoral, intertidal, and subtidal regions of estuaries, beaches, and continental shelf are highly complex ecosystems with "noisy" rhythmic configurations of multiple tidal, solar, and lunar frequencies. A final set of environments is the aphotic zone of caves or the great ocean depths, where little or no rhythmic information is available. In all but the last of these environments, the capacity for rhythmic anticipatory response by organisms presumably equips them better for individual survival, as well as for increased genetic fitness by programming the appropriate event at an advantageous place at the right time. The intervention of a clock allows not only an alarm system to alert the organism in advance of need as in wake-up timers, but also a precision measuring instrument for continuous consulting as in migration of many species.

In suggesting directions for future research, one warning should first be raised. A common ecological practice has in the past been to correlate biological response with environmental stimuli and to attribute a causal relationship to the factor most strongly correlated. This paper suggests that such a simplistic view, which neglects possible programming by an endogenous pacemaker, will not give a complete answer. Ecologists cannot assume that simple correlation between animal function and environmental variables will reveal causal relationships (Enright, 1970).

In the future, the comparative ecological study of clock-controlled functions in crustaceans will give ever increasing understanding of the survival value of time structuring and of how it has evolved in animals. Inferences on the probable fitness of precise events such as larval release or sun-compass orientation should be followed up by showing underlying endogenous events and the corresponding timing mechanism to the temporal features of the niche. Studies should thus include the function of the rhythmic event, the limiting temporal conditions in the environment, and the specific attributes of the timer, such as period, precision, PRC, and the synchronizing agents.

Finally, crustaceans will provide an excellent model for studying the

150

Patricia J. DeCoursey

physiology of clocks. Along with the *Aplysia* eye, the bird SCN–pineal complex, and the mammal SCN, the crustacean brain–eyestalk complex is rapidly becoming a favored system for analysis. The relative simplicity of neural–hormonal structure and accessibility for manipulation in crustaceans such as the crayfish have attracted an increasing number of scientists in the past two decades, and the field promises to continue yielding important new insights and breakthroughs in the future.

ACKNOWLEDGMENTS

This paper is Contribution Number 414 of the Belle W. Baruch Institute; partial support was provided by an AAUW Senior Postdoctoral Fellowship. I thank Beth Stewart Martin for technical assistance; Kitty Harper and Pat Lambert typed manuscript drafts. Dr. Michael Menaker and his students provided a subbatical haven and gave many helpful suggestions and comments. Dr. James Enright, Dr. Joseph Takahashi, Dr. William Herrnkind, and Ron Lipcius provided invaluable suggestions for improvement of the manuscript.

REFERENCES

Al-Adhub, A. H. Y., and Naylor, E. (1975). Emergence rhythms and tidal migrations in the brown shrimp, *Crangon crangon* (L.). *J. Mar. Biol. Assoc. U. K.* **55**, 801–810.
Al-Adhub, A. H. Y., and Naylor, E. (1976). Daily variation in *Dichelopandulus bonnieri* (Caullery) as a component of the epibenthos. In "Biology of Benthic Organisms" (B. F. Keegan, P. O. Ceidigh, and P. J. S. Boaden, eds.), pp. 1–6. Pergamon, New York.
Alheit, J., and Naylor, E. (1976). Behavioural basis of intertidal zonation in *Eurydice pulchra* Leach. *J. Exp. Mar. Biol. Ecol.* **23**, 135–144.
Allen, J. A. (1966). The rhythms and population dynamics of decapod Crustacea. *Oceanogr. Mar. Biol.* **4**, 247–266.
Allen, J. A. (1972). Recent studies on rhythms of post-larval decapod Crustacea. *Oceanogr. Mar. Biol.* **10**, 415–436.
Aréchiga, H. (1977). Circadian rhythmicity in the nervous system of crustaceans. *Fed. Proc., Fed. Am. Soc. Exp. Biol.* **36**, 2036–2041.
Aréchiga, H., and Mena, F. (1975). Circadian variations of hormonal content in the nervous system of the crayfish. *Comp. Biochem. Physiol.* **52A**, 581–584.
Aréchiga, H., and Wiersma, C. A. G. (1969). Circadian rhythm of responsiveness in crayfish visual units. *J. Neurobiol.* **1**, 71–85.
Arnold, D. C. (1970). A tidal rhythm in the response of the barnacle, *Balanus balanoides*, to water of diminished salinity. *J. Mar. Biol. Assoc. U. K.* **50**, 1040–1055.
Arudpragasam, K. D., and Naylor, E. (1964). Gill ventilation volumes, oxygen consumption, and respiratory rhythms in *Carcinus maenas* (L.) *J. Exp. Biol.* **41**, 309–321.
Aschoff, J. (1951). Messung der lokomotorischen Aktivität von Mäusen mittels der mechanischer Gleichrichter. *Pfluegers Arch.* **254**, 262–266.
Aschoff, J. (1960). Exogenous and endogenous components in circadian rhythms. *Cold Spring Harbor Symp. Quant. Biol.* **25**, 11–28.
Aschoff, J., ed. (1965a). "Circadian Clocks." North-Holland Publ., Amsterdam.

Aschoff, J. (1965b). Response curves in circadian periodicity. In "Circadian Clocks" (J. Aschoff, ed.), pp. 95–111. North-Holland Publ., Amsterdam.
Aschoff, J., ed. (1981). "Handbook of Behavioral Neurobiology. IV. Biological Rhythms." Plenum Press, New York.
Aschoff, J., and Homa, K. (1950). Art- und Individual-Muster der Tagesperiodik. Z. Physiol. **42**, 383–392.
Aschoff, J., and Meyer-Lohmann, J. (1954). Angeborene 24-Stunden Periodik beim Kücken. Pfluegers Arch. **260**, 170–176.
Aschoff, J., Fatransk, M., Giedke, H., Doerr, P., Staman, D., and Wisser, H. (1971a). Human circadian rhythms in continuous darkness: Entrainment by social clues. Science **171**, 213–215.
Aschoff, J., Gerecke, U., Kureck, A., Pohl, H., Rieger, P., von Saint-Paul, U., and Wever, R. (1971b). Interdependent parameters of circadian activity rhythms in birds and man. In "Biochronometry" (M. Menaker, ed.), pp. 3–29. Natl. Acad. of Sci., Washington, D.C.
Atkinson, R. J. A., and Naylor, E. (1976). An endogenous activity rhythm and the rhythmicity of catches of Nephrops norvegicus (L.). J. Exp. Mar. Biol. Ecol. **25**, 95–108.
Bainbridge, R. (1961). Migrations. In "Physiology of Crustacea" (T. H. Waterman, ed.), vol. II, pp. 431–463. Academic Press, New York.
Ball, E. E. (1968). Activity patterns and retinal pigment migration in Pagurus (Decapoda, Paguridae). Crustaceana **14**, 302–306.
Barham, E. G. (1963). Siphonophores and the deep scattering layer. Science **140**, 826–828.
Barham, E. G. (1966). Deep scattering layer migration and composition: Observation from a diving saucer. Science **151**, 1399–1402.
Barnes, H. (1963). Light, temperature, and breeding of Balanus balanoides. J. Mar. Biol. Assoc. U. K. **43**, 717–728.
Barnwell, F. H. (1966). Daily and tidal patterns of activity in individual fiddler crabs (Genus Uca) from the Woods Hole region. Biol. Bull. (Woods Hole, Mass.) **130**, 1–17.
Barnwell, F. H. (1968). The role of rhythmic systems in the adaptation of fiddler crabs to the intertidal zone. Am. Zool. **8**, 569–583.
Barnwell, F. H. (1976). Variation in the form of the tide and some problems it poses for biological timing systems. In "Biological Rhythms in the Marine Environment" (P. J. DeCoursey, ed.), pp. 161–188. Univ. of South Carolina Press, Columbia, South Carolina.
Baylor, E. R., and Smith, F. E. (1957). Diurnal migration of planktonic crustaceans. In Recent Advances in Invertebrate Physiology" (B. Sheer, ed.), pp. 21–35. Univ. of Oregon Publications, Eugene, Oregon.
Bennett, M. F. (1974). "Living Clocks in the Animal World." Thomas, Springfield, Illinois.
Bennett, M. F., Shriner, J., and Brown, R. A. (1957). Persistent tidal cycles of spontaneous motor activity in the fiddler crab, Uca pugnax. Biol. Bull. (Woods Hole, Mass.) **112**, 267–275.
Benson, J. A., and Lewis, R. D. (1976). An analysis of the activity rhythm of the sand beach amphipod, Talorchestia quoyana J. Comp. Physiol. **105**, 339–352.
Benzie, V. (1968). Some ecological aspects of the spawning behavior and early development of the common whitebait Galaxias maculatus attenuatus (Jenyns). Proc. N. Z. Ecol. Soc. **15**, 31–39.
Bergin, M. E. (1978). Hatching rhythms in Uca pugilator (Bosc) (Decapoda, Barchyura) from North Inlet, South Carolina. M.S. Thesis, 46 pp. Univ. of South Carolina, Columbia, South Carolina.
Bergin, M. E. (1981). Hatching rhythms in Uca pugilator (Decapoda: Brachyura). Mar. Biol. **63**, 151–158.
Berry, A. J., and Chew, E. (1973). Reproductive systems and cyclic release of eggs in Littorina

melanostoma from Malayan mangrove swamps (Mollusca: Gastropoda). *J. Zool.* **171,** 333–344.

Bierhuizen, J. F., Organizing Chairman (1972). "Circadian Rhythmicity: Proceedings of the Symposium on Circadian Rhythmicity." Center for Agricultural Publishing and Documentation, Wageningen.

Binkley, S. (1976). Computer methods of analysis for biorhythm data. *In* "Biological Clocks in the Marine Environment" (P. J. DeCoursey, ed.), pp. 53–62. Univ. of South Carolina Press, Columbia, South Carolina.

Blaxter, J. H. S., and Currie, R. I. (1967). The effect of artificial lights on acoustic scattering layers in the ocean. *Symp. Zool. Soc. London* **19,** 1–14.

Boden, B. P., and Kampa, E. M. (1967). The influence of natural light on the vertical migrations of an animal community in the sea. *Symp. Zool. Soc. London* **19,** 15–26.

Bosch, H. F., and Taylor, W. R. (1973). Diurnal vertical migration of an estuarine cladoceran *Podon polyphemoides* in Chesapeake Bay. *Mar. Biol. (Berlin)* **19,** 172–181.

Brady, J. (1967). Control of the circadian rhythm of activity in the cockroach. *J. Exp. Biol.* **47,** 165–178.

Brady, J. (1979). "Biological Clocks." Camelot Press, Ltd., Southhampton.

Branford, J. R. (1978). The influence of daylength, temperature, and season on the hatching rhythm of *Homarus gammarus. J. Mar. Biol. Assoc. U. K.* **58,** 639–658.

Bregazzi, P. K. (1972). The effects of low temperature upon the locomotor activity rhythm of *Talitrus saltator* (Montague) (Crustacea: Amphipoda). *J. Exp. Biol.* **57,** 393–399.

Bregazzi, P. K., and Naylor, E. (1972). The locomotor activity rhythm of *Talitrus saltator* (Montague) (Crustacea: Amphipoda). *J. Exp. Biol.* **57,** 375–591.

Bright, T. F., Ferrari, F., Martin, D., and Franceschini, G. A. (1972). Effects of a total solar eclipse on the vertical distribution of certain oceanic zooplankters. *Limnol. Oceanogr.* **17,** 296–301.

Brown, F. A., Jr. (1972). The "clocks" timing biological behavior. *Am. Sci.* **60,** 756–766.

Brown, F. A., Jr., Fingerman, M., Sandeen, M., and Webb, H. M. (1953). Persistent diurnal and tidal rhythms of color change in the fiddler crab, *Uca pugnax. J. Exp. Zool.* **123,** 29–60.

Brown, F. A., Jr., Bennett, M. F., and Webb, H. M. (1954). Daily and tidal rhythms of O_2-consumption in fiddler crabs. *J. Cell. Comp. Physiol.* **44,** 477–506.

Brown, F. A., Jr., Webb, H. M., and Bennett, M. F. (1955). Proof for an endogenous component in persistent solar and lunar rhythmicity in organisms. *Proc. Natl. Acad. Sci. U.S.A.* **41,** 93–100.

Bünning, E. (1936). Die endogene Tagesrhythmik als Grundlage der photoperiodische Reaktion. *Ber. Dtsch. Bot. Ges.* **54,** 590–607.

Bünning, E. (1959). "Die physiologische Uhr." Springer-Verlag, Berlin and New York.

Caruthers, J. W., Thompson, R. C., Novarini, J. C., and Francischini, G. A. (1970). The response of deep scattering layers in the Gulf of Mexico to a total solar eclipse. *Deep Sea Res.* **19,** 337–338.

Cederwall, H. (1979). Diurnal oxygen consumption and activity of two *Pontoporeia* (Amphipoda: Crustacea) species. *In* "Cyclic Phenomena in Marina Plants and Animals" (E. Naylor and R. G. Hartnoll, eds.), pp. 309–316. Pergamon, New York.

Chandrashekaran, M. K. (1965). Persistent tidal and diurnal rhythm of locomotor activity and oxygen consumption in *Emerita asiatica. Z. Physiol.* **50,** 137–150.

Chauvnick, A., ed. (1960). "Biological Clocks." Waverly, Baltimore, Maryland.

Christy, J. H. (1978). Adaptive significance of reproductive cycles in the fiddler crab *Uca pugilator:* An hyopthesis. *Science* **199,** 453–455.

Christy, J. H., and Stancyk, S. E. (1982). Timing of larval production and flux of invertebrate larvae in a well-mixed estuary. In "Estuarine Comparisons" (V. S. Kennedy, ed.), pp. 489–503. Academic Press, New York.

Corey, S. (1970). The diurnal vertical migration of some Cumacea (Crustacea, Pericarida) in Kames Bay, Isle of Cumbrae, Scotland. *Can. J. Zool.* **48,** 1385–1388.

Corliss, J. B., Dymond, J., Gordon, L. I., Edmond, J. M., von Herzen, R. P., Ballard, R. D., Green, K., Williams, D., Bainbridge, A., Crane, K., and von Andel, T. H. (1979). Submarine thermal springs on the Galápagos Rift. *Science* **203,** 1073–1083.

Coull, B. C., and Vernberg, W. B. (1975). Reproductive periodicity of meiobenthic copepods: Seasonal or continuous? *Mar. Biol. (Berlin)* **32,** 289–293.

Cronin, T. W., and Forward, R. B., Jr. (1979). Tidal vertical migration: An endogenous rhythm in estuarine crab larvae. *Science* **205,** 1020–1022.

Currie, R. I., Boden, B. P., and Kampa, E. M. (1969). An investigation on sonic scattering layers: The RRS DISCOVERY SOND cruise 1965. *J. Mar. Biol. Assoc. U. K.* **49,** 489–514.

Cushing, D. (1951). Vertical migration of planktonic Crustacea. *Biol. Rev.* **26,** 158–192.

Cushing, D. H. (1955). Some experiments on the vertical migration of zooplankton. *J. Anim. Ecol.* **24,** 137–166.

Daan, S., and Pittendrigh, C. S. (1976a). A functional analysis of circadian pacemakers in nocturnal rodents, III. Heavy water and constant light: Homeostasis of frequency? *J. Comp Physiol.* **106,** 267–290.

Daan, S., and Pittendrigh, C. S. (1976b). A functional analysis of circadian pacemakers in nocturnal rodents. II. The variability of phase response curves. *J. Comp. Physiol.* **106,** 253–266.

Davis, F. C., and Menaker, M. (1980). Hamsters through time's window: Temporal structure of hamster locomotor rhythmicity. *Am. J. Physiol.* **239,** 149–155.

DeCoursey, G., and DeCoursey, P. J. (1964). Adaptive aspects of activity rhythms in bats. *Biol. Bull. (Woods Hole, Mass.)* **126,** 14–27.

DeCoursey, P. J. (1961). Effect of light on the circadian activity rhythm of the flying squirrel, *Glaucomys volans. Z. Physiol.* **44,** 331–354.

DeCoursey, P. J. (1964). Function of light response rhythm in hamsters. *J. Cell. Comp. Physiol.* **63,** 189–196.

DeCoursey, P. J. (1973). Free-running rhythms and patterns of circadian entrainment in three species of diurnal rodents. *J. Interdiscp. Cycle Res.* **4,** 67–77.

DeCoursey, P. J., ed. (1976a). "Biological Rhythms in the Marine Environment," 283 pp. Univ. of South Carolina Press, Columbia, South Carolina.

DeCoursey, P. J. (1976b). Vertical migration of larval *Uca* in a shallow estuary. *Am. Zool.* **16,** 244.

DeCoursey, P. J., ed. (1977). "PRC Atlas." Hopkins Marine Station, Pacific Grove.

DeCoursey, P. J. (1979). Egg-hatching rhythms in three species of fiddler crabs. *In* "Cyclic Phenomena in Marine Plants and Animals" (E. Naylor and R. G. Hartnoll, eds.), pp. 399–406. Pergamon, New York.

Dieleman, J. (1977). Circatidal activity rhythms and the annual migration cycle in an estuarine population of *Gammarus zaddachi* Sexton, 1912. *Crustaceana Suppl.* **4,** 81–87.

Dietz, R. (1962). The seas deep scattering layers. *Sci. Am.* **207,** 44–50

Donaldson, H. A. (1975). Vertical distribution and feeding of sergestid shrimps (Decapoda: Natantia) collected near Bermuda. *Mar. Biol.* **31,** 37–50.

Donaldson, H. A., and Pearcy, W. G. (1972). Sound scattering layers in the northeastern Pacific. *J. Fish. Res. Board Can.* **29,** 1419–1423.

Dudley, D. L., and Judy, M. H. (1973). Seasonal abundance and distribution of juvenile blue crabs in Core Sound, North Carolina, 1965–1968. *Chesapeake Sci.* **14,** 51–55.

Dumont, H. J. (1972). A competition-based approach of the reverse vertical migration in zooplankton and its implication, chiefly based on a study of the interactions of the rotifer *Asplanchna priodota* with several crustacean Entomostraca. *Int. Rev. Gesamten Hydrobiol.* **57,** 1–38.

Elliott, J. A. (1976). Circadian rhythms and photoperiodic time measurement in mammals. *Fed. Proc., Fed. Am. Soc. Exp. Biol.* **35,** 2339–2347.

Elliott, J. A. (1981). Circadian rhythms, entrainment, and photoperiodism in the Syrian hamster. *In* "Biological Clocks in Seasonal Reproductive Cycles" (B. K. Follett and D. E. Follett, eds.), pp. 203–217. John Wright and Sons, Ltd., Bristol.

Ennis, G. P. (1973a). Behavioral responses to changes in hydrostatic pressure and light during larval development of the lobster *Homarus gammarus. J. Fish Res. Board Can.* **30,** 1349–1360.

Ennis, G. P. (1973b). Endogenous rhythmicity associated with larval hatching in the lobster *Homarus gammarus. J. Mar. Biol. Assoc. U. K.* **53,** 531–538.

Enright, J. T. (1963). The tidal rhythms of activity of a sand-beach amphipod. *Z. Physiol.* **46,** 276–313.

Enright, J. T. (1965a). Entrainment of a tidal rhythm. *Science* **147,** 864–867.

Enright, J. T. (1965b). Accurate geophysical rhythms and frequency analysis. *In* "Circadian Clocks" (J. Aschoff, ed.), pp. 31–42. North-Holland Publ., Amsterdam.

Enright, J. T. (1965c). The search for rhythmicity in biological time-series. *J. Theor. Biol.* **8,** 426–468.

Enright, J. T. (1965d). Synchronization and ranges of entrainment. *In* "Circadian Clocks" (J. Aschoff, ed.), pp. 112–125. North-Holland Publ., Amsterdam.

Enright, J. T. (1970). Ecological aspects of endogenous rhythmicity. *Annu. Rev. Ecol. Syst.* **1,** 221–238.

Enright, J. T. (1971a). Heavy water slows biological timing processes. *Z. Physiol.* **72,** 1–16.

Enright, J. T. (1971b). The internal clock of drunken isopods. *Z. Physiol.* **75,** 332–346.

Enright, J. T. (1972). A virtuoso isopod: Circa-luna rhythms and their tidal fine structure. *J. Comp. Physiol.* **77,** 141–162.

Enright, J. T. (1975). Orientation in time: Endogenous clocks. *Mar. Ecol.* **2,** pt. 2, 917–944.

Enright, J. T. (1976a). Resetting a tidal clock: A phase response curve for *Excirolana. In* "Biological Rhythms in the Marine Environment" (P. J. Decoursey, ed.), pp. 103–114. Univ. of South Carolina Press, Columbia, South Carolina.

Enright, J. T. (1976b). Plasticity in an isopod's clockworks: Shaking shapes form and affects phase and frequency. *J. Comp. Physiol.* **107,** 13–37.

Enright, J. T. (1977a). Diurnal vertical migration: Adaptive significance and timing. Part I. Selective advantage: A metabolic model. *Limnol. Oceanogr.* **22,** 856–872.

Enright, J. T. (1977b). Copepods in a hurry: Sustained high-speed upward migration. *Limnol. Oceanogr.* **22,** 118–125.

Enright, J. T. (1978). Migration and homing of marine invertebrates: A potpourri of strategies. *In* "Animal Migration, Naviation, and Homing" (K. Schmidt-Koenig and W. T. Keeton, eds.), pp. 440–446. Springer-Verlag, Berlin and New York.

Enright, J. T., and Hamner, W. M. (1967). Vertical diurnal migration and endogenous rhythmicity. *Science* **157,** 937–941.

Enright, J. T., and Honegger, H.-W. (1977). Diurnal vertical migration: Adaptive significance and timing. Part II. Test of the model: Details of timing. *Limnol. Oceanogr.* **22,** 873–886.

Enright, J. T., Newman, W. A., Hessler, R. R., and McGowan, J. A. (1981). Deep-ocean hydrothermal vent communities. *Nature (London)* **289,** 219–221.

Eskin, A. (1971). Some properties of the system controlling the circadian activity rhythm of sparrows. *In* "Biochronometry" (M. Menaker, ed.), pp. 55–80. Natl. Acad. of Sci., Washington, D.C.

Eskin, A. (1979). Circadian system of the *Aplysia* eye: Properties of the pacemaker and mechanisms of its entrainment. *Fed. Proc., Fed. Am. Soc. Exp. Biol.* **38**, 2573–2579.

Fincham, A. A. (1970a). Rhythmic behavior of the intertidal amphipod *Bathyporeia pelagica*. *J. Mar. Biol. Assoc. U. K.* **50**, 1057–1068.

Fincham, A. A. (1970b). Amphipods in the surf plankton. *J. Mar. Biol. Assoc. U. K.* **50**, 177–198.

Fingerman, M., and Lowe, M. (1957). Twenty-four hour rhythm of distal pigment migration in the dwarf crayfish. *J. Cell. Comp. Physiol.* **50**, 371–379.

Fingerman, M., and Yamamoto, Y. (1964). Daily rhythm of color change in eyestalkless fiddler crabs, *Uca pugilator*. *Am. Zool.* **4**, 334.

Fish, J. D., and Fish, S. (1972). The swimming rhythm of *Eurydice pulchra* Leach and a possible explanation of intertidal migration. *J. Exp. Mar. Biol. Ecol.* **8**, 195–200.

Follett, B. K. (1973). Circadian rhythms and photoperiodic time measurement in birds. *J. Reprod. Fert., Suppl.* **19**, 5–18.

Follett, B. K., Robinson, J. E., Simpson, S. M. and Harlow, C. R. (1981). Photoperiodic time measurement and gonadotrophin secretion in quail. *In* "Biological Clocks in Seasonal Reproduction Cycles" (B. K. Follett, ed.). John Wright & Sons, Ltd., Bristol. (In press.)

Forward, R. B., Jr. (1974). Negative phototaxis in crustacean larvae: Possible functional significance. *J. Exp. Mar. Biol. Ecol.* **16**, 11–17.

Forward, R. B., Jr. (1976). Light and diurnal vertical migration: Photobehavior and photophysiology of plankton. *Photochem. Photobiol. Rev.* **1**, 157–209.

Forward, R. B., Jr. (1980). Phototaxis of a sand-beach amphipod: Physiology and tidal rhythms. *J. Comp. Physiol.* **135**, 243–250.

Forward, R. B., Jr., and Costlow, J. D., Jr. (1974). The ontogeny of phototaxis by larvae of the crab *Rhithropanopeus harrisii*. *Mar. Biol.* **26**, 27–33.

Forward, R. B., Jr. and Cronin, T. W. (1980). Tidal rhythms of activity and phototaxis of an estuarine crab larva. *Biol. Bull. (Woods Hole, Mass.)* **158**, 295–303.

Foulds, J. B., and Roff, J. C. (1976). Oxygen consumption during simulated vertical migration in *Mysis relicta* (Crustacea: Mysidacea). *Can. J. Zool.* **54**, 377–385.

Foxton, P. (1970). The vertical distribution of pelagic decapods (Crustacea: Nantantia) collected on the SOND Cruise 1965. I. The Caridae. II. The Penaeidea and general discussion. *J. Mar. Biol. Assoc. U. K.* **50**, 939–960, 961–1000.

Gibson, R. N. (1970). The tidal rhythm of *Coryphoblennius galerita* (L.) (Teleostei, Blenniidae). *Anim. Behav.* **18**, 539–543.

Gifford, C. A. (1962). Some observations on the general biology of landcrabs, *Cardisoma guanhumi* (Latreille), in South Florida. *Biol. Bull. (Woods Hole, Mass.)* **123**, 207–223.

Grindley, J. R. (1964). Effect of low salinity water on the vertical migration of estuarine plankton. *Nature (London)* **203**, 781–782.

Guerderat, J. A., and Friess, R. (1971). The extent of the diurnal migration of bathypelagic copepods. *Cah. ORSTOM, Ser. Oceanogr.* **9**, 187–196.

Gwinner, E. (1966). Entrainment of a circadian rhythm in birds by species specific song cycles (Aves, Fringillidae: *Carduelis spinus, Serinus serinus*). *Experientia* **22**, 765.

Hagerman, L. (1970). Locomotory activity patterns of *Crangon vulgaris* (Fabricius) (Crustacea, Natantia). *Ophelia* **8**, 255–266.

Hamann, A. (1974). Die neuroendokrine Steuerung tagesrhythmischer Blutzuckerschwankungen durch die Sinusdrüse beim Flusskrebs. *J. Comp. Physiol.* **89**, 197–214.

Hammond, R. D., and Fingerman, M. (1975). Entrainment analysis of the circadian locomotor activity rhythm in specimens of the crayfish, *Faxonella clypeate*, having activity peaks at different times of the solar day. *Chronobiologia (Milan)* **2**, 119–132.

Hammond, R. D., and Naylor, E. (1977). Effects of dusk and dawn on locomotor activity rhythms in the Norway lobster *Nephrops norvegicus. Mar. Biol. (Berlin)* **39,** 253–260.

Haney, J. F., and Hall, D. J. (1975). Diel vertical migration and filter-feeding activities of *Daphnia. Arch. Hydrobiol.* **75,** 413–441.

Hardy, A. C., and Paton, W. N. (1947). Experiments on the vertical migration of plankton animals. *J. Mar. Biol. Assoc. U. K.* **26,** 467–526.

Harker, J. E. (1964). "The Physiology of Diurnal Rhythms." Cambridge Univ. Press, London and New York.

Harris, J. E. (1963). The role of endogenous rhythms in vertical migration. *J. Mar. Biol. Assoc. U. K.* **43,** 153–166.

Harris, R. P. (1972). Seasonal changes in population density and vertical distribution of harpacticoid copepods on an intertidal sand beach. *J. Mar. Biol. Assoc. U. K.* **52,** 493–505.

Hartwick, R. F. (1976). Aspects of celestial orientation behavior in talitrid amphipods. In "Biological Rhythms in the Marine Environment" (P. J. DeCoursey, ed.), pp. 189–198. Univ. of South Carolina Press, Columbia, South Carolina.

Hauenschild, C. (1960). Lunar periodicity. *Cold Spring Harbor Symp. Quant. Biol.* **25,** 491–497.

Herman, S. S. (1963). Vertical migration of the opossum shrimp *Neomysis americana* Smith. *Limnol. Oceanogr.* **8,** 228–238.

Herrnkind, W. F. (1969). Queuing behavior of spiny lobsters. *Science* **164,** 1425–1427.

Herrnkind, W. F. (1970). Migration of the spiny lobster. *Nat. Hist.* **79,** 36–43.

Herrnkind, W., and Kanciruk, P. (1978). Mass migration of spiny lobster, *Panulirus argus* (Crustacea: Palinuridae:) Synopsis and orientation. In "Animal Migration, Navigation, and Homing" (K. Schmidt-Koenig and W. T. Keeton, eds.), pp. 430–439. Springer-Verlag, Berlin and New York.

Hoffmann, K. (1955). Aktivitätsregistrierungen bei frisch geschlüpften Eidechsen. *Z. Physiol.* **37,** 253–262.

Hoffmann, K. (1965a). Overt circadian frequencies and circadian rule. In "Circadian Clocks" (J. Aschoff, ed.), pp. 87–94. North-Holland Publ., Amsterdam.

Hoffmann, K. (1965b). Clock-mechanisms in celestial orientation of animals. In "Circadian Clocks" (J. Aschoff, ed.), pp. 426–441. North-Holland Publ., Amsterdam.

Hoffmann, K. (1968). Synchronisation der circadianen Aktivitätsperiodik von Eidechsen durch Temperaturcyclen verschiedener Amplitude. *Z. Physiol.* **58,** 225–228.

Hoffmann, K. (1969). Die relativ Wirksamkeit von Zeitgebern. *Oecologia* **3,** 184–206.

Honegger, H.-W. (1973a). Rhythmic motor activity responses of the California fiddler crab *Uca crenulata* to artificial light conditions. *Mar. Biol. (Berlin)* **18,** 19–31.

Honegger, H.-W. (1973b). Rhythmic activity responses of the fiddler crab *Uca crenulata* to artificial tides and artificial light. *Mar. Biol. (Berlin)* **21,** 196–212.

Honegger, H.-W. (1976). Locomotor activity in *Uca crenulata,* and the response to two Zeitgebers, light-dark and tides. In "Biological Rhythms in the Marine Environment" (P. J. DeCoursey, ed.), pp. 93–102. Univ. of South Carolina Press, Columbia, South Carolina.

Hu. V. J. H. (1978). Relationships between vertical migration and diet in four species of eupbausids. *Limnol. Oceanogr.* **23,** 296–306.

Hughes, D. A. (1972). On the endogenous control of tide-associated displacements of pink shrimp, *Penaeus duodarum. Biol. Bull. (Woods Hole, Mass.)* **142,** 271–280.

Hulbert, E. M. (1957). The distribution of *Neomysis americana* in the estuary of the Delaware River. *Limnol. Oceanogr.* **2,** 1–11.

Huntsman, A. G. (1948). *Odontosyllis* at Bermuda and lunar periodicity. *J. Fish. Res. Board Can.* **7,** 363–369.

Jacklet, J. W. (1969). Circadian rhythm of optic nerve impulses recorded in darkness from isolated eye of *Aplysia*. *Science* **164**, 562–563.

Jegla, T. C., and Poulson, T. L. (1968). Evidence of circadian rhythms in a cave crayfish. *J. Exp. Zool.* **168**, 273–282.

Jones, D. A., and Naylor, E. (1970). The swimming rhythm of the sand beach isopod *Eurydice pulchra*. *J. Exp. Mar. Biol. Ecol.* **4**, 188–199.

Kampa, E. M. (1975). Observations of a sonic-scattering layer during the total solar eclipse, 30 June 1973. *Deep Sea Res.* **22**, 417–423.

Kampa, E. M. (1976). Photoenvironment and vertical migrations of mesopelagic marine animal communities. *In* "Biological Rhythms in the Marine Environment" (P. J. DeCoursey, ed.), pp. 257–272. Univ. of South Carolina Press, Columbia, South Carolina.

Kampa, E. M., and Boden, B. P. (1954). Submarine illumination and the twilight movements of a sonic scattering layer. *Nature (London)* **174**, 869–973.

Kanciruk, P., and Herrnkind, W. (1978). Mass migration of spiny lobster, *Panulirus argus* (Crustacea: Palinuridae): Behavior and environmental correlates. *Bull. Mar. Sci.* **28**, 601–623.

Kerfoot, W. B. (1970). Bioenergetics of vertical migration. *Am. Nat.* **104**, 529–546.

Klapow, L. A. (1972). Natural and artificial rephasing of a tidal rhythm. *J. Comp. Physiol.* **79**, 233–258.

Klapow, L. A. (1976). Lunar and tidal rhythms of an intertidal crustacean. *In* "Biological Rhythms in the Marine Environment" (P. J. DeCoursey, ed.), pp. 215–224. Univ. of South Carolina Press, Columbia, South Carolina.

Knudsen, J. W. (1960). Reproduction, life history, and larval ecology of the California Xanthidae, the pebble crabs. *Pac. Sci.* **14**, 3–17.

Korringa, P. (1957). Lunar periodicity. *Mem. Geol. Soc. Am.* **67**, 917–934.

Kramer, G. (1951). Eine neue Methode zur Erforschung der Zugorientierung und die bisher damit erzielten Ergebnisse. *Proc. Int. Ornithol. Congr.* **10**, 271–280.

Lance, J. (1962). Effect of reduced salinity on the vertical migration of zooplankton. *J. Mar. Biol. Assoc. U. K.* **42**, 131–154.

Landry, M. R. (1975). Dark inhibition of egg hatching of the marine copepod *Acartia clausi*. *J. Exp. Mar. Biol. Ecol.* **20**, 43–47.

Larimer, J. L., and Smith, J. T. F. (1980). Circadian rhythm of retinal sensitivity in crayfish: modulation by the cerebral and optic ganglia. *J. Comp. Physiol.* **136**, 313–326.

Lehmann, U. (1976). Interpretation of entrained and free-running locomotor activity patterns of *Uca*. *In* "Biological Rhythms in the Marine Environment" (P. J. DeCoursey, ed.), pp. 77–92. Univ. of South Carolina Press, Columbia, South Carolina.

Lehmann, U., Neumann, D., and Kaiser, H. (1974). Gezeitenrhythmische und spontane Aktivitätsmuster von Winkerkrabben. I. Ein neuer Ansatz zur quantitativen Analyse von Lokomotionsrhythmen. *J. Comp. Physiol.* **91**, 187–221.

Lincoln, R. J. (1970). Laboratory investigation of effects of hydrostatic pressure on vertical migration of planktonic Crustacea. *Mar. Biol. (Berlin)* **6**, 5–11.

Lincoln, R. J. (1971). Observations on effects of changes of hydrostatic pressure and illumination on behavior of planktonic Crustacea. *J. Exp. Biol.* **54**, 677–688.

Lipcius, R., Winter, M., Andrea, S., and Herrnkind, W. (1979). Endogenous control of behavioral rhythms in solitary and grouped western Atlantic spiny lobster, *Panulirus argus*. *Am. Zool.* **19**, 889.

Little, G. (1968). Induced winter breeding and larval development in the shrimp, *Palaemonetes pugio* Holthuis (Caridea, Palaemonidae). *Crustaceana Suppl.* **2**, 19–26.

Loher, W. (1972). Circadian control of stridulation in the cricket *Teleogryllus commodus* Walker. *J. Comp. Physiol.* **79**, 173–190.

Longhurst, A. R. (1976). Vertical migration. In "The Ecology of the Seas" (D. H. Cushing and J. J. Welsh, eds.), pp. 116–137. Saunders, Philadelphia, Pennsylvania.

Londsale, D. J., and Coull, B. C. (1977). Composition and seasonality of zooplankton of North Inlet, South Carolina. Chesapeake Sci. **18**, 272–283.

Lough, R. G. (1976). Larval dynamics of the dungeness crab, Cancer magister, off the central Oregon coast, 1970–71. Fish Bull. **74**, 353–376.

McDowell, R. M. (1976). Lunar rhythms in aquatic animals, a general review. Tuatara **17**, 133–144.

McLaren, I. A. (1974). Effects of temperature on growth of zooplankton, and the adaptive value of vertical migration. J. Fish. Res. Board Can. **20**, 685–727.

Marlow, C. J., and Miller, C. B. (1975). Patterns of vertical distribution and migration of zooplankton at ocean station "P." Limnol. Oceanogr. **20**, 824–844.

Marzolf, G. R. (1965). Vertical migration of Pontoporeia affinis (Amphipoda) in Lake Michigan. Great Lakes Res. Div., Univ. Mich., Pub. No. 13, pp. 133–140.

Menaker, M., ed. (1971). "Biochronometry." Natl. Acad. of Sci., Washington, D.C.

Menaker, M., and Eskin, A. (1966). Entrainment of circadian rhythms by sound in Passer domesticus. Science **154**, 1579–1581.

Menaker, M., Takahashi, J. S., and Eskin, A. (1978). The physiology of circadian pacemakers. Annu. Rev. Physiol. **40**, 501–526.

Mercer, D. M. A. (1965). The limitations of detection of periodicities in random noise. In "Circadian Clocks" (J. Aschoff, ed.), pp. 23–30. North-Holland Publ., Amsterdam.

Middaugh, D. P. (1981). Reproductive ecology and spawning periodicity of the Atlantic silverside, Menidia menidia (Pisces: Atherinidae). Copeia (In press.)

Moller, T. H., and Branford, J. R. (1979). A circadian hatching rhythm in Nephrops norvegicus (Crustacea: Decapoda). In "Cyclic Phenomena in Marine Plants and Animals" (E. Naylor and R. G. Hartnoll, eds.), pp. 391–398. Pergamon, New York.

Moller, T. H., and Jones, D. A. (1975). Locomotory rhythms and burrowing habits of Penaeus semisulcatus (de Haan) and P. monodon (Fabricius) (Crustacea: Penaeidae). J. Exp. Mar. Biol. Ecol. **18**, 61–77.

Moore, R. Y. (1978). Central neural control of circadian rhythms. In "Frontiers in Neuroendocrinology" (W. F. Ganong and L. Martini, eds.), Vol. 5, pp. 185–206. Raven, New York.

Moore-Ede, M. C., Sulzman, F. M., and Fuller, C. A. (1982). "The Clocks that Time Us." Harvard University Press, Cambridge, Mass.

Morgan, E. (1965). The activity rhythm of the amphipod Corophium volutator (Pallas) and its possible relationship to changes in hydrostatic pressure associated with tides. J. Anim. Ecol. **34**, 731–746.

Naylor, E. (1958). Tidal and diurnal rhythms of locomotory activity in Carcinus maenas (L.). J. Exp. Biol. **35**, 602–610.

Naylor, E., and Hartnoll, R. G., eds. (1979). "Cyclic Phenomena in Marine Plants and Animals." Pergamon, New York.

Naylor, E., Atkinson, R. J. A., and Williams, B. G. (1971). External factors influencing the tidal rhythm of shore crabs. J. Interdiscipl. Cycle Res. **2**, 173–180.

Neumann, D. (1976a). Entrainment of a semilunar rhythm. In "Biological Rhythms in the Marine Environment" (P. J. DeCoursey, ed.), pp. 115–127. Univ. of South Carolina Press, Columbia, South Carolina.

Neumann, D. (1976b). Mechanismen für die zeitliche Anpassung von Verhaltenund Entwicklungsleistungen an den Gezeitenzyklus. Verh. Dtsch. Zool. Ges. 1976 pp. 9–28.

Neumann, D. (1981). Tidal and lunar rhythms. In "Handbook of Behavioral Neurobiology. IV. Biological Rhythms" (J. Aschoff, ed.), pp. 351–380. Plenum Press, New York.

Page, T. L., and Larimer, J. L. (1972). Entrainment of the circadian locomotor activity rhythm in crayfish: The role of the eyes and caudal photoreceptor. *J. Comp. Physiol.* **78,** 107–120.

Page, T. L., and Larimer, J. L. (1975a). Neural control of circadian rhythmicity in the crayfish. I. The locomotor activity rhythm. *J. Comp. Physiol.* **97,** 59–80.

Page, T. L., and Larimer, J. L. (1975b). Neural control of circadian rhythmicity in the crayfish. II. The ERG amplitude rhythm. *J. Comp. Physiol.* **97,** 81–96.

Page, T. L., and Larimer, J. L. (1976). Extraretinal photoreception in entrainment of crustacean circadian rhythms. *Photochem. Photobiol.* **23,** 245–251.

Palmer, J. D. (1964). A persistent, light-preference rhythm in the fiddler crab, *Uca pugnax,* and its possible adaptive significance. *Am. Nat.* **98,** 431–434.

Palmer, J. D. (1967). Daily and tidal components in the persistent rhythmic activity of the crab, *Sesarma. Nature (London)* **215,** 64–66.

Palmer, J. D. (1971). Comparative studies of circadian locomotory rhythms in four species of terrestrial crabs. *Am. Midl. Nat.* **85,** 97–107.

Palmer, J. D. (1973). Tidal rhythms: The clock control of the rhythmic physiology of marine organisms. *Biol. Rev.* **48,** 377–418.

Palmer, J. D. (1974). "Biological Clocks in Marine Organisms." Wiley (Interscience), New York.

Palmer, J. D. (1976). "An Introduction to Biological Rhythms." Academic Press, New York.

Papi, F. (1960). Orientation by night: The moon. *Cold Spring Harbor Symp. Quant. Biol.* **25,** 475–480.

Pardi, L. (1960). Innate components in the solar orientation of littoral amphipods. *Cold Spring Harbor Symp. Quant. Biol.* **25,** 395–401.

Pardi, L., and Papi, F. (1952). Die Sonne als Kompass bei *Talitrus saltator* (Montague) (Amphipoda-Talitridae). *Naturwissenschaften* **39,** 262–263.

Pavlidis, T. (1973). "Biological Oscillators: Their Mathematical Analysis." Academic Press, New York.

Pearcy, W. G. (1970). Vertical migration of the ocean shrimp *Pandalus jordani:* A feeding and dispersal mechanism. *Calif. Fish Game* **56,** 77–140.

Pearse, J. S. (1975). Lunar reproductive rhythms in sea urchins: A review. *J. Interdiscip. Cycle Res.* **6,** 47–52.

Pengelley, E. T., ed. (1974). "Circannual Clocks." Academic Press, New York.

Pittendrigh, C. S. (1954). On temperature independence in the clock system controlling emergence time in Drosophila. *Proc. Natl. Acad. Sci. U.S.A.* **40,** 1018–1029.

Pittendrigh, C. S. (1960). Circadian rhythms and the circadian organization of living systems. *Cold Springs Harbor Symp. Quant. Biol.* **25,** 159–184.

Pittendrigh, C. S. (1973). Circadian oscillations in cells and the circadian organization of multicellular systems. *In* "The Neurosciences; Third Study Program" (F. O. Schmidtt, ed.), pp. 437–458. MIT Press, Cambridge, Massachusetts.

Pittendrigh, C. S., and Daan, S. (1974). Circadian oscillations in rodents: A systematic increase of their frequency with age. *Science* **186,** 548–550.

Pittendrigh, C. S., and Daan, S. (1976a). A functional analysis of circadian pacemakers. I. The stability and lability of spontaneous frequency. *J. Comp. Physiol.* **106,** 223–252.

Pittendrigh, C. S., and Daan, S. (1976b). A functional analysis of circadian pacemakers in nocturnal rodents. V. Pacemaker structure: A clock for all seasons. *J. Comp. Physiol.* **106,** 333–355.

Poolsanguan, B., and Uglow, R. F. (1974). Quantitative changes in blood sugar levels in *Crangon vulgaris. J. Comp. Physiol.* **93,** 1–6.

Powell, B. L. (1962a). Types, distribution, and rhythmical behavior of the chromatophores of juvenile *Carcinus maenas. J. Anim. Ecol.* **31,** 251–261.

Powell, B. L. (1962b). Studies on rhythmical behavior in crustacea. I. Persistent locomotor activity in juvenile *Carcinus maenas* (L.) and *Ligia oceanica* (L.). *Crustaceana* **4**, 42–46.

Powell, B. L. (1966). The control of the 24-hour rhythm of colour change in juvenile *Carcinus maenas*. *Proc. R. Ir. Acad.* **64B**, 379–399.

Rajan, K. P., Kharout, H. H., and Lockwood, A. P. M. (1979). Rhythmic cycles of blood sugar concentrations in the crab *Carcinus maenas*. In "Cyclic Phenomena in Marine Plants and Animals" (E. Naylor and R. G. Hartnoll, eds.), pp. 451–458. Pergamon, New York.

Rawson, K. S. (1960). Effects of tissue temperature on mammalian activity rhythms. *Cold Spring Harbor Symp. Quant. Biol.* **25**, 105–113.

Rawson, K. S., and DeCoursey, P. J. (1976). A comparison of the rhythms of mice and crabs from intertidal and terrestrial habitats. In "Biological Rhythms in the Marine Environment" (P. J. DeCoursey, ed.), pp. 32–52. Univ. of South Carolina, Columbia, South Carolina.

Raymont, J. E. G. (1963). "Plankton and Productivity in the Oceans," pp. 418–466. Pergamon, New York.

Reaka, M. L. (1976). Lunar and tidal periodicity of molting and reproduction in stomatopod crustacea: a selfish herd hypothesis. *Biol. Bull.* *(Woods Hole Mass.)* **150**, 468–490.

Ringelberg, J. (1964). The positively phototactic reaction of *Daphnia magna* Straus, a contribution to the understanding of diurnal vertical migration. *Neth. J. Sea. Res.* **2**, 319–406.

Roberts, S. K. (1960). Circadian activity rhythms in cockroaches. I. The free-running rhythm in steady state. *J. Cell. Comp. Physiol.* **55**, 99–110.

Roberts, S. K. (1962). Circadian activity rhythms in cockroaches. II. Entrainment and phase shifting. *J. Cell. Comp. Physiol.* **59**, 175–186.

Roberts, S. K. (1974). Circadian rhythms in cockroaches. Effects of the optic lobe lesions. *J. Comp. Physiol.* **88**, 21–30.

Robertson, A. I., and Howard, R. K. (1978). Diel trophic interactions between vertically-migrating zooplankton and their fish predators in an eelgrass community. *Mar. Biol.* *(Berlin)* **48**, 207–213.

Rodriquez, G., and Naylor, E. (1972). Behavioural rhythms in littoral prawns. *J. Mar. Biol. Assoc. U. K.* **52**, 81–95.

Roe, H. S. J. (1972). The vertical distributions and diurnal migrations of calanoid copepods collected on the SOND Cruise, 1965. *J. Mar. Biol. Assoc. U. K.* **52**, 277–314, 314–343, 525–552, 1021–1044.

Rudjakov, J. A. (1970). The possible causes of diel vertical migrations of planktonic animals. *Mar. Biol.* *(Berlin)* **6**, 98–105.

Rudjakov, Y. A., and Veronina, N. M. (1973). Diurnal vertical migrations of copepod *Metridia gerlachei* in the Scotia Sea. *Okeanologiya* *(Moscow)* **13**, 512–514.

Rusak, B., and Zucker, I. (1975). Biological rhythms and animal behavior. *Annu. Rev. Psychol.* **26**, 137–171.

Rusak, B., and Zucker, I. (1979). Neural regulation of circadian rhythms. *Physiol. Rev.* **59**, 449–526.

Russell, F. S. (1927). The vertical distribution of plankton in the sea. *Biol. Rev.* **2**, 213–62.

Russell-Hunter, W. D., Apley, M. L., and Hunter, P. D. (1972). Early life-history of *Melampus* and the significance of semilunar synchrony. *Biol. Bull.* *(Woods Hole, Mass.)* **143**, 623–656.

Saigusa, M. (1980). Entrainment of a semilunar rhythm by a simulated moonlight cycle in the terrestrial crab, *Sesarma haematocheir*. *Oecologia* 46, 38–44.

Saigusa, M., and Hidaka, T. (1978). Semilunar rhythm in the zoea-release activity of the land crabs *Sesarma*. *Oecologia* **37**, 163–176.

Sandifer, P. A. (1973). Distribution and abundance of decapod crustacean larvae in the York River Estuary and adjacent lower Chesapeake Bay, Virginia, 1968–1969. *Chesapeake Sci.* **14**, 235–257.

Saunders, D. S. (1977). "An Introduction to Biological Rhythms." Blackie, Glasgow and London.

Spindler, M., Hemleben, C., Bayer, U., Bé, A. W. H., and Anderson, O. R. (1979). Lunar periodicity of reproduction in the plankton Foraminifera *hastigerina pelagica*. *Mar. Ecol: Prog. Ser.* **1**, 61–64.

Stephens, G. C. (1962). Circadian melanophore rhythms of the fiddlecrab: Interaction between animals. *Ann. N.Y. Acad. Sci.* **98**, 926–939.

Stickney, R. R., and Knowles, S. C. (1975). Summer zooplankton distribution in a Georgia estuary. *Mar. Biol. (Berlin)* **33**, 147–154.

Struhsaker, J. W. (1966). Breeding, spawning, spawning periodicity, and early development in the Hawaiian *Littorina 1. pintado* (Wood), *L. picta* Philippi and *L. scabra* (Linné). *Proc. Malac. Soc. London* **37**, 137–166.

Sulkin, S. D. (1973). Depth regulation of crab larvae in the absence of light. *J. Exp. Mar. Biol. Ecol.* **13**, 73–82.

Suter, R. B., and Rawson, K. S. (1968). Circadian activity rhythm of the deer mouse, *Peromyscus:* Effect of deuterium oxide. *Science* **160**, 1011–1014.

Swade, R. H., and Pittendrigh, C. S. (1967). Circadian locomotor rhythms of rodents in the Arctic. *Am. Nat.* **101**, 431–466.

Takahashi, J., and Menaker, M. (1979). Physiology of avian circadian pacemakers. *Fed. Proc., Fed. Am. Soc. Exp. Biol.* **38**, 2583–2588.

Takahashi, J. S., and Zatz, M. (1982). Regulation of circadian rhythmicity. *Science* **217**, 1104–1111.

Taylor, A. C., and Naylor, E. (1977). Entrainment of the locomotor rhythm of *Carcinus* by cycles of salinity change. *J. Mar. Biol. Assoc. U. K.* **57**, 273–277.

Thompson, D. A., and Muench, K. A. (1976). Influence of tides and waves on the spawning behavior of the Gulf of California grunion, *Leuresthes sardina* (Jenkins and Evermann). *Bull. S. Calif. Acad. Sci.* **75**, 198–203.

Thorpe, J. E., ed. (1978). "Rhythmic Activity of Fishes." Academic Press, New York.

Thorson, G. (1964). Light as an ecological factor in the dispersal and settlement of larvae of marine bottom invertebrates. *Ophelia* **1**, 167–208.

Truman, J. W. (1972). Physiology of insect rhythms. II. The silkworm brain as the location of the biological clock controlling eclosion. *J. Comp. Physiol.* **18**, 99–114.

Truman, J. W. (1974). Physiology of insect rhythms. IV. Role of the brain in the regulation of the flight rhythm of the giant silkmoth. *J. Comp. Physiol.* **95**, 281–296.

Tsalkina, A. V. (1970). Vertical distribution and diurnal migration of some Cyclopoida (Copepoda) in the tropical region of the Pacific Ocean. *Mar. Biol. (Berlin)* **5**, 275–282.

Tsalkina, A. V. (1972). Vertical distribution and diurnal migrations of Cyclopoida (Copepoda) in the northeastern part of the Indian Ocean. *Okeanologiya (Moscow)* **12**, 677–688.

Walker, B. (1952). A guide to the grunion. *Calif. Fish Game* **38**, 409–420.

Webb, H. M. (1971). Effects of artificial 24-hour cycles on the tidal rhythm of activity in the fiddler crab, *Uca pugnax*. *J. Interdiscp. Cycle Res.* **2**, 191–198.

Webb, H. M. (1972). Phasing of locomotor activity in the fiddler crab, *Uca pugnax*. *J. Interdiscp. Cycle Res.* **3**, 179–185.

Welsh, J. H. (1941). The sinus glands and 24-hour cycles of retinal pigment migration in the crayfish. *J. Exp. Zool.* **86**, 35–49.

Wheeler, D. E. (1978). Semilunar hatching periodicity in the mud fiddler crab *Uca pugnax*. *Estuaries* **1**, 268–269.

162   Patricia J. DeCoursey

Wheeler, J. F. G. (1937). Further observations on lunar periodicity. *J. Linn. Soc. London, Zool.* **40**, 325–345.

Wildish, D. J. (1970). Locomotory activity rhythms in some littoral *Orchestia* (Crustacea: Amphipoda). *J. Mar. Biol. Assoc. U. K.* **50**, 241–252.

Williams, A. B. (1971). A ten-year study of meroplankton in North Carolina estuaries: Annual occurrence of some brachyuran developmental stages. *Chesapeake Sci.* **12**, 53–61.

Williams, B. G. (1969). The rhythmic activity of *Hemigrapsus edwardsi. J. Exp. Mar. Biol. Ecol.* **3**, 215–223.

Williams, B. G., and Naylor, E. (1969). Synchronization of the locomotor tidal rhythm of *Carcinus. J. Exp. Biol.* **51**, 715–725.

Williams, J. A. (1979). A semi-lunar rhythm of locomotor activity and moult synchrony in the sand beach amphipod *Talitrus saltator.* In "Cyclic Phenomena in Marine Plants and Animals" (E. Naylor and R. G. Hartnoll, eds.), pp. 407–414. Pergamon, New York.

Williams, J. A., Pullin, R. S. B., Naylor, E., Smith, G., and Williams, B. G. (1979). The role of Hanström's organ in clock control in *Carcinus maenas.* In "Cyclic Phenomena in Marine Plants and Animals" (E. Naylor and R. G. Hartnoll, eds.), pp. 459–466. Pergamon, New York.

Withrow, R. B., ed. (1959). "Photoperiodism and Related Phenomena in Plants and Animals." A. A. A. S., Washington, D.C.

Wooldridge, T. (1976). The zooplankton of Msikabe Estuary. *Zool. Afr.* **11**, 23–44.

Zaret, T. M., and Suffern, J. S. (1976). Vertical migration in zooplankton as a predator avoidance mechanism. *Limnol. Oceanogr.* **21**, 804–813.

Zucker, N. (1976). Behavioral rhythms in the fiddler crab *Uca terpsichores.* In "Biological Rhythms in the Marine Environment" (P. J. DeCoursey, ed.), pp. 145–159. Univ. of South Carolina Press, Columbia, South Carolina.

Note in press: Since the completion of this chapter in 1980, a number of relevant publications have appeared. Particularly noteworthy are two major volumes summarizing biological timing (Aschoff, 1981; Moore-Ede et al., 1982), a review of tidal and lunar rhythms (Neumann, 1981), a review article on physiology of biological timers (Takahashi and Zatz, 1982), and an important work on decapod hatching rhythms (Christy and Stancyk, 1982).

4

Symbiotic Relations

D. M. ROSS

I. INTRODUCTION

Symbiosis is a recurrent theme in the biological literature. In recent years it has received perhaps more attention than ever before, e.g., the works edited by Henry (1966), Cheng (1971), Vernberg (1974), and Jennings and Lee (1975), along with numerous review articles, e.g., Davenport (1955), Dales (1957), Yonge (1957), Cheng (1967), Ross (1974), and Bruce (1976a). The reasons for this continuing interest are easily understood. Adaptations generate curiosity, particularly the adaptations that enable pairs of organisms to live in, on, or with each other, such as a fish living safely amid the

THE BIOLOGY OF CRUSTACEA, VOL. 7
Copyright © 1983 by Academic Press, Inc.
All rights of reproduction in any form reserved.
ISBN 0-12-106407-7

tentacles of a sea anemone. But some symbioses are vitally important for human welfare and for preserving the balance of nature; these pose research problems for every type of biologist.

Current opinion has reestablished de Bary's original meaning of symbiosis as any situation where organisms live together (see Henry's Foreword and Preface to "Symbiosis," Vol. 1, 1966). This definition disregards any advantages or disadvantages for the participants and embraces parasitism, commensalism, mutualism, phoresy, and inquilinism. The notion that symbiosis refers to nonparasitic associations still persists in popular writing and to some extent also in the scientific literature, perpetuated in part by Caullery's (1952) otherwise invaluable work "Parasitism and Symbiosis." Contemporary work has shown that there are many associations with no obvious consequences to the participants, so that definitions based on such considerations cannot apply generally. There are also cases where the boundary between parasitism and predation is unclear, reflecting only the relative sizes of host and symbiont. For these reasons, a term is needed for all cases of organisms living together as the term "symbiosis" is now used.

Among the environmental adaptations of Crustacea, symbioses are particularly important because these animals have more symbioses with members of other phyla and with one another than perhaps any major group of invertebrates. Over one century ago, Verrill (1869) commented on the numbers of crustaceans that live with other animals as parasites. Using Kükenthal's (1926–1927) "Handbuch der Zoologie" (Vol. 3, Part 1), I have tried to estimate how many crustaceans live in symbioses. That work names about 1800 crustacean genera. From the descriptions of families and from the text, about 500 genera, or about 28%, are involved in symbioses, either as symbionts or as hosts. Waterman and Chace (1960) estimated that there were approximately 26,000 extant crustacean species at that time. Assuming the percentage of symbiotic species to be the same as symbiotic genera, we arrive at a rough estimate of over 7000 symbiotic species. Admittedly far from accurate, this shows that living in symbiosis is one of the most common environmental adaptations in the Crustacea.

Symbioses occur in all the major groups of Crustacea: copepods, cirripedes, isopods, amphipods, and decapods. Most species in the smaller groups, the Branchiopoda, Ostracoda, Leptostraca, Mysidacea, and Euphausiacea, live independently. This suggests a strong trend toward symbiosis in the actively evolving Crustacea, stronger than in other major coelomate phyla, the annelids, echinoderms, and mollusks. Symbioses are relatively rare in those phyla, and parasitic forms are almost nonexistent. The polychaete parasite *Ichthyotomus*, the bivalve *Entovalva*, and about 30 prosobranch species that parasitize echinoderms (Caullery, 1952) are isolated exceptions. Leeches are considered ectoparasitic annelids, but they are better described as temporary or partial predators; typical ectosymbionts remain with their hosts, and leeches do not.

Another feature of crustacean symbioses is the number of cases in which both symbiont and host are crustaceans. Copepods, cirripedes, and isopods have evolved many lines, in some cases whole families, which live on or in other crustaceans. Insects also have established many symbioses with other insects (Hartzell, 1967; Steffan, 1967). Evidently this is a direction that arthropods have followed in their evolution and one to be noted. In this review, the dual roles of crustaceans as hosts and symbionts will be stressed by considering first those symbioses in which crustaceans serve as hosts and later those where they serve as symbionts with non-crustacean hosts.

II. SYMBIOSES WITH CRUSTACEAN HOSTS

Crustaceans serve as hosts for both ectosymbionts and endosymbionts; some of these associations are standard textbook examples, such as the dromiid crabs, which invest themselves with sponges, and the brachyuran crabs, which are parasitized by the cirripede *Sacculina*. It is not surprising that we know most about symbioses in which the large and familiar decapods serve as hosts and that we know more about ectosymbiosis than endosymbiosis. In this section we shall examine typical symbioses involving crustacean hosts, including some where the symbiont does not live in or on the host itself, but on a shell carried by the host or in a burrow or shelter occupied by the host.

Most of the earlier work in this field was descriptive and taxonomic (see relevant sections of Caullery, 1952; Dales, 1966). These continue to be the dominant themes at the present time. Attempts have been made, however, to turn to analytical and experimental studies of symbiosis, largely due to the influence of Davenport (1955, 1966a,b).

A. Ectosymbionts of Decapods

1. ON ECTOSKELETON OF HOST

a. With Crustacean Symbionts. Epizootics often live on decapod exoskeletons or in the gill chambers in associations that are presumed to be sometimes commensal and sometimes parasitic, but rarely mutually beneficial. The decapod exoskeleton would seem to offer any settling organism an advantageous location. Yet the external surfaces of most decapods are relatively free of epizootics; older and more heavily calcified individuals may be encrusted with barnacles and bryozoans, but most of these are unspecific and random settlers. No doubt ecdysis does much to keep the crustacean exoskeleton free of ectosymbionts during earlier growth stages.

Decapods are hosts to other animals in many species-specific symbioses. Bopyrine epicaridean isopods were well known as parasites of various decapods nearly one century ago from the work of Giard and Bonnier (see Caullery, 1952). Later work has added more examples, especially with natantian hosts (see Overstreet, Chapter 4, Volume 6 of this treatise). In collections of shrimps of the Subfamily Pontoniinae from Indo-Pacific locations, Bruce (1972a,b, 1973, 1974) found four new genera and species. These parasites cling to various places on the host's exterior, the location being characteristic for each species pair. The pontoniines are often commensals living with actinians, holothurians, echinoids, or corals, so these associations with bopyrids are frequently hypersymbioses.

The ectoparasitic family of isopods, Bopyridae [Phryxidae], is found on the full range of decapod hosts, but with characteristic geographical and species distributions and much variation in levels of infestation, features that are largely unexplained. Allen (1966) and later Warren (1974) found the bopyrid *Hemiarthrus abdominalis* (Krøyer) on 2.2 and 0.4%, respectively, of *Pandalus montagui* Leach in two North Sea locations on the English Coast. They found none on *P. borealis* Krøyer in the same collections. Warren confirmed earlier observations that on the Canadian Atlantic coast the opposite situation exists. *Hemiarthrus abdominalis* is found on *P. borealis* but not on *P. montagui*. Other collections in which new species and genera have been discovered came from Indo-Pacific locations (Danforth, 1971, 1976) and from the western Atlantic (Markham, 1973, 1974, 1975a,b).

Table I lists some of the decapods on which symbiotic isopods have been found. From this sample, one sees that the bopyrids have attached themselves to species in all the main groups of decapods. No doubt further investigation will uncover numerous other examples. In the course of evolution, the decapods seem to have become the targets for the radiation of ectosymbionts within this isopod line.

Decapods are hosts for ectosymbionts from many other groups. Among the copepods, some harpacticoids, most of which are free-living, occur in large numbers in the branchial chambers of decapods. Soyer (1973) collected 700 specimens of a new species, *Paramphiascopsis paromolae*, in the gill chambers of the homolid crab, *Paromola cuvieri* (Risso). Humes (1947, 1958, 1971a) has described harpacticoids from the gill chambers of tropical land crabs and hermit crabs. Hamond (1972) has named four new species of harpacticoids collected in Australia from *Diogenes senex* Heller with infestations of 7 and 29%. Earlier records of the harpacticoid *Sunaristes paguri*

TABLE I
Decapod Hosts of Ectosymbiotic Isopods

Decapod host	Family	Geographical location	Isopod symbiont	Position on host
Parapenaeus fissurus Bate[a]	Penaeidae	Mozambique	*Epipenaeon fissurae* Kensley	Branchial chamber
Alpheus crassimanus Heller[a]	Alpheidae	Natal	*Bopyrella hodgarti* Chopra	Branchial chamber
Alpheus malabaricus Fabricius[b]	Alpheidae	South India	*Stegoalpheon kempi* Chopra	Branchial chamber
Pandalus borealis Krøyer[c]	Pandalidae	Eastern North America	*Hemiarthrus abdominalis* (Krøyer)	Ventral abdomen
Pandalus montagui Leach[c]	Pandalidae	United Kingdom	*Hemiarthrus abdominalis* (Krøyer)	Ventral abdomen
Heterocarpus ensifer A. Milne Edwards[d]	Pandalidae	Guam	*Zonophryxus dodecapus* Holthuis	Dorsal carapace?
Palaemon serratus (Pennant)[e]	Palaemonidae	Europe	*Bopyrus fougerouxi* Giard and Bonnier	Unknown
Palaemonetes pugio Holthuis[f]	Palaemonidae	Eastern North America	*Probopyrus pandalicola* (Packard)	Branchial chamber
Coralliocaris superba (Dana)[h]	Palaeomonidae	Indo-West Pacific	*Allophryxus malindiae* Bruce	Ventral abdomen
Periclimenes hertwigi Balss[g]	Palaeomonidae	Australia	*Filophryxus dorsalis* Bruce	Dorsal abdomen
Callianassa stebbingi[e,n]	Callianassidae	Europe	*Ione thoracica* Montagu	Branchial chamber
Pagurus longicarpus Say[f]	Paguridae	North Carolina	*Asymmetrione desultor* Markham	Branchial chamber
Clibanarius tricolor (Gibbes)[m]	Diogenidae	Bermuda	*Stegias clibanarii* Richardson	Branchial chamber
Clibanarius misanthropus (Risso)[l]	Diogenidae	Europe	*Pseudione fraissei*	Branchial chamber
Dardanus arrosor (Herbst)[l]	Diogenidae	Morocco	*Asymmetrione dardani* Bourdon	Branchial chamber
Galathea rostrata A. Milne Edwards[j]	Galatheidae	Europe/North America/West Pacific	*Pleurocrypta floridana* Markham	Branchial chamber
Petrolisthes armatus (Gibbes)[k]	Porcellanidae	North West Atlantic	*Aporobopyrus curtatus* (Richardson)	Branchial chamber
Munida iris A. Milne Edwards[i]	Galatheidae	Western Atlantic	*Anuropodione carolinensis* Markham	Branchial chamber?
Stenopus hispidus (Olivier)[a]	Stenopodidae	Mozambique	*Argeiopsis inhacae* Kensley	Branchial chamber
Paratergatis longimanus Sakai[a]	Xanthidae	Mozambique	*Gigantione sagamiensis* Shiino	Branchial chamber
Pilumnus hirtellus (L.)[e]	Xanthidae	Europe	*Cepon elegans*[o]	Branchial chamber?
Nikoides danae Paulson[a]	Processidae	Natal	*Nikione natalensis* Kensley	Branchial chamber

[a] From Kensley (1974). [b] From Kannupandi (1976). [c] From Warren (1974). [d] From Danforth (1976). [e] From Caullery (1952). [f] From Anderson (1977). [g] From Bruce (1972b). [h] From Bruce (1974). [i] From Markham (1973). [j] From Markham (1974). [k] From Markham (1975a). [l] From Markham (1975b). [m] From Markham (1975c). [n] Now known as *Cancricepon elegans*. [o] Now known as *C. tyrrhena*.

Hesse linked this symbiont with *Diogenes pugilator* (Roux) and *Pagurus bernhardus* (L.) from Norway through the Mediterranean and into the Black Sea (Hamond, 1972). Raibaut (1968) found an unknown harpacticoid heavily infesting the gill lamellae in three out of forty *Maja squinado* (Herbst) and named it *Paralaophonte ormieresi*. Seeing no lesions associated with this infestation, he considered it an example of inquilinism. Perhaps many branchial infestations should be so described as sheltering rather than parasitic.

Some of the cirripedes that live on decapod crustaceans are true ectosymbionts. Bowers (1968) has studied the interactions between the pedunculate barnacle, *Trilasmis fissum* (Pilsbry), and its hosts, *Panulirus japonicus* (De-Siebold) and *P. penicillatus* (Olivier), in Hawaii. This symbiont settles on the mouthparts whereas other more or less naked pedunculates (*Octolasmis* spp.) are found in the gill chambers (Newman, 1960, 1961a,b). From his own and earlier studies Walker (1974) noted that *Octolasmis mülleri* (Coker) in North Carolina has many hosts. They settle on those gills where they can best avoid being cleaned off by the maxillipeds and where the respiratory current is strong. In heavy infestations they are found also on the chamber walls. The cyprids settle themselves as Dinamani (1964) stated, so that the cirral net meets the current.

A few other decapods are hosts to crustacean ectoparasites. Williams and Brown (1972) reported that 10% of *Munida iris* A. Milne Edwards housed a species of the bopyrid, *Anuropodione*, in the gill chamber. Occasionally the symbiont is an amphipod, e.g., *Gitanopsis paguri* Myers in the gills of *Dardanus megistos* (Herbst) (Myers, 1974). An unusual case is Baker's (1969) report of an association between the ostracod *Ankylocythere sinuosa* (Rioja) and the crayfish *Procambarus simulans* (Faxon). Finally, Bruce (1969) reported the rare circumstance of a decapod symbiont on another decapod, the alphaeid shrimp *Areotopsis amabilis* DeMan on pagurids in the Seychelles.

With Non-Crustacean Symbionts. Decapod crustaceans which have so many crustacean ectosymbionts might be expected to be targets also for the settlement of many non-crustacean ectosymbionts. However, there have been few reports of non-crustacean ectosymbionts seeking out a decapod as a host. Perhaps this is because the other invertebrate phyla which have symbiotic tendencies have produced mostly endosymbionts. Cases of non-crustacean ectosymbionts mentioned in recent papers include the following: the small polychaete, *Histriobdella homari* van Beneden, living in the gill chambers of lobsters and feeding on microorganisms in the current and not on the host's tissues (Jennings and Gelder, 1976); another polychaete, *Stratodrilus* spp., living in the same way on various freshwater and marine

Crustacea in the southern hemisphere; unidentified oligochaetes in the family Branchiobdellidae living under the rostrum and attaching cocoons to the abdomen of freshwater crayfish in central Canada (Bishop, 1968); eggs and hatched larvae of the liparid fish, *Careproctus* sp., in the gills of *Lopholithodes foraminatus* (Stimpson), possibly impairing respiration (Peden and Corbett, 1973).

In contrast with the examples cited above in which the symbionts find their hosts, the remaining examples consist of non-crustacean symbionts which are actively acquired by certain brachyuran crabs, mostly in the families Dromiidae and Majidae. That *Dromia vulgaris* Milne Edwards actively invests itself with *Suberites* by cutting out portions of the sponge and shaping them to fit its carapace has been known for many years (Fenizia, 1935) and has been well documented in films (Schöne, 1976). This dromiid–sponge association is clearly a symbiosis, but the behavior that produces it is not exclusively linked to sponges, since analogous behavior can occur with nonliving substitutes: The crab cuts and shapes cardboard and holds it above its carapace by the rear pereiopods (Dembowska, 1926). However, living material seems to be preferred (usually the sponge *Suberites* or an ascidian, e.g., *Ascidia mentula* O. F. Müller) and is held in position and placed in order to become fixed to the cuticle (Balss, 1924; Carlisle, 1953).

Many majids mask themselves by cutting off pieces of algae and sponge, which they "plant" on limbs, rostrum, and carapace, and which can grow to cover the crab completely. Dr. René Catala (personal communication) has described a special case in which a spider crab in New Caledonia transplanted pieces of an alcyonacean, which grew into healthy colonies on the carapace, whereas his own attempts to perform such transplants always failed. Some majids display remarkable behavior patterns in acquiring actinians as specific ectosymbionts. Cutress *et al.* (1970) described how the Caribbean spider crab *Stenocionops furcata* palpates the anemone *Calliactis tricolor* (Lesueur) until it releases its pedal disk and then, after much manipulation, uses its cheliped like a crane to hoist the relaxed anemone onto its carapace (Fig. 1). This crab even shows memory-storing behavior by testing the carapace beforehand for empty areas on which the symbiont can be placed (Fig. 1E).

Some calappid crabs also carry actinians, but with less direct participation in their acquisition. *Hepatus epheliticus* scratches *C. tricolor* briefly and then presents its carapace to the anemone, which transfers to the crab by itself (Cutress *et al.*, 1970). In a much earlier account (Bürger, 1903), the ectosymbiont *Antholoba reticulata* transferred to its host, *Hepatus chiliensis*, without any assistance. Both of these examples from the calappids bear further investigation.

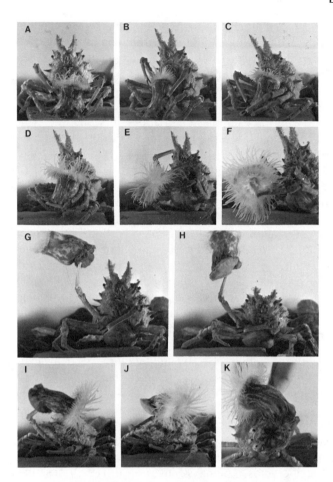

Fig. 1. *Stenocionops furcata* and *Calliactis tricolor*. Detachment and transfer of anemone from tile to carapace of crab in sequence A to K. Approximately 1.0 min between each photo, except G to H (10 sec) and J to K (4 min).

More specific brachyuran–actinian symbioses are found in the Indo-Pacific boxer crabs in the family Xanthidae. *Lybia tesselata* (Latreille) and *Polydectus cupilifera* (Latreille) carry single actinians, species of *Bunodeopsis* and *Phellia*, respectively, on their chelipeds. Duerden (1905) described this association as a protective adaptation, and he observed how the crabs acquired the anemones. This association has long awaited further study.

The pagurid *Diogenes edwardsi* (DeHaan) also bears a single actinian, *Sagartiomorphe* sp., on the propodus of its cheliped so that it fills the aperture when the crab withdraws into the shell. *Diogenes edwardsi* acquires the anemone in a special behavior pattern differing from its behavior with other actinians that are placed on the shell (Ross, 1975). Some other cases are known only from preserved specimens. *Paguropsis typica* Henderson has no shell, but covers its abdomen with an actinian, *Mammillifera* sp., using modified claws on the fourth pereiopods and stilleto-like processes on the uropods to hold the actinian in position (Balss, 1924). The recently re-discovered association between an unnamed actinian and *Munidopagurus macrocheles* (A. Milne Edwards) in the West Central Atlantic may be similar (Provenzano, 1971). Thus, the active acquisition of ectosymbionts is a fairly general adaptation among brachyurans and anomurans, since it occurs in diverse species in several scattered families. The habit must have evolved independently a number of times, a point that will be reinforced by facts to be presented in the next section.

2. ON SHELLS CARRIED BY HOSTS

The shells carried by hermit crabs and some brachyurans provide settling sites and living quarters for animals from all the major phyla. These are to be treated as symbiotic associations only when there is evidence of specificity and when significant interactions take place between the symbiont and the host or its shell. Sometimes beneficial or damaging consequences for the host may be demonstrated, but in most cases the effects are problematical and remain to be determined.

Certain sponges, hydroids, actinians, polychaetes, copepods, amphipods, and even shrimps live in or on the shells carried by pagurids and on some dorippid crabs. These symbioses are not directly with the crab; the essential bond in the association is the shell on whose surface or in whose cavity the symbiont lives.

a. Symbioses Involving Actinians. Among symbioses of this kind those between sea anemones and pagurids have been studied most. Gosse (1860) was one of the first to draw attention to the fact that certain species of pagurids and actinians are always found living together. Collections and expeditions produced more examples of these species-pairs, and the association became a model for studies on the interactions between ectosymbionts and their hosts (Brunelli, 1910, 1913; Faurot, 1910, 1932; Cotte, 1922; Balss, 1924; Brock, 1927; Carlgren, 1928; Ross, 1960 et seq.). In earlier reviews (Ross, 1967, 1974), I have discussed interactions between pagurids and actinians with the emphasis on the role of the actinian symbiont. In this article the symbioses will be considered with the emphasis on the role of the pagurid host.

Several dozen pagurid species are known to carry actinians on their shells, most of them living in warm temperate and tropical seas. Interactions of the participants have been studied in about 20 of these (McLean and Mariscal, 1973; Ross, 1974; Hand, 1975).

i. PAGURID HOSTS. It is impossible at this stage to produce a comprehensive world list of pagurid hosts of actinians. In recent decades, the taxonomy of the pagurids has undergone extensive revision. The familiar species present no problem, but many of the older records of unfamiliar species can only be related to current species lists by careful investigation and comparisons of types to determine synonyms. Even so many uncertainties remain. The chief difficulty is that the reports of many of the collections make no reference to the presence or absence of actinians on their shells, e.g., the "Calypso" expedition on the Atlantic Coast of South America from 1961 to 1962 (Forest and Saint Laurent, 1967). In order to extract the maximum amount of information from collections of this kind, records and specimens of symbionts should be kept along with data of the main collections.

The occurrence of symbiosis with actinians within the Superfamily Paguroidea reflects the distribution of the actinian symbionts (see below), as well as the ranges of the host genera. The species pairs display a geographical distribution of their own, some hosts and some symbionts living beyond the range of their usual partners. In general, this symbiotic relationship occurs in warm temperate and tropical seas. Host genera are found in both the families Diogenidae and Paguridae. Caution is necessary in appraising reports of the occasional occurrence of actinians in shells of pagurids (*sensu lato*) not usually found living with actinian symbionts. Such a host might have acquired the anemones by moving into a shell formerly occupied by another pagurid with the symbiotic habit.

Among the Diogenidae, the genus *Dardanus* Paulson (about 30 species worldwide, two thirds in the Indo-West Pacific) is the most consistent actinian carrier. To date the only nonhost reported is *D. cressimanus* H. Milne Edwards in Japan (Ross, 1975). Species of the huge genus *Paguristes* Dana (>100 species) carry actinians in the Mediterranean and New Zealand, but information is lacking about actinians on the shells of most of these. Single species of *Petrochirus* Stimpson (three species in the Amphi-American Tropics and in Africa), *Clibanarius* Dana (50 species worldwide, warm, temperate, and tropics), and *Diogenes* Dana (fewer and scattered species) are known as actinian hosts. Among the Paguridae, the genus *Pagurus* Fabricius (>140 species worldwide at all latitudes) has species here and there that are frequently found with actinian symbionts, but the vast majority live in waters where such associations are unknown, e.g., the North Pacific with 25 species (McLaughlin, 1974). A new family has recently been proposed to accommodate the genus *Parapagurus* Smith (40 species worldwide in deeper waters) (Saint Laurent, 1972). Older records (Balss, 1924; Carlgren, 1928) report *P. pilosimanus* (Smith) as hosts of actinians. Haig (1974) has recently described a new species, *Pagurus imarpe,* whose shells were completely covered by anemones. No doubt many other examples of such associations with *Pagurus* spp. remain to be discovered.

ii. ACTINIAN SYMBIONTS. Most of the identified symbiotic actinians of pagurid hosts are in three genera of the family Hormathiidae, namely, *Calliactis* Verrill, *Paracalliactis* Carlgren, and *Adamsia* Forbes. Species of *Sagartiomorphe* Kwietniewski of the family Sagartiomorphidae have also been identified as symbionts of hermit crabs in the West Pacific and Caribbean. The number of actinian species is small compared to the crustacean hosts, and many of these symbionts are found on a large number of pagurids, e.g., *Calliactis polypus* Forskål, which may turn up on the shell of ten or more pagurid hosts in the Indo-Pacific region. *Calliactis parasitica* (Couch) is found on several hermits in the Mediterranean and on *Pagurus bernhardus* in the Atlantic, although this pagurid has an extensive northern and trans-atlantic range beyond that of the actinian which stops short at a latitude of 51°. It is rare for a symbiotic anemone to be restricted to a single host, but it happens with the obligate partnership between *Adamsia palliata* (Bohadsch) and *P. prideauxi* Leach. These organisms live throughout the Mediterranean and along the European coast to northern Norway and south to the Cape Verde Islands.

iii. ROLES OF HOSTS AND SYMBIONTS IN ESTABLISHING SYMBIOSES. That some pagurids actively transfer actinians to their shells probably became known soon after these animals could be observed in aquaria. The sequence of papers describing this behavior was as follows: Brunelli (1910, 1913) on *Dardanus arrosor* and *C. parasitica* in Naples; Faurot (1910) on *P. prideauxii* and *A. palliata,* with a short account of *D. arrosor* and *C. parasitica* at Banyuls-sur-Mer; Cowles (1919) on some pagurid–anemone associations in the Far East; and Brock (1927) on the sensory cues involved in the behavior of *D. arrosor* toward *C. parasitica* in Naples.

Years later I found (Ross, 1960) that *Pagurus bernhardus,* unlike other pagurid hosts, does not transfer *Calliactis* to its shell. Instead, the anemone transfers itself in response to a substance of molluscan origin within the shell (Ross and Sutton, 1961a). The association between *P. bernhardus* and *C. parasitica* was the only one in which the pagurid seemed to be completely passive. In fact, *P. bernhardus* may not be completely passive; Ross (1979b) found that some young *P. bernhardus* make brief but active contact with *Calliactis* and then present the shell to the anemone; this interaction seems to facilitate the transfer of *C. parasitica* to the shells when contact is made.

In *Dardanus* spp. in Hawaii and Japan, the pagurid's palpations are the

trigger for the detachment and transfer of *C. polypus;* apparently this actinian is relatively unresponsive to shells and unable to transfer by itself (Ross,
1975). In some other associations, the pagurid's palpations and the response
of *Calliactis* to the shell contribute in varying degrees to the mutual behavior
patterns required for successful transfers (Ross and Sutton, 1961b; Cutress *et
al.,* 1970). Apparently there is no standard formula for the partition of the
behavioral contributions of pagurid and actinian in these partnerships. Each
species pair seems to have evolved behavior patterns in which a greater or
lesser share of the activity is assigned to the host and to the symbiont.

iv. THE ROLE OF THE SHELL IN PAGURID–ACTINIAN SYMBIOSES.   The lives of all
pagurids revolve around their shells. The periodic need to move into bigger
shells to accommodate growth is a recurring crisis for the pagurid. In-
terestingly, pagurids demonstrate preferences for certain types of shells:
their capacities for selecting shells for size, gastropod species, weight,
shape, and soundness are one of the most impressive examples of complex
behavior among the invertebrates (Reese, 1962). There is evidence of information services of a high order when a hermit crab tests an empty shell. It
makes certain quantitative and qualitative tests and often makes the final
decision whether to transfer or remain, only after several movements back
and forth.

There is no evidence that the pagurid hosts of actinians are influenced in
their choice of shells by the presence or absence of actinians. When *Dardanus* without *Calliactis* were offered empty shells with *Calliactis,* they tested the empty shells but rejected them, although later they transferred most of
the *Calliactis* to their own shells. In the reverse experiment, two *Dardanus*
out of nine tested left shells with *Calliactis* for shells without. The anemones
from the deserted shells were transferred to the new shells only when they
were encountered later in the pagurids' random movements. These observations (D. M. Ross, unpublished) point to the overriding importance of the
shell in pagurids, including those pagurids which carry actinians. Because
of the shell-changing habit, some pagurids acquire actinians, not through
their own or the actinian's behavior, but when they enter a shell with an
actinian already on it.

v. INTRASPECIFIC AND INTERSPECIFIC COMPETITION FOR ACTINIANS.   Once it is
on a shell, a symbiotic anemone can remain on that shell indefinitely. Hand
(1975) saw the New Zealand *Calliactis conchicola* Parry transferring spontaneously from one pagurid-occupied shell to another, something which I
have never seen. He adds, however, that typically the anemone does not
leave the shell even when the shell has been vacated. This corresponds to
my experience and is contrary to many published accounts, e.g., Gotto

(1969). The manipulations of an active pagurid like *Dardanus* spp. are required to release the pedal disk and transfer the anemone. Thus pagurid hosts can only acquire anemones in nature as a result of encounters, intraspecific and interspecific, between individual hermits.

It seems that *Calliactis* are rarely stolen in encounters between two *Dardanus arrosor* (Ross, 1979a). By contrast, *D. arrosor* lacking *Calliactis* are extremely active in acquiring them from *Paguristes oculatus* (Fabricius) and from *Pagurus alatus* Fabricius. After a few hours of such interspecific encounters, almost all the actinians on these two species were stolen by *D. arrosor;* in reciprocal trials, none were taken from *D. arrosor.* Apparently *D. arrosor* is a dominant species in the competition for actinians among pagurids, which may account for its cosmopolitan distribution (Mediterranean, South Atlantic, Indo-Pacific) (Ross, 1979a) (Figs. 2 and 3).

Dominance within the species is also a factor in pagurid–actinian symbioses. Mainardi and Rossi (1969) have shown that a dominant–subordinate relationship exists in pairs of *D. arrosor.* When *C. parasitica* were offered to such pairs, they were always taken by the dominant individual; the subordinate showed no activity toward the actinian until the dominant individual was removed.

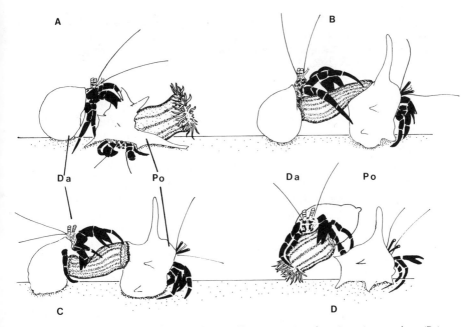

Fig. 2. *Dardanus arrosor* (Da) "stealing" *Calliactis parasitica* from *Paguristes oculatus* (Po). Drawn from frames of 16-mm film. *Paguristes oculatus* neither resists nor reacts in any way to this activity of *D. arrosor.* Times between A and B: 6 min; B and C: 8 min; C and D: 11 min.

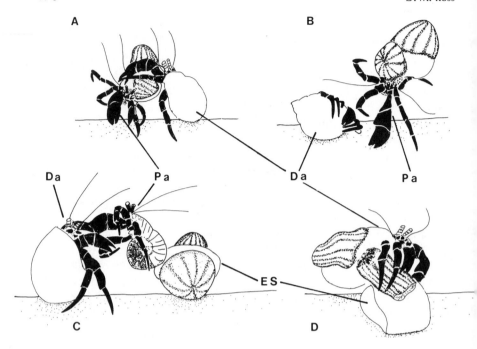

Fig.3. *Dardanus arrosor* (Da) "stealing" *Calliactis parasitica* from *Pagurus alatus* (Pa). Drawn from frames of 16-mm film. *Pagurus alatus* resists and attempts to flee; encounters end by *P. alatus* deserting or being pulled out of the shell as in Fig. 3C, whereupon *D. arrosor* either occupies the empty shell (ES) with the *Calliactis* or transfers the *Calliactis* to its own shell as in Fig. 3D. Times between A and B: 2 min; B and C: 1 min; C and D: 18 min.

vi. BENEFITS TO PAGURIDS FROM ACTINIAN SYMBIONTS. The activities of pagurids in acquiring actinians suggest that actinians bring important benefits to their hosts. The benefits may differ in different species. In the case of *Pagurus prideauxii* and *Adamsia palliata,* the actinian provides a shelter which grows with the pagurid and thus eliminates the need to change shells. It also gives the hermit greater mobility (Faurot, 1910). It has always been thought, however, that actinians may protect pagurids from predators, and evidence to support this now exists. Boycott (1954) observed that octopus retreated after contact with *C. parasitica.* Ross (1971) confirmed this and showed in trials that *D. arrosor* without *Calliactis* on the shells were always taken as food by *Octopus vulgaris* Lamarck, whereas *D. arrosor* with *Calliactis* on their shells were never taken. Later, McLean and Mariscal (1973) showed that *Pagurus pollicaris* Say is also protected from its predator, the oxystomatid crab *Calappa flammea* (Herbst), by its symbiont, *Calliactis tricolor.* Hand (1975) has shown that a New Zealand spider crab *Leptomithrax longipes* is also protected from *Octopus maorum* by its *Calliactis.* No doubt the nematocysts on the tentacles and the acontia of the actinian deter these would-be predators.

Balasch and Mengual (1974) observed that when *D. arrosor* were kept for many months in an aquarium not containing octopus, the pagurids lost their drive to acquire actinians. However, when the octopus *Eledone moschata* (Lamarck) was introduced into the aquarium, the activity of *Dardanus* toward *Calliactis* returned. Ross and von Boletzky (1979) have further shown that this renewed activity is due to chemical, not visual, cues. All of these facts point to the extraordinary complexity and diversity of the relationships and interactions that are involved in the special environment of the molluscan shells occupied by pagurids and their partners.

vii. THE ONTOGENY AND PHYLOGENY OF PAGURID–ACTINIAN SYMBIOSES. Only in the case of *Pagurus prideauxii* is the early history of a symbiosis with an actinian known with any certainty. Faurot (1910) described how the young *P. prideauxii* picks up a tiny cylindrical *Adamsia palliata* and settles it on its shell, establishing a partnership between two individuals that is henceforth virtually inseparable. The more typical and less intimate pagurid– actinian symbioses as between species of *Dardanus* and *Calliactis* are known from late juvenile and adult stages only. It is not known where the early postmetamorphic *Calliactis* settle or how they are eventually acquired by the hermits which are always found with adult *Calliactis* only. A recent discovery (D. M. Ross, unpublished observation) of some early stages of *C. parasitica* on *Paguristes oculatus* at Banyuls, and not on other species in the area, gives a hint of what might happen. Preferential settling and early development on certain species of pagurid may provide the pool of young individuals from which juveniles and adults are recruited in the interspecific encounters described above. An intensive search for early actinian symbionts could be rewarding.

The most plausible explanation for the phylogenetic origin of pagurid–actinian symbioses is that certain actinians developed a habit of attaching themselves to shells of living gastropods, thereby gaining the advantage of being transported (Balss, 1924; Ross, cited in Vernberg, 1974). A number of actinians live as passengers on gastropods (Hand, 1975; Ross and Kikuchi, 1976). The extension of this shell-climbing habit to the shells of hermit crabs is a short step. The development, by certain pagurids, of a tendency to acquire these partners would be another short step once the selective advantage of the additional protection against predators made itself felt. This would also explain why there are so few gastropod hosts of actinians at the present time; as the pagurids developed active behavior patterns toward actinians, the shell-climbing habit of actinians became less important in establishing these partnerships (Ross, 1974).

b. Non-Actinian Ectosymbionts. Shells of hermit crabs are settlement sites for all kinds of symbionts. Closely related to the actinian–pagurid symbioses are those in which zoanthids are the symbionts. Balss (1924) tabulated the published reports on ten zoanthids named as associates of about 15 hermit crabs. It seems that no work has been done on these symbioses since the original systematic descriptions.

Familiar hydrozoan colonies of *Hydractinia echinata* (Fleming) and *Podocoryne* sp. interact with their pagurid hosts in a number of ways. Jensen (1970), in two-choice tests, demonstrated a preference by *P. bernhardus* for shells with *H. echinata* and showed that the choice was based on chemoreception. Conover (1976) reported that *P. pollicaris* and *P. longicarpus* selected shells with *H. echinata* much more often than they chose naked shells. Earlier, Wright (1973) had shown that the aggressive and dominant species, *Clibanarius vittatus* (Bosc), rejects shells with *Hydractinia* and reacts to contact with it as if stung. *Pagurus longicarpus* and *P. pollicaris* were not affected by *H. echinata* and prefer shells with the hydroid to clean shells, which offsets the advantage of behavioral dominance enjoyed by *C. vittatus* in the competition for shells.

Species of the sponge *Suberites* occur not only on dromiid crabs, but also occasionally invest and partially replace the shell of a pagurid. There is no information on how this relationship is established.

Other creatures that utilize the shells to live in symbioses with hermit crabs include the small burrowing acrothoracic barnacles, *Trypetesa* spp. (Tomlinson, 1955; Turquier, 1972), and possibly a few rare bryozoans (Balss, 1924). Although not involving a hermit crab, the analogous use of bivalve shells as shelters by some dorippids is utilized by some actinians as their preferred locations. *Dorippe granulata* DeHaan carries the half-shell of a bivalve, *Macoma* sp., above itself by special claws on the fourth pereiopod. A small anemone, *Carcinactis ichikawai* Uchida, not found elsewhere, lives on the shell (Uchida, 1960).

In concluding this section, it is noteworthy that most of the work cited is purely descriptive and systematic. The study of the behavior of actinian symbionts in relation to their hosts has been a rewarding field of investigation for physiologists (McFarlane, 1973; Ross, 1973). The experimental study of the behavior of crustacean hosts in relation to their symbionts might be equally rewarding, but so far little use has been made of the opportunities that seem to exist.

B. Co-habitants of Burrows and Shelters

Animals occupying permanent burrows or cavities invariably share their living speces with cohabitants. Among crustaceans, the mud shrimps share their burrows with several other species, and the pagurids always have other animals living inside their shells.

The intertidal burrowing genera *Callianassa* and *Upogebia* are regularly accompanied in their burrows by worms and other crustaceans. *Callianassa californiensis* Dana lives with the scaleworm, *Hesperonoë complanata* (Johnson), the peacrab, *Scleroplax granulata* Rathbun, and the copepods, *Clausidium vancouverense* (Haddon) and *Hemicyclops callianassae* Wilson (MacGinitie, 1934). *Upogebia pugettensis* (Dana) shares its burrow with the same peacrab, *S. granulata*, another small bivalve, *Pseudopythina* sp., another species of *Hesperonoë*, a copepod, an ascidian, and sometimes with an itinerant goby, *Clevelandia ios* (Jordan and Gilbert) (MacGinitie, 1930).

Sleeter and Coull (1973) found nematodes, the archiannelid *Dinophilus* sp., the tubiculous polychaete *Polydora* sp., several copepods, two gammarids, and an acoelous turbellarian in the tunnels of the wood-boring isopod, *Limnoria tripunctata* Menzies. Apparently an entire community uses the shelter and the food created by the activities of this borer. Studies on burrows of land crabs have revealed many mosquitoes in these habitats (Hogue and Bright, 1971). The squillids (order Stomatopoda) are all burrow dwellers, but surprisingly they have few co-dwellers. With more than 100 species in the order, only a scale worm, *Lepidasthenia digueti* Gravier, which lives with a Californian species, *Lysiosquilla digueti* Coutière, in the tube of a balanoglossid (Balss, cited in Kükenthal, 1926–1927) and *Pseudopythina subsinuata* (Lischke) with a squillid in Hong Kong (Morton, 1972) have been named.

Among the co-dwellers found in shells occupied by hermit crabs are the amphipods *Podoceropsis nitida* (Stimpson) (family Photidae)[1] and several species in the families Lysianassidae and Stenothoidae (Vader, 1971). More intimate co-dwellers are found in shells inhabited by hermit crabs. The best known examples are *Nereis fucata* (Savigny) in Europe and *Cheilonereis cyclurus* (Harrington) on the west coast of North America. Gilpin-Brown (1969) showed that this symbiosis is established in two phases: a "searching" phase in which the worms leave their burrows and lash about in the direction of the pagurid; and an "entering" phase which occurs when the worm contacts the shell and immediately slides up into the interior. The searching phase is triggered by the vibrations set up by the shell as the hermit crab moves about; chemoreception is not involved. The entering phase follows contact with the shell; the pagurid is not necessary, since an empty shell will be entered as readily as an occupied one. Worms enter shells of any pagurid of the appropriate size, but hermit crabs are the only creatures living on the bottom dragging a heavy object about, so the response is highly adaptive. The benefits in this relation seem to be all on the

[1] Included by Bowman and Ahele in the family Corophiidae.

side of the worm. It occupies the upper whorls of the shell and emerges to take food from the crab's mouthparts whenever it feeds. Yet the pagurid's tolerance of the worm suggests that the symbiosis has some benefit for it, too. The relationship invites further research.

C. Ectosymbionts on Non-Decapod Crustaceans

There are not many reports of ectosymbionts living on non-decapod crustaceans. No doubt many are still to be discovered. The small size of some large classes of crustaceans, especially branchiopods and copepods, dictates that sessile protozoans will be common ectosymbionts on them. Delamare-Deboutteville and Nunes (1951) described epizootic hydroids that live on some parasitic copepods.

Among the non-decapod Malacostraca, Dales (1966) cited two turbellarian species, *Hypotrichina tergestina* Calandr. and *H. marseilensis* Calandr., as active parasites of *Nebalia* (Leptostraca). Field (1969) has studied *Notophryxus lateralis* G. O. Sars, an ectosymbiotic isopod, on a euphausiacean *Nematoscelis difficilis* Hansen, which was found infesting 0.24% of the host species off California. However, other than the decapods, isopods seem to be the chief target for ectosymbionts. Thus, Hastings (1972) found 15 specimens of the barnacle *Conchoderma virgatum* (Spengler) on the parasitic isopod *Nerocila acuminata* Schioedte and Meinert, which in turn was on the pectoral fin of the orange file-fish. Ritchie (1975) studied two parasitic copepods collected on two deep-sea asellotes. Several papers report isopod ectosymbionts on other isopods: *Cabirops* sp. on *Pseudione reverberii* (Restivo, 1971); *Iais californica* (Richardson) associating with *Sphaeroma quoyanum* H. Milne Edwards (Rotramel, 1975); and *Clypeoniscus hanseni* Giard and Bonnier on *Idotea pelagica* Leach (Sheader, 1977). Future research will undoubtedly turn up many more examples.

D. Endosymbionts of Crustacean Hosts

Compared with ectosymbionts, knowledge about endosymbionts in Crustacea is meager. Attention has been focused on the endosymbionts of economically important crustaceans (see Volume 6). In the brief statement appropriate here, a few examples will be noted in which the interactions between endosymbionts and their crustacean hosts contribute to our knowledge of crustacean biology.

1. NON-CRUSTACEAN ENDOSYMBIONTS

Records of the occurrence of protozoan and helminth endosymbionts still cover only a small minority of crustaceans. Infestations of sporozoans, described as heavy (8–41%), occur in crangonids (Breed and Olson, 1977). Turbellarian infestations in an amphipod *Ampelisca macrocephala* Liljeborg at times reached 60% (Christensen and Kanneworff, 1965) and in some hermit crabs averaged 68% (Lytwyn and McDermott, 1976). Jennings (1974) tabulated data on turbellarian species occurring in symbioses in major animal groups. Of the 57 species described as "entocommensals," none were associated with crustacean hosts, whereas of the 53 species described as "ectocommensals," 17 were associated with crustaceans. On the other hand, 11 forms were listed as "parasitic," and eight of these had crustacean hosts. Jennings pointed out that turbellarian flatworms, a mixture of free-living and symbiotic species, were instructive indicators of symbiotic and parasitic trends. The figures suggest that there are basic differences between the Crustacea and other major groups, such as mollusks and echinoderms, in the opportunities that they present for the establishment of symbioses.

Certain Crustacea are well known as intermediate hosts in the life-cycles of endoparasites, e.g., waterfowl parasites using ostracods and cyclopoid copepods as intermediates. A variant on this was reported by Dobrohotova (1975), who found that calanoid copepods are more important as the hosts of the hymenolepid cysticercoids that parasitize waterfowl in Kazakhstan, a fact that she links to semidesert and high saline conditions.

Decapod crustaceans are intermediate hosts of parasitic nematodes of the spirurid genus, *Ascarophis*, which live as adults in marine fishes (Uspenskaya, 1953; Uzmann, 1967). Poinar and Kuris (1975) and Poinar and Thomas (1976) showed that juvenile *Ascarophis* occur in some decapod species and not in others on the Californian coast, so there are obvious preferences in the intermediate hosts. The physiological and ecological interactions between endosymbionts and their hosts are now recognized as the most challenging areas for research. This approach has been extended recently to studies on both intermediate and adult crustacean hosts of helminths and bopyrids (Rumpus and Kennedy, 1974; Anderson, 1975, 1977). A promising direction in such studies opened up when changes in behavior were detected in amphipods harboring cystacanths of acanthocephalans (Hindsbo, 1972; Holmes and Bethel, 1972). In later papers, Bethel and Holmes (1973, 1974) described how *Gammarus lacustris* Sars, infected with cystacanths of *Polymorphus paradoxus* Connell and Corner, became attracted to light, whereas uninfected individuals were photophobic and negatively phototactic. As a result of this behavioral change, infective gammarids became much more vulnerable to predation by mallards and muskrats, and definitive hosts.

2. CRUSTACEAN ENDOSYMBIONTS

Crustacean endosymbionts with crustacean hosts are found in the endoparasitic entoniscid isopods and the rhizocephalan cirripedes. These endosymbionts probably began as ectosymbionts, since they enter the host, not through the digestive system, but from the point at which they settle on the cuticle.

There has been no recent work on the entoniscids. Caullery (1952) recounted the history of research on the group up to the work of Drach (1941) and Veillet (1945), which proved that the entoniscids penetrate the body cavities of their hosts.

The elucidation of the life history of the rhizocephalan parasite *Sacculina carcini* Thompson is one of the classics of zoological discovery. Delage (1884) reported observations so extraordinary that they were disputed for 20 years (Potts, 1915a). His is still the definitive paper on the subject. Caullery (1952) gives an authoritative summary of the parasitic adaptations in the Rhizocephala. The different genera of rhizocephalans are associated with certain families and superfamilies among their decapod hosts, e.g., *Sacculina* on cancroid and portunid crabs, *Parthenopea* on the callianassids, and *Peltogaster* on anomurans. One of the directions in which evolution proceeded in the rhizocephalans was from single hermaphroditic individuals, such as *Sacculina*, to gregarious infestations produced asexually, such as *Thompsonia* Kossmann (Potts, 1915a).

New rhizocephalans continue to be described. Boschma (1968) named a parasite of the crab *Anasimus latus* Rathbun as a new species, *Loxothylacus engeli*. A species described earlier, *Briarosaccus callosus* Boschma, was found on 14 of 21 *Lithodes aequispina* in Alaska (McMullen and Yoshihara, 1970). Rhizocephalans are not confined to the decapods; two new species have been discovered living on balanomorph cirripedes: *Chthamalophilus delagei* (Bocquet-Védrine) on *Chthamalus stellatus* (Poli), and *Boschmaella* (=*Microgaster*) *balani* Bocquet-Védrine on *Balanus improvisus* Darwin (Bocquet-Védrine, 1967). In one of the few experimental studies, Herberts (1978) compared the hemolymph from healthy and parasitized crabs, confirming the presence of a specific protein fraction and demonstrating an antisacculin precipitant reaction in the infested crabs.

The phenomenon of "parasitic castration," first observed in crabs parasitized by rhizocephalans, has been actively investigated for many years. The term refers to the loss of gonads and secondary sexual characters in parasitized male crabs. In females, too, ovaries are often reduced and reproductive activity blocked. Most of the work in the field has been done with rhizocephalans such as *Sacculina*. However, parasites of many kinds have similar effects, varying in degree, upon their hosts, e.g., bopyrids on shrimps.

The loss of male gonads and the assumption of some female features in parasitized males has been explained by linking the genetic theory of sex determination with concepts of balancing male and female hormones. Reinhard (1956) reviewed earlier work on the subject and stressed the genetic viewpoint. Later work, mainly centered in France, swung opinion in favor of hormonal influence as the primary factor, not surprising in view of the expansion of knowledge about crustacean endocrine systems (Charniaux-Cotton, 1960; Hartnoll, 1967). More recently, Kuris (1974) has reviewed parasitic castration as a specialized form of predation comparable to insect parasitoids in that only one host is reproductively killed. Baudoin (1975) has examined the phenomenon from the standpoint of strategic advantages in evolution, noting the ways in which the loss of reproductive capacity in the host benefits the parasite, while at the same time it serves as a partial defensive strategy for the host.

We have now touched on the various situations in which crustaceans serve as hosts for symbionts. Apart from the widespread occurrence and diversity of adaptations in these symbioses, the most noteworthy general observation is the number and diversity of the symbioses in which both symbiont and host are crustaceans. There are some obvious reasons for this. The planktonic larvae of crustaceans usually persist long enough to have well-developed sensory and locomotory powers. They are well equipped for host-finding while swimming in the plankton or in the nekton or later peram-bulating in the benthos. Also, no other animals are as well equipped as Crustacea for becoming attached to crustaceans; with only minor adapta-tions the appendages can possess claws, spines, hairs, and other processes which fit them to cling to and to crawl about on the articulated and often spiny surfaces of other crustaceans. But this remarkable array of symbioses of crustacean with crustacean only serves to emphasize the adaptational resources which have made the crustaceans a major group. Their capacity to evolve creatures with structural, developmental, and behavioral adapta-tions to other creatures of their own kind is impressive and instructive.

III. SYMBIOSES WITH NON-CRUSTACEAN HOSTS

Crustacean symbionts have hosts in almost every phylum: sponges, cnidarians, annelids, mollusks, echinoderms, ascidians, and fishes. Indeed, they rival helminths as the most widespread and diverse symbionts in the aquatic environment. Some complete their life cycles as grotesquely modi-fied parasites, but all other kinds of symbiotic relationships occur, some beneficial, others harmful, and many with unknown consequences. Copepods, cirripedes, isopods, and a few decapods make up the majority. Their symbioses will be reviewed here as case histories in crustacean evolu-tion and environmental adaptation.

A. Copepods

Copepod symbionts live in, on, or with members of every major class of aquatic metazoans. They come from all copepod orders except the calanoids, so presumably the habit has arisen independently many times.

Some of the conspicuous copepod parasites of fish were known to Linnaeus and before that to many generations of fishermen as "anchor worms." By the end of the nineteenth century, the main groups of symbiotic copepods had been discovered so that the chief activity since then has been the exploration of the copepod fauna in host after host. Most of the recent papers consist of systematic descriptions of new species, sometimes with information about life histories and ecology with an occasional major work revising classificatory systems and faunal lists. Some of this work has been routine, e.g., the identification of material collected in the international Indian Ocean expedition at various locations, including Mauritius (Humes, 1975) and Madagascar (Stock and Humes, 1970). This work has added knowledge about the biogeography of many symbioses, but it has added little to the understanding of the nature of these relationships in living animals.

Symbiotic copepods will be reviewed here by their hosts in view of the differing opinions about names and relationships at the levels of order and suborder (Bocquet and Stock, 1963; Yamaguti, 1963; Fryer, 1968). Although it is reasonable to treat copepod symbionts of non-crustaceans as "cyclopoids" or "caligoids," our purpose is best served by using the family divisions whose names have persisted when those of some higher taxa have become extinct or rearranged. Among the copepods, families are generally associated with hosts in certain phyla or classes. Thus, the Notodelphyidae are associated with tunicates, the Clausidiidae with mollusks and tunicates, the Monstrillidae with polychaetes as larval endoparasites, etc. In surveying the hosts of symbiont copepods, one notes that none of the copepod families that parasitize fishes have become associated with invertebrate hosts.

New species of symbiotic copepods are being described as quickly as new hosts are examined. The following examples can be cited from the recent literature: sponges (Yeatman, 1970; Humes, 1973a); hydroids (Humes, 1966); medusae (Reddiah, 1968; Humes, 1970); octocorals (Bouligand, 1966; Stock and Humes, 1970; Humes and Stock, 1973); zoanthids (Humes and Ho, 1966); actinians (Vader, 1970c,d); scleractinian corals (Humes, 1974a); sabellid and polynoid polychaetes (Carton, 1971, 1974a); gastropods (Laubier and Bouchet, 1976); pteropods (Stock, 1971a); holothurians (Humes, 1967, 1973b, 1974b; Ho and Perkins, 1977); asteroids (Humes, 1971b; Carton, 1974b); ophiuroids (Stock, 1971b; Humes and Hendler, 1972; Bartsch, 1975); and crinoids (Stock, 1967; Humes, 1972); ascidians (Illg, 1970; Illg and Humes, 1971; Laubier and Lafargue, 1974).

Additions have been made to the following families of fish parasites: Ergasilidae (Burris and Miller, 1972; Johnson and Rogers, 1972; Fernando and Hanek, 1973; Cressey, 1976); Lernaeopodidae (Kabata, 1967, 1969);

Chondracanthidae (Ho, 1972a,b, 1974); and Caligidae (Hameed and Pillai, 1973; Kabata, 1974).

The descriptive literature on parasitic copepods is immense, but limited in scope. Classical monographs describe their morphology and their life histories where known. Analytical studies have been delayed by disagreements and misconceptions about nomenclature and relationships. More rational schemes seem to be emerging that will assist the turn to more profitable studies on the interactions between host and symbiont, such as Dudley's (1968) investigation of the tissue exchanges between a parasitic copepod and its ascidian host, using both light and electron microscopes.

The symbioses in which copepods participate provide examples of the ultimate in transformation of the basic body plan for which Kabata (1970) coined the phrase "morphological exuberance." Yet our knowledge of the biology of these transformed organisms and their interactions with their hosts as living animals is meager. The papers of Boxshall (1974a,b, 1976, 1977) on *Lepeophtheirus pectoralis* (Müller) are important in showing a turn to more biological studies. His observation that larvae from *L. pectoralis*, a species which is found on several different fish, prefer to attach to the same species from which they were released is very interesting. The culturing of larvae of parasitic copepods undertaken by Kabata and Cousens (1973) and Kabata (1976) is particularly important. This can open up a route to the study of host specificity and host recognition, key problems in understanding how these animals come together, which is the way to intelligent control.

B. Cirripedes

Both the balanomorph and lepadomorph barnacles have evolved species living as ectosymbionts in phoretic associations. The whale barnacles of the genera *Coronula, Tubicinella,* and *Xenobalanus* are balanomorphs which penetrate the integument and become firmly lodged there by ridges on their plates. The lepadomorph whale barnacles, *Conchoderma auritum* (L.), settle on hard surfaces and sometimes on the plates of the balanomorph *Coronula diadema* (L). (MacGinitie and MacGinitie, 1968). Another species, *Conchoderma virgatum,* only occasionally lives on other animals and then only on hard structures such as the lateral spine of the fish *Diodon hystrix* L. (Balakrishnan, 1969). Another specialized lepadomorph, *Anelasma squalicola* (Lovén), becomes implanted on a small shark and forms an implantation cavity where one or two additional *A. squalicola* may also settle (Hickling, 1963).

Other aquatic tetrapods are also hosts of symbiotic barnacles. *Platylepas hexastylus* (O. Fabricius) occurs on the dugong and on the flippers of marine turtles; *Chelonebia testudinaria* (L.) is found on the shells of the same turtle.

The barnacles, *Platylepas ophiophilus* Lanchester, on sea snakes in south-east Asia and Australasia were known to Darwin. Zann (1978) has noted how remarkable it is that the larvae of *P. ophiophilus* should find such mobile and solitary hosts and that pairs of *Platylepas* should settle close enough to reproduce sexually by internal fertilization. Although attached by ribs which penetrate the snake's skin, this ectosymbiont usually does not survive the shedding of the skin. About 50% of the population is infested, an unusually high figure. *Platylepas ophiophilus* lives on several species, so the symbiosis is not species-specific (Zann, 1978). Lepadomorphs also occur on various invertebrates. Utinomi (1970) found a new scalpellid species, *Calantica pusilla*, on a gorgonian at Amakusa, Japan. Another species, *Rhizolepas annelidicola* Day, is an obligatory symbiont on errant polychaetes (Day, 1939). It sends out invading rootlets into the host's tissues; these absorb nutrients so that the animal lacks a digestive system.

Among the balanids, species in the family Pyrgomatidae live in various relationships with corals. *Boscia* (=*Pyrgoma*) *anglicum* (Sowerby) is an obligate symbiont on several different corals (*Caryophyllia smithi* Stokes in Britain). Anderson (1978) has shown that coral is only a perch and that the barnacle feeds independently, extending its cirri to capture microscopic food which is taken by the maxillipeds. *Hoekia* (=*Pyrgoma*) *monticulariae* (Gray), on the other hand, feeds directly on the tissues of its coral host, since ingested nematocysts are found in the gut. Moreover, the coral coenenchyme covers the cirripede, leaving only a small aperture for the mouth (Ross and Newman, 1969).

Acrothoracic symbionts are not known with non-crustacean hosts, but some ascothoracic cirripedes are ectosymbionts on cnidarians or echinoderms (the genera *Synagoga*, *Laura*, and *Petrarca*); others are endosymbionts on echinoderms (*Dendrogaster*). These organisms were featured in the classical discoveries of Lacaze-Duthiers, A. M. Norman, and G. H. Fowler (see Caullery, 1952). Recent papers report the occasional discovery of new ascothoracic cirripedes, e.g., a new species of *Dendrogaster* living on the starfish *Asterina burtoni* Gray in the Gulf of Elat (Achituv, 1970). Finally, a group long considered as a cirripede suborder, the so-called Apoda, with the single genus *Proteolepas* Darwin, has been critically reviewed and recommended for transfer. According to Bocquet-Védrine (1972), these animals belong among the cryptoniscan isopods, animals which are found on certain pedunculid barnacles.

C. Isopods

Isopod larvae were found by Temnikow (1974) in washings from the mantle of *Mytilus californianus* Conrad. This is the only report of an epicaridean isopod on a non-crustacean host. Since the possibility that the isopod came from a pinnotherid crab in the mussel was not explored, there is some uncertainty about the observation.

Other families of isopods are associated with non-crustaceans. The Gnathiidae are well known from Monod's (1926) monograph as ectoparasitic larvae which pierce the skin and suck the blood of fishes and later become free-living adults. The Cymothoidae also include some species that live ectoparastically on fish, e.g., *Ichthyoxenus* spp., penetrating deeply into the body wall.

Glynn (1968) investigated the associations of sphaeromid isopods with chitons in the Caribbean area. He found six isopod–chiton symbioses with *Dynamenella perforata* (Moore) on *Acanthopleura granulata* Gmelin often occurring together. In many places, isopods were living in the pallial grooves of all chitons, with numbers ranging from 1 to 77 (average 30 per individual). Other species of *Dynamenella* live with limpets (*Patella* spp.) in South Africa (Branch, 1975).

The most surprising relationship involving an isopod was discovered by Barham and Pickwell (1969). In a submersible off San Diego at 723 m they collected a large scyphozoan, subsequently named *Deepstaria enigmatica* Russell, to which a giant (8-cm long) isopod was clinging. It was later identified as *Anuropus bathypelagicus* Menzies and Dow. The hooklike appendages of the isopod and the fragments of medusoid material in the collection indicate that the host is a medusa.

D. Amphipods and Other Non-Decapod Symbionts

Amphipods are generally associated with other animals as shelter seekers and occasionally as commensals. Thus, *Hyperia medusarum* (Müller) is found in immense numbers under the umbrella of *Rhizostoma pulmo* (Machrie). Another species, *H. galba* Montagu, is found with *R. octopus* L. and *Chrysaora hysoscella* (L.). Harbison et al. (1977) showed that symbiotic amphipods associate with particular coelenterate hosts. Thus, five families occurred only with siphonophores, three with medusae, and another with ctenophores. They concluded that all hyperiid amphipods live with "gelatinous zooplankton" at some stage in their lives and that their highly specific relationships explain the explosive evolution of the group.

Other hyperiid amphipods have recently been described as specific and obligate symbionts of salps (Madin and Harbison, 1977). In 381 scuba dives, they collected 18 species of hyperiids from salps in six different genera. They showed that the amphipods are not only obtaining shelter, but also in several species of *Vibilia*, they are feeding on the mucous string which carries food to the mouth of the salp. In species of *Lycaea*, the amphipod feeds directly on the salp's tissues. Here the boundary between symbiosis and predation is unclear.

Examples of amphipods living as shelter-seekers also occur in limpets, sponges, and starfish. Branch (1975) showed that *Calliopiella michaelseni* has a preference for certain limpets (*Patella* spp.) in South Africa. Species of *Hyale* live with various limpets in several continents (see Branch, 1975, for references), but only as juveniles. When under water they emerge and only retreat under the shell for shelter when the tide retreats (Branch, 1975). Connes et al. (1971) found the amphipod *Perrierella audouiniana* (Bate) in the diverticula of the canals of the sponge, *Suberites,* presumably sheltering in the food-rich environment.

Caprellids are equipped to live as ectosymbionts, but many presumed relationships are random and without significance as symbioses. However, Patton (1968) found that *Caprella grahami* Wigley and Shave never left its host, *Asterias forbesi* (Desor), and was never seen elsewhere. Surprisingly, they are not disturbed by the pedicellariae as described earlier by Jennings (1907) in other starfish.

Associations between amphipods and Norwegian sea anemones were reviewed by Vader (1967); in the sea anemone *Bolocera tuediae* (Johnston), in particular, he found infestations of *Onesimus normani* Sars in 67–80% of specimens of the anemone with as many as 20 amphipods in a single anemone (Vader, 1970b). Vader and Lönning (1973) later showed that the amphipods living symbiotically with anemones were immune to toxic substances from anemones which quickly killed free-living closely related amphipods in the same habitat. Some of these associations are not specific as to hosts. Thus, *Aristias neglectus* Hansen, which is associated with *Bolocera,* also occurs in sponges, echinoderms, and tunicates (Vader, 1970b), and surprisingly with at least two brachiopod species (Vader, 1970a). Published records of associations between amphipods and mollusks, and medusae and echinoderms have been reviewed by Vader (1972c,d, 1978).

The Cyamidae, a family close to the Caprellidae, are the only bona fide amphipod parasites. These whale lice, equipped with extremely sharp terminal claws on their legs and with a greatly reduced abdomen, attain populations of tens and sometimes hundreds of thousands of individuals per whale (Gotto, 1969).

Mysids are essentially pelagic filter-feeders, but several species of *Heteromysis* S. I. Smith are now known to live with specific invertebrate hosts: e.g., *H. actiniae* Clarke among the tentacles of the anemone, *Bartholomea annulata* (LeSueur); *H. harpax* Hilgendorf with hermit crabs, *Dardanus* spp.; *H. zeylanica* W. M. Tattersall in the central cavity of a tubular sponge; and *H. gymnura* W. M. Tattersall on a giant ophiuroid near Zanzibar. Apparently all of these species feed on wastes from their hosts and gain semiprotected locations, and *H. actiniae* also enjoys an immunity from the anemone's toxins, as is usual for all actinian commensals (Tattersall, 1962).

E. Caridean Decapods

1. SYMBIOTIC SHRIMPS

The caridean shrimps comprise over 20 families, three of which—Palaemonidae, Gnathophyllidae, and Alphaeidae—have numerous symbiotic species. These are species of shallow tropical waters, many on coral reefs, that are generally described as commensal: the shrimps have no apparent bad effects on their hosts and sometimes gain obvious benefits from the association. These are not random or occasional associations, but true symbionts. The shrimps are not found apart from their hosts, and the associations are generally specific, in some cases species-specific.

The subfamily Pontoniinae of the Palaemonidae has been the subject of much recent work. The pontoniines were known from earlier studies to be obligate commensals, and monographs were devoted to them by Kemp (1922) and Holthuis (1952). Contemporary work has added many new species and some new genera to the subfamily (Bruce, 1975a,b), and firm data have been obtained on hosts and distributions. As a result, this subfamily of relatively small and inconspicuous decapods is now recognized as one of the most successful and ubiquitous elements in the tropical marine fauna. A review by Bruce (1976a), based on his exceptional contribution to the subject, gives a full account which can be treated only briefly here.

The hosts of pontoniines are usually sponges, actinians, corals, bivalves, nudibranchs, and all five classes of echinoderms. Associations with hydroids, scyphozoans, and tunicates are rare and with annelids, nonexistent. Of 41 genera on Bruce's (1976a) list, half had been named since 1950 and 39 of these were composed of commensal species only. Only one species, *Eupontonia noctalbata* Bruce, is described as noncommensal. Figures 4 and 5 (from Bruce, 1976a) show how different genera link up with hosts in certain taxa and how groups of species in the genus *Periclimenes* are restricted to certain kinds of hosts. The behavioral physiology of host detection and selection in these cases is a challenging problem on which almost no work has been done. Probably larvae of different stages must be selectively collected and reared before progress can be made on such behavioral problems.

One aspect of these symbioses that has been investigated experimentally is that between certain shrimps in the genus *Periclimenes* and their actinian hosts. Levine and Blanchard (1980) worked with *P. rathbunae* Schmitt and *P. anthophilus* Holthuis and Eibl-Eibesfeldt, which are obligate commensals

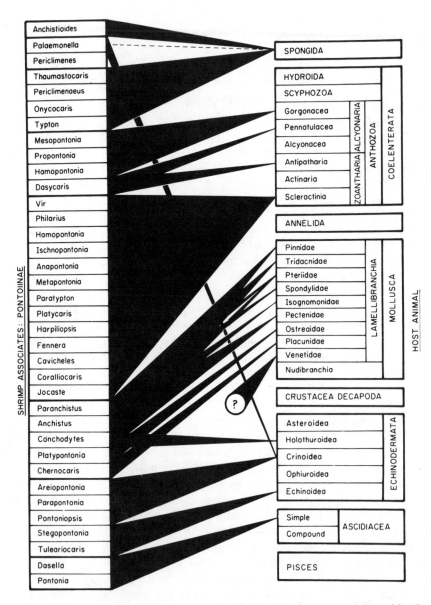

Fig. 4. The relationship of the commensal Indo-West Pacific genera of the subfamily Pontoniinae (excluding *Periclimenes*) to their host types. (From Bruce, 1976a.)

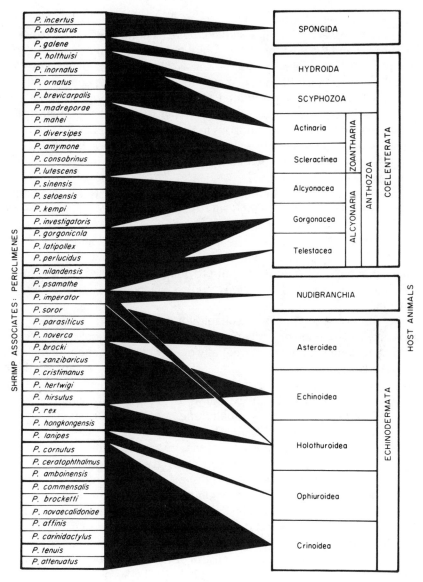

Fig. 5. The relationship of the commensal Indo-West Pacific species of the genus *Periclimenes* to their host types. (From Bruce, 1976a.)

191

of the anemones *Stoichactis helianthus* (Ellis) and *Condylactis gigantea* (Weinland), respectively. They studied the behavioral interactions between the shrimps and the anemones and found that shrimps which had been out of contact with the anemones for as long as 24 hr were no longer protected from the nematocysts. Periods of from 1 to 5 hr of cautious contact with the anemone's tentacles were required to acclimate the shrimps so that the actinian's tentacles no longer stung them and stuck to them. Their results showed a situation closely analogous to the relationship that exists between anemone-fishes and their hosts. Possibly the mechanism is the same, viz., the coating of the symbiont with anemone mucus so that it becomes chemically camouflaged.

Strict rules cannot be formulated about host specificity for pontoniine species or genera. Some widely distributed species in the Indo-Pacific (ranges from Hawaii to East Africa are not unusual) occur on different hosts in different marine provinces. Equally common is the situation in which a shrimp species is associated with a single host and tied to its distribution. In a field in which collections are still providing new species for description and new facts about distributions and associations, it is premature to generalize.

The array of special niches available for groups adopting a commensal habit is reflected in the diversity of form exhibited by the pontoniines. Small size and flattening, either dorsoventrally or laterally, are features of different genera adapted to different corallites. Those living inside bivalves tend to have smooth, swollen bodies, e.g., *Conchodytes* spp. living in *Pinna* sp. (Hipeau-Jacquotte, 1973); an exception to this generalization is the flat bivalve *Placuna* spp., whose symbiont *Chernocaris placunae* n.gen., n.sp., is also flattened (Johnson, 1967). Shrimps usually have many spines and processes, and the pontoniines offer many examples of the reduction and loss of rostral teeth, of the reduction of carapace and abdominal spines, and of the smoothing of the appendages, presumably to assist the movements on and in the host. For example, those pontoniines which inhabit the canals of sponges are generally smooth and cylindrical, e.g., *Onycocaris zanzibarica* Bruce (Bruce, 1971). Special adaptations on the appendages enable shrimps to cling to different hosts and settle in various sites. Feeding appendages have many specialized features, most of which cannot be interpreted because so little is known about feeding habits. The second pereiopods are adapted in many ways for feeding and for predation or defense, e.g., by enlarged chelae (often unequal) equipped with special teeth or scoops. Many pontoniines produce sounds by means of devices incorporated into the chelae of the first pereiopod. The sounds are often louder than those of the more familiar snapping shrimp in the Alpheidae (for details, see Bruce, 1976a).

Behavior patterns and postures of the pontoniines are closely related to the nature and habits of their hosts. Shrimps that live with corals generally cling to the branches in characteristic locations, sometimes near the tips (*Periclimenes amymone* DeMan), sometimes on the intermediate branches (*P. lutescens*), and in other cases at the base of branches on the main stem (*Coralliocaris* spp.). Species that live among the spines of sea urchins (*Tuleariocaris neglecta* Chace and *T. zanzibarica* Bruce) swim constantly among the spines with their heads directed downward toward the test (Castro, 1974). Fricke and Hentschel (1971) reported that the shrimps showed a preference for dark surroundings, and for dark spines in a vertical position and in dense clusters.

2. CLEANING SYMBIOSES

The subject of cleaning symbiosis in fishes and in shrimps was covered so effectively by the pioneer investigator and diver, the late Conrad Limbaugh (Limbaugh, 1961; Limbaugh et al., 1961) and so well reviewed by Feder (1966) that there is little need to do more here than to refer to those earlier works. The shrimps involved in cleaning symbioses are *Periclimenes pedersoni* Chase and *P. yucatanicus* (Ives); two species of hippolytid shrimps, *Lysmata grabham* (Gordon) and *H. californica* Stimpson; and two species of *Stenopus*, *S. hispidus* (Olivier) and *S. scutellatus* Rankin. With the exception of *H. californica*, all of these shrimps occur in the western Atlantic and Caribbean regions; two of the species found there, *H. grabhami* and *S. hispidus*, also occur in Indo-Pacific locations.

Cleaning symbiosis is surely one of the most remarkable of all ecological and behavioral adaptations. That certain shrimps as well as certain fishes should independently have acquired the habit of establishing stations to which fishes come to have their parasites removed and their wounds cleaned up must be one of the most striking examples of convergence. Moreover, the habit occurs in three different families, and therefore it must have been acquired at least three times independently in different shrimps. For the purposes of this review, the point to be emphasized is that crustaceans can react to such subtle signals and can carry out a complex behavior pattern in their relations with their living environment.

F. Brachyuran Decapods

1. THE PINNOTHERIDS

Pea crabs were probably the first crustacean symbionts to be recognized as such. Aristotle was aware of them, but no doubt so were most Mediterranean peoples, who were always consumers of shellfish, where pinnotherids are invariably to be seen. Not only bivalves, but also holothurians, echinoids, a few gastropods, opistobranchs, chitons, and ascidians are hosts of pinnotherids. Others are cohabitants of the tubes and burrows of polychaetes and callianassids. Associations with bivalves are fairly specific, some highly so; those with other hosts are less so. Infestation rates often exceed 50% with certain hosts, but these vary greatly according to local circumstances, e.g., depth (Gray et al., 1968; Kruczynski, 1974).

The genus *Pinnotheres* has over 100 species, all symbiotic. The most familiar species, *P. pisum* (Linnaeus), is associated typically with *Mytilus edulis* L. in Europe. In eastern North America, *P. ostreum* Say and *P. maculatus* Say are linked with oysters and scallops. On the west coast of North America, the genus *Pinnixa* has many species whose most common hosts are the holothurian, *Molpadia intermedia* (Ludwig), with *P. barnharti* Rathbun; the gaper clam, *Tresus capax* (Gould), with *P. faba* (Dana); the cockle, *Clinocardium* spp., with *P. littoralis* Holmes; the lugworm, *Arenicola cristata* Stimpson, with *P. eburna* Wells; terebellid worms with *P. tubicola* Holmes; and the mud shrimps, *Callianassa* and *Upogebia*, and the "inn-keeper," *Urechis caupo* Fisher and MacGinitie, with *P. franciscana* Rathbun (Ricketts and Calvin, 1968).

Species of *Dissodactylus* are associated with irregular echinoids and with sand dollars on both coasts of North America (Gray et al., 1968). Another species from California, *Opisthopus transversus* Rathbun, occurs in many molluscan hosts, including *Cryptochiton stelleri* (Middendorff), the prosobranches, *Megathura crenulata* (Sowerby), *Astraea undosa* (Wood), and *Polinices lewisii* Gould, the opisthobranchs, *Bulla gouldiana* Pilsbry, *Aplysia vaccaria* Winkler, and *Navanax inermis* Cooper, the bivalves, *Mytilus edulis*, *Tresus nuttallii* (Conrad), and *Modiolus* sp., and the holothurian, *Parastichopus californicus* (Stimpson) (Beondé, 1968).

Symbiont–host relations and reproduction have been the focal points of investigations on pinnotherids. Orton (1921) cut windows in mussel shells and showed that *P. pisum* feeds on the mucous strings by which the bivalve conveys its filtered food to its mouth. Pearce (1966) confirmed this for *Fabia subquadrata* Dana living in *Modiolus modiolus* L. He also confirmed that *P. ostreum* in oysters and mussels damages the host's gills. It is not known whether this is incidental damage due to the food taken by the crab or by direct feeding off the host's tissues. The damage may be severe, and therefore pinnotherids are generally regarded as parasites, not commensals. Kruczynski (1974) fed scallops and mussels with labeled diatoms, which *P. maculatus* ingested. Some of the labeled material became incorporated, but it is not known whether it came directly from the plankton or from the tissues or feces of the host.

Experiments to ascertain what attracts pea crabs to their hosts have shown that chemical substances from the host are involved. Sastry and Menzel

(1962) showed that *P. maculatus* in a choice-apparatus were attracted to their hosts, *Aequipecten irradians* Say and *Atrina rigida* Solander. But Gray *et al.* (1968), following up earlier conflicting reports on reactions of *Dissodactylus mellitae* to "host-water" effluent from the sand dollar, *Mellita quinquiesperforata* (Leske), observed a response in a Y-tube apparatus to stronger currents as well as to host-water. In an olfactometer with equalized flow rates, *D. mellitae* demonstrated a highly positive response to host-water.

Confusion about the life history of pinnotherids was cleared up when Christensen and McDermott (1958) discovered that the first invasive stage in *P. ostreum* was not the hard crab stage but a first crab stage, which they found both in plankton and in the host, *Crassostrea virginica* (Gmelin). The hard crab stage, previously thought to be the first invasive stage, is in fact a mating stage. In some of the species that infest bivalves, females remain within the shells of their hosts, and the males enter to copulate. In others, both sexes leave the bivalves after a few molts and engage in copulatory swimming in the sea, e.g., *Fabia subquadrata* in Puget Sound (Pearce, 1966) and *Tritodynamia horvathi* Nobili in Japan (Sakai, 1939). It has been shown in *P. pisum* in some locations that there is a change in the host species after copulatory swimming; the invasive first crab stages are passed in the clam, *Spisula solida* (L.), and the post-hard crab stages in mussels (Christensen and McDermott, 1958). Presumably in different species there are many variants on this general theme. Thus, the adaptations for symbioses in the pinnotherids are extensive and, judging from the numbers of species and individuals, they have been successful.

2. OTHER BRACHYURAN SYMBIONTS

The Hapalocarcinidae, the gall crabs of coral reefs, is the only brachyuran family besides the Pinnotheridae that is typically symbiotic. Verrill (1869) discovered the nature of the coral galls, and Potts (1915b) wrote the definitive paper on the group. Little information has since been added. As a member of the Carnegie expedition to the Torres Strait, Potts described the growth of the branching coral genera, *Pocillopora* and *Seriatopora,* in response to the female *Hapalocarcinus marsupialis* Stimpson located at one of the bifurcations of the colony. Starting with a cup, then with an upper chamber with a slit, and eventually with pores only, the female crab becomes an imprisoned filter-feeder, reproducing after being impregnated by a much smaller male. Potts held the opinion that the currents generated by the crab produced the growth of coral in the shape of a gall, but this is still an open question. Later work has increased the number of described species (see Patton, 1976). Only a few hapalocarcinids produce the typical galls; others live in more or less enclosed pits or recesses of various kinds. A high degree of host-specificity is evident from the data on distributions, but nothing is known about host recognition and settlement in these animals.

Some symbiotic species occur in other brachyuran and some anomuran families, especially the xanthids. Most of the associations are with corals. *Domecia hispida* Eydoux and Souleyet in the Indo-Pacific and *D. acanthophora* (Desbonne and Schramm) in the Caribbean live on various species of *Acropora,* inhabiting sheltering slits and crevices which apparently are due to modified growth in the coral in response to the crab's presence (Patton, 1967).

Species of *Trapezia* were collected in the coral *Pocillopora damicornis* (L.) on the Great Barrier Reef by Patton (1974), along with other xanthids, *Chlorodiella nigra* (Forskål), *Phymodius ungulatus* (H. Milne Edwards), and *Cymo andreossyi* (Audouin). At Eniwetok, Knudsen (1967) found that *Trapezia* is an obligate commensal of pocilloporids, whereas *Tetralia* was found only on acroporids. When *Tetralia* was transferred to *Pocillopora* and *Trapezia* to *Acropora,* each crab returned to its normal host. Knudsen showed that *Trapezia* scraped its food from the coral's surface and argued that these crabs are "obligate ectoparasites." In such a case, "predation" or "grazing" are almost equally acceptable in defining the relationship. Castro (1978a) stressed the obligate nature of this symbiosis. He found that *Trapezia ferruginea* Latreille, when placed with separate colonies of *Pocillopora damicornis* of different sizes, moved among the colonies and assumed special positions, especially at night. Heterosexual pairs were formed, and larger crabs occupied the bigger colonies.

Lamberts and Garth (1977) reported a novel relationship between living corals and two xanthid crabs in Samoa. They found 25 specimens of *Actumnus digitalis* (Rathbun) and *A. antelmei* (Ward) living underneath pieces of coral of 12 different species ranging from 10 to 150 g in weight. The crabs excavate cavities in the coral to fit themselves somewhat like the dromiid sponge crabs mentioned earlier. In an aquarium, these pieces move about in jerks; the situation also is analogous to hermit crabs in their shells. The corals remained alive, and the authors suggest that they are disseminated by this association.

Glynn (1976) tested the suggestion that the symbiotic crabs and shrimps on corals reduced predation by *Acanthaster planci* (L.). He showed that aggressive reactions of *Trapezia ferruginea* and of the snapping shrimp, *Alpheus lottini* Guérin, frequently prevented *Acanthaster* from mounting and feeding on the colonies of their host, *Pocillopora* spp.

Symbionts in the family Parthenopidae are associated with echinoids. Bouvier and Seurat (1905) noted such an association between *Eumedon convictor* (=*Echinoecus pentagonus* A. Milne Edwards) and *Echinothrix turcanum.* Castro (1971, 1978b) has studied the symbiosis between *E. pentagonus* and *Echinothrix calamaris* (Pallas). In this case, small males and immature females live as ectosymbionts and feed on the surface tissues of the urchin; large mature females enter the rectum and feed on feces. The symbiosis is established at the megalopa stage. A response to attractant from the host, together with responses to gravity and a negative response to light, combine to direct the larvae to *E. calamaris,* a dark-colored benthic organism.

Spider crabs have previously been noted as hosts of symbionts, but there are also cases where they are symbionts on hosts in other phyla, especially with sea anemones. Davenport (1962) described the peculiar relationship between the sea anemone, *Tealia felina* (L.), and *Hyas araneus* (L.). This majid is a facultative symbiont of the actinian, and it survives contact with the anemone's tentacles when other crabs are quickly paralyzed and ingested. Another majid, *Mithrax commensalis* Manning, has a close association with its host, *Stoichactis* sp. It is not found elsewhere, and like other commensals, e.g., shrimps and fish, it no doubt gains protection from predators among the anemone's tentacles, having acquired its own immunity from the host's nematocysts.

Weinbauer *et al.* (1982) recently studied a symbiosis between the majid *Inachus phalangium* (Fabricius) and *Anemonia sulcata* Pennant *in situ* in the Mediterranean. They identified crabs individually and removed them from their anemone hosts. All crabs returned to these anemones within 24 hr, but not necessarily to the same *Anemonia* from which they had been removed. Protection from ingestion by *Anemonia* was removed by treating *Inachus* with acetone. Multiple infestations rarely occurred. Thirteen *Inachus,* placed on a large *Anemonia* already occupied by a single crab, gradually deserted this host until only the single crab remained.

Finally, a few anomuran species in the family Porcellanidae live in symbioses with ophiuroids and with the asteroid, *Luidia* spp. (Gore and Shoup, 1968). Another species, *Porcellana sayana* Leach, has been found in *Strombus gigas* (L.) shells occupied by the large pagurid, *Petrochirus diogenes* (L.) (Telford and Daxboeck, 1978). These crabs are also found near the operculum on the foot of the living *S. gigas.*

This review of brachyuran symbionts shows that even at this peak of crustacean evolution, a symbiotic existence is one of the chief modes of life. Nature has endowed crustaceans with a potential for linking their lives with those of other animals in a multitude of diverse ways.

IV. COMMENTS AND PERSPECTIVES

This review of a representative sample of the thousands of symbioses in the Crustacea has reinforced the statements made earlier about the number and diversity of their symbiotic relationships. It seems that the Crustacea are basically adapted for a symbiotic existence both as symbionts and hosts, and this helps to explain why there are so many symbioses in which both participants are crustaceans.

A special feature of crustacean symbioses that is not found on other groups outside the insects is that the symbioses do not depend wholly on the searching and settling activities of the symbiont. We have documented a number of crustacean hosts, especially among the hermits and the decorating crabs, that reverse the usual procedure and seek out certain symbionts as partners. Such activities reveal a sophisticated type of environmental adaptation which requires that these crustaceans possess elaborate neurosensory equipment.

One is struck by the fact that with the exception of cleaning symbioses, all the basic symbiotic situations in the Crustacea were discovered long ago; most of them have been known for more than a century. This shows how completely the Victorian zoologists and collectors surveyed the fauna of the globe, including the depths of the sea. At the visual level they left little for future generations to discover. The limitation was that few of the symbioses were studied alive, although with the advent of marine laboratories and marine aquaria this limitation was partially overcome. Later, with the introduction of scuba, underwater observation of marine animals in their natural habitat could take place, and the discovery of cleaning symbiosis was one of the first results of that activity.

Even now, describing new examples of symbiosis and naming new species and genera to be added to the list still make up a large part of contemporary work. One recognizes that such work has to be done, but nothing new in principle can come out of it since it only adds new examples to classes of phenomena that are already well known. Even the impressive data about the symbioses among the pontoniine shrimps, both as symbionts and hosts, do little to extend a horizon that was already well in view.

In the book that was a landmark in studies of symbiosis, Caullery (1952) stressed that answers to the basic questions that arose from symbiotic relationships could only come from experimental studies. Studies of symbioses involving microorganisms already follow that direction (see Jennings and Lee, 1975). Research on symbiosis among crustaceans has not made a general move in that direction. For instance, the emphasis in the behavioral studies on actinian–pagurid symbioses has been on the actinian rather than on the pagurid (Ross, 1974). One line of work that is beginning to write a new chapter is to be seen in papers by Anderson (1975, 1977), who has used a symbiosis between a bopyrid and a shrimp to obtain data on the metabolic interaction between host and symbiont. But the most promising direction that studies on crustacean symbioses could take would involve a general move to study more examples in the living state. An explosion of knowledge has taken place in recent years in crustacean neurophysiology and behavior. Applied to such problems as host recognition and the interactions in crustacean symbioses, this new knowledge can move the subject far beyond the descriptive level at which it now stands.

This is not a new perspective. It is a viewpoint expressed many times during the past quarter century by Demorest Davenport, who has been not only a contributor to, but also the progenitor of, much of the work that has gone on in recent years on marine invertebrate symbioses. Most of the good experiments that have been done have either been carried out or inspired by Davenport. Like most convincing experiments, his had the elegance of simplicity. Beginning with his pioneer study on polychaete symbionts which saw the introduction of the invaluable Y-tube and trough olfactometers, and continuing with the clarification of the anemone–fish symbiosis through to the development of a method for the study of host location in small organisms, Davenport's work has broken new ground. By his example, personality, and ideas, "Dav," who is always looking for triggers to explain behavioral events, has himself been a trigger for the activities of his colleagues. It is fitting to close this article by quoting from one of his early reviews (Davenport, 1955), which pointed out a direction that my own work took a few years later:

> There remains the basic investigation of the control of adaptive behavior in the numerous fascinating partnerships in the sea that have aroused the interest of naturalists since earliest times. Examples wonderfully suited to critical study are such symbioses as those between hermit-crabs and anemones . . . many of these associations have been studied and described; in some, a few simple experiments on specificity have been conducted. But in every case without exception the behavior concerned is in need of precise experimental analysis by modern techniques. In each case we must identify stimuli eliciting special behavior and determine precisely the manner in which the partners are first brought together. As one may observe in many ecology textbooks, these associations have acquired a reputation of being "unique," "unusual," or "extraordinary." Certainly their adaptive "reason-for-being" is no more extraordinary than that of any other individual natural adaptation. Actually, this reputation depends largely upon the fact that two organisms are linked together, and the behavior concerned in the linkage is not understood. It is unusual behavior only because, without analysis, it seems to demand reason or, at least, learning. That it is not unusual or inexplicable at all will certainly be disclosed by the critical analysis to which it is so clearly susceptible.

This statement has a special relevance for studies on symbiosis in Crustacea. It points to the very areas which this account has shown to be thinly developed. It reminds us that it is not enough to describe and to wonder at the remarkable adaptations for symbiosis that exist in the Crustacea. The primary aim now should be to analyze and to understand these relationships. Fortunately, the basic knowledge and the essential tools for this task now exist.

ACKNOWLEDGMENTS

Thanks are due to my secretary, Mrs. Kay Baert, not only for careful typing of the manuscript, but also for invaluable and painstaking help with the bibliography and for her personal interest in all phases of the project. Miss Lynne Metcalfe's assistance with the literature search is acknowledged with thanks. Some of the expenses of the bibliographical work were paid out of operating grant NSERC A-1445 to the author. Dr. A. J. Bruce and Academic Press permitted reproduction of Fig. 4 and 5, and the National Research Council of Canada granted permission for the reproduction of certain photographs in Fig. 1 that were previously published in the *Canadian Journal of Zoology* **48,** 371–376, 1970.

REFERENCES

Achituv, Y. (1970). *Dendrogaster asterinae* n.sp., an ascothoracid (Cirripedia) parasite of the starfish *Asterina burtoni* of the Gulf of Elat. *Crustaceana* **21,** 1–4.

Allen, J. A. (1966). Notes on the relationship of the bopyrid parasite *Hemiarthrus abdominalis* (Krøyer) with its hosts. *Crustaceana* **10,** 1–6.

Anderson, D. T. (1978). Cirral activity and feeding in the coral-inhabiting barnacle *Boscia anglicum* (Cirripedia). *J. Mar. Biol. Assoc. U. K.* **58,** 607–626.

Anderson, G. (1975). Metabolic response of the caridean shrimp *Palaemonetes pugio* to infection by the adult epibranchial isopod parasite *Probopyrus pandalicola. Comp. Biochem. Physiol.* **52A,** 201–297.

Anderson, G. (1977). The effects of parasitism on energy flow through laboratory shrimp populations. *Mar. Biol. (Berlin)* **42,** 239–251.

Baker, J. H. (1969). On the relationship of *Ankylocythere sinuosa* (Rioja 1942) (Ostracoda, Entocytheridae) to the crayfish *Procambarus simulans* (Faxon, 1884). *Trans. Am. Microsc. Soc.* **88,** 293–294.

Balakrishnan, K. P. (1969). Observations on the occurrence of *Conchoderma virgatum* (Spengler) (Cirripedia) on *Diodon hystrix* Linnaeus (Pisces). *Crustaceana* **16,** 101–102.

Balasch, J., and Mengual, V. (1974). The behavior of *Dardanus arrosor* in association with *Calliactis parasitica* in artificial habitat. *Mar. Behav. Physiol.* **2,** 251–260.

Balss, H. (1924). Über Anpassungen und Symbiose der Paguriden. Eine Zusammenfassende Übersicht. *Z. Morph. Oekol. Tiere* **1,** 752–792.

Barham, E. G., and Pickwell, G. V. (1969). The giant isopod, *Anuropus:* A scyphozoan symbiont. *Deep-Sea Res.* **16,** 525–529.

Bartsch, I. (1975). *Cancerilla oblonga* n.sp., ein cyclopoider copepode auf *Amphiura capensis* Ljungman (Ophiuroidea, Echinodermata). *Crustaceana* **29,** 290–294.

Baudoin, M. (1975). Host castration as a parasitic strategy. *Evolution* **29,** 335–352.

Beondé, A. C. (1968). *Aplysia vaccaria,* a new host for the pinnotherid crab *Opisthopus transversus. Veliger* **10,** 375–378.

Bethel, W. M., and Holmes, J. C. (1973). Altered evasive behavior and responses to light in amphipods harboring acanthocephalan cystacanths. *J. Parasitol.* **59,** 945–956.

Bethel, W. M., and Holmes, J. C. (1974). Correlation of development of altered evasive behavior in *Gammarus lacustris* (Amphipoda) harboring cystacanths of *Polymorphus paradoxus* (Acanthocephala) with the infectivity to the definitive host. *J. Parasitol.* **60,** 272–274.

Bishop, J. E. (1968). An ecological study of the branchiobdellid commensals (Annelida-Branchiobdellidae) of some mid-western Ontario crayfish. *Can. J. Zool.* **46,** 835–843.

Bocquet, C., and Stock, J. H. (1963). Some recent trends in work on parasitic copepods. *Oceanogr. Mar. Biol.* **1,** 289–300.

Bocquet-Védrine, J. (1967). Un nouveau Rhizocéphale parasite de Cirripède: *Microgaster balani* n.gen., n.sp. *C. R. Hebd. Seances Acad. Sci., Ser.* D **265,** 1630–1632.

Bocquet-Védrine, J. (1972). Suppression de l'ordre des Apodes (Crustacés Cirripedès) et rattachment de son unique représentant, *Proteolepas bivincta,* à la famille des Crinoniscidae (Crustacés Isopodes, Cryptonisciens). *C. R. Hebd. Seances Acad. Sci., Ser.* D **275,** 2145–2148.

Boschma, H. (1968). *Loxothylacus engeli* nov. spec., a rhizocephalan parasite of the crab *Anasimus latus* Rathbun. *Beaufortia* **15,** 21–26.

Bouligand, Y. (1966). Recherches recentes sur les copépodes associés aux anthozoaires. *Symp. Zool. Soc. London* **16,** 267–306.

Bouvier, L., and Seurat, G. (1905). *Eumedon convictor,* crabe commensal d'un Oursin. *C. R. Hebd. Seances Acad. Sci., Ser.* D **140,** 629–631.

Bowers, R. L. (1968). Observations on the orientation and feeding behavior of barnacles associated with lobsters. *J. Exp. Mar. Biol. Ecol.* **2,** 105–112.

Boxshall, G. A. (1974a). *Lepophtheirus pectoralis* (O. F. Müller, 1776); a description, a review, and some comparisons with the genus *Caligus* Müller, 1785. *J. Nat. Hist.* **8,** 445–468.

Boxshall, G. A. (1974b). The developmental stages of *Lepeophtheirus pectoralis* (Müller, 1776) (Copepoda: Caligidae). *J. Nat. Hist.* **8,** 681–700.

Boxshall, G. A. (1976). The host specificity of *Lepeophtheirus pectoralis* (Müller, 1776) (Copepoda: Caligidae). *J. Fish Biol.* **8,** 255–264.

Boxshall, G. A. (1977). The histopathology of infection by *Lepeophtheirus pectoralis* (Müller) (Copepoda: Caligidae). *J. Fish Biol.* **10,** 411–415.

Boycott, B. B. (1954). Learning in *Octopus vulgaris* and other cephalopods. *Pubbl. Stn. Zool. Napoli* **25,** 67–93.

Branch, G. M. (1975). The ecology of *Patella* from the Cape Peninsula, South Africa. 5. Commensalism. *Zool. Afr.* **10,** 133–162.

Breed, G. M., and Olson, R. E. (1977). Biology of the microsporidan parasite *Pleistophora crangoni* n.sp. in three species of crangonid sand shrimps. *J. Invertebr. Pathol.* **30,** 387–405.

Brock, F. (1927). Des Verhaltens des Einsidlerkrebses *Pagurus arrosor* Herbst während des Aufsuchens, Ablösens und Aufpflanzens seiner Seerose *Sagartia parasitica* Gosse. *Arch. Entwicklungsmech. Org.* **112,** 205–238.

Bruce, A. J. (1969). *Aretopsis amabilis* de Man, an alpheid shrimp commensal of pagurid crabs in the Seychelle Islands. *J. Mar. Biol. Assoc. India* **11,** 175–181.

Bruce, A. J. (1971). *Onycocaris zanzibarica* sp. nov., a new pontoniid shrimp from East Africa. *J. Nat. Hist.* **5,** 293–298.

Bruce, A. J. (1972a). *Orophryxus shiinoi* gen. nov., sp. nov., an unusual phryxid (Crustacea, Isopoda, Bopyridae) parasitic upon a pontoniid shrimp from Zanzibar. *Parasitology* **64,** 445–450.

Bruce, A. J. (1972b). *Filophryxus dorsalis* gen. nov., sp. nov., an unusual bopyrid parasite from eastern Australia. *Parasitology* **65,** 351–358.

Bruce, A. J. (1973). *Mesophryxus ventralis* gen. nov., sp. nov., a phryxid bopyrid parasitic upon the pontoniinid shrimp *Harpiliopsis beaupresi* (Audouin). *Parasitology* **66,** 515–523.

Bruce, A. J. (1974). *Allophryxus malindiae* gen. nov., sp. nov., a hemiarthrinid bopyrid parasitic upon the pontoniinid shrimp *Coralliocaris superba* (Dana). *Parasitology* **68,** 127–134.

Bruce, A. J. (1975a). Notes on some Indo-Pacific Pontoniinae. XXV. Further observations upon *Periclimenes noverca* Kemp, 1922, with the designation of a new genus *Zenopontonia,*

202   D. M. Ross

and some remarks upon *Periclimenes parasiticus* Borradaile (Decapoda Natantia, Palaemonidae). *Crustaceana* **29**, 275–285.

Bruce, A. J. (1975b). Notes on some Indo-Pacific Pontoniinae. XXVI. *Neoanchistus cardiodytes* gen. nov., sp. nov., a new mollusc-associated shrimp from Madagascar (Decapoda, Palaemonidae). *Crustaceana* **29**, 149–165.

Bruce, A. J. (1976a). Shrimps and prawns of coral reefs, with special reference to commensalism. In "Biology and Geology of Coral Reefs" (O. A. Jones and R. M. Dean, eds.), Vol. 3, pp. 37–94. Academic Press, New York.

Bruce, A. J. (1976b). A report on some pontoniid shrimps collected from the Seychelle Islands by the F.R.V. Manihine, 1972, with a review of the Seychelles pontoniid shrimp fauna. *J. Linn. Soc. London, Zool.* **59**, 89–153.

Brunelli, G. (1910). Osservazioni ed esperienze sulla simbiosi dei Paguridi e delle Attinnie. *Atti Accad. Naz. Lincei, Rend., Ser.* **5, 19**, 77–82.

Brunelli, G. (1913). Ricerche etologiche. Osservazioni ed esperienze sulla simbiosi dei Paguridi e delle Attinie. *Zool. Jahrb., Abt. Allg. Zool. Physiol. Tiere* **34**, 1–26.

Bürger, O. (1903). Ueber das Zusammenleben von *Antholoba reticulata* Couth und *Hepatus chilensis* M.E. *Biol. Zentralbl.* **23**, 678–679.

Burris, K. W., and Miller, C. G. (1972). Parasitic copepods of some freshwater fishes from North Carolina. *J. Elisha Mitchell Sci. Soc.* **88**, 18–20.

Carlgren, O. (1928). Zur Symbiose zwischen Actinien und Paguriden. *Z. Morphol. Oekol. Tiere* **12**, 165–173.

Carlisle, A. I. (1953). Observations on the behaviour of *Dromia vulgaris* Milne Edwards with simple ascidians. *Pubbl. Stn. Zool. Napoli* **24**, 142–151.

Carton, Y. (1971). Copépodes parasites de Madagascar. I. Description de *Sabellacheres aenigmatopygus* n.sp., copépode parasite de *Potamilla reniformis* (Polychètes, Sabellidae). *Crustaceana* **21**, 145–152.

Carton, Y. (1974a). Description de *Selioides guineensis* sp.n., copépode cyclopoide parasite d'Aphroditidae. *Arch. Zool. Exp. Gen.* **115**, 129–139.

Carton, Y. (1974b). Copépodes parasites de Madagascar. II. Description de *Botulosoma endoarrhenum* n.gen., n.sp. (Lichomolgidae) parasite d'*Othilia purpurea* (Echinodermata, Asteridae); étude de ses relations anatomiques avec l'hôte. *Crustaceana* **26**, 65–79.

Castro, P. (1971). The natantian shrimps (Crustacea, Decapoda) associated with invertebrates in Hawaii. *Pac. Sci.* **25**, 395–403.

Castro, P. (1974). A new host and notes on the behavior of *Tuleariocaris neglecta* Chace, 1969 (Decapoda, Palaemonidae, Pontoniinae), a symbiont of diadematid sea urchins. *Crustaceana* **26**, 318–320.

Castro, P. (1978a). Movements between coral colonies in *Trapezia ferruginea* (Crustacea: Brachyura), an obligate symbiont of scleractinian corals. *Mar. Biol. (Berlin)* **46**, 237–245.

Castro, P. (1978b). Settlement and habitat selection in the larvae of *Echinoecus pentagonus* (A. Milne Edwards), a brachyuran crab symbiotic with sea urchins. *J. Exp. Mar. Biol. Ecol.* **34**, 259–270.

Caullery, M. (1952). "Parasitism and Symbiosis." Sidgwick and Jackson Limited, London.

Charniaux-Cotton, H. (1960). Sex determination. *Physiol. Crustacea* **1**, 411–447.

Cheng, T. C. (1967). Marine molluscs as hosts for symbioses with a review of known parasites of commercially important species. *Adv. Mar. Biol.* **5**, 1–424.

Cheng, T. C., ed. (1971). "Aspects of the Biology of Symbiosis." Univ. Park Press, Baltimore, Maryland.

Christensen, A. M., and Kanneworff, B. (1965). Life history and biology of *Kronborgia amphipodicola* Christensen and Kanneworff (Turbellaria, Neorhabdocoeia). *Ophelia* **2**, 237–252.

Christensen, A. M., and McDermott, J. J. (1958). Life-history and biology of the oyster crab *Pinnotheres ostreum* Say. *Biol. Bull. (Woods Hole, Mass.)* **114,** 146–179.

Connes, R., Paris, J., and Sube, J. (1971). Réactions tissulaires de quelques démosponges vis-à-vis de leurs commensaux et parasites. *Naturaliste Can.* **98,** 923–935.

Conover, M. R. (1976). The influence of some symbionts on the shell-selection behaviour of the hermit crabs, *Pagurus pollicaris* and *Pagurus longicarpus. Anim. Behav.* **24,** 191–194.

Cotte, J. (1922). Études sur le comportement et les réactions des actinies. *Bull. Inst. Oceanogr.* **410,** 1–44.

Cowles, R. P. (1919). The habits of tropical Crustacea. III. Habits and reactions of hermit crabs associated with sea anemones. *Philipp. J. Sci.* **15,** 81–90.

Cressey, R. F. (1976). *Shiinoa elegata,* a new species of parasitic copepod (Cyclopida) from *Elagatus* (Carangidae). *Proc. Biol. Soc. Wash.* **88,** 433–438.

Cutress, C., Ross, D. M., and Sutton, L. (1970). The association of *Calliactis tricolor* with its pagurid, calappid, and majid partners in the Caribbean. *Can. J. Zool.* **48,** 371–376.

Dales, R. P. (1957). Commensalism. *In* "Treatise on Marine Ecology and Paleoecology" (J. W. Hedgpeth, ed.), Vol. 1, pp. 391–412. Memoir 67Geol. Soc. Am., New York.

Dales, R. P. (1966). Symbiosis in marine organisms. *In* "Symbiosis" (S. M. Henry, ed.), Vol. 1, pp. 299–327. Academic Press, New York.

Danforth, C. G. (1971). New bopyrids (Isopoda) from the Indian and Pacific Oceans. *Micronesica* **7,** 163–177.

Danforth, C. G. (1976). Epicaridea (Isopoda) of Guam. *Crustaceana* **31,** 78–80.

Davenport, D. (1955). Specificity and behavior in symbioses. *Q. Rev. Biol.* **30,** 29–46.

Davenport, D. (1962). Physiological notes on actinians and their associated commensals. *Bull. Inst. Oceanogr.* **50,** No. 59, 1–15.

Davenport, D. (1966a). Cnidarian symbioses and the experimental analysis of behavior. *Symp. Zool. Soc. London* **16,** 361–372.

Davenport, D. (1966b). The experimental analysis of behavior in symbioses. *In* "Symbiosis" (S. M. Henry, ed.), Vol. 1, pp. 381–429, Academic Press, New York.

Day, J. H. (1939). A new cirripede parasite, *Rhizolepas annelidicola,* nov. gen. et sp. *Proc. Linn. Soc. London* **51,** 64–79.

Delage, Y. (1884). Evolution de la Sacculine (*Sacculina carcini* Thomps.), Crustacé endoparasite de l'ordre nouveau de Kentrogonides. *Arch. Zool. Exp. Gen. Ser.* **2,** 417–736.

Delamare-Deboutteville, C., and Nunes, L. P. (1951). Sur le comportement d'*Octopicola superba* Humes, n.g., n.sp. parasite de la Pieuvre *Octopus vulgaris* Lamarck. *C. R. Hebd. Seances Acad. Sci.* **244,** 504–506.

Dembowska, W. S. (1926). Study on the habits of the crab *Dromia vulgaris* (M. (M.E.). *Biol. Bull. (Woods Hole, Mass.)* **50,** 162–178.

Dinamani, P. (1964). Variation in form, orientation and mode of attachment of the cirriped, *Octolasmis stella* (Ann.), symbiotic on the gills of lobster. *J. Anim. Ecol.* **33,** 357–362.

Dobrohotova, O. V. (1975). Copepod crustaceans of the suborder Calanoida, the intermediate hosts of *Hymenolepididae (Cestoda)* in Kazakhstan. *Acta Parasitol. Pol.* **23,** 237–242.

Drach, P. (1941). Nouvelle conception sur les rapports ethologiques des Entonisciens et de leurs hôtes. Critique de la theorie ectoparasitaire. *C. R. Hebd. Seances Acad. Sci.* **213,** 80–82.

Dudley, P. L. (1968). A light and electron microscopic study of tissue interactions between a parasitic copepod, *Scolecodes huntsmani* (Henderson), and its host ascidian, *Styela gibbsii* (Stimpson). *J. Morphol.* **124,** 263–281.

Duerden, J. E. (1905). On the habits and reactions of crabs bearing actinians in their claws. *Proc. Zool. Soc. London* **2,** 494–511.

Faurot, L. (1910). Étude sur les associations entre les Pagures et les Actinies; *Eupagurus prideauxi* Heller et *Adamsia palliata* Forbes, *Pagurus striatus* Latreille et *Sagartia parasitica* Gosse. *Arch. Zool. Exp. Gen.* **5,** 421–486.

Faurot, L. (1932). Actinies et pagures. Étude de psychologie animale. *Arch. Zool. Exp. Gen.* **74,** 139–154.

Feder, H. M. (1966). Cleaning symbiosis in the marine environment. *In* "Symbiosis" (S. M. Henry, ed.), Vol. 1, pp. 327–380. Academic Press, New York.

Fenizia, G. (1935). La *Dromia vulgaris* (M. Edwards) e le sue abitudini. *Arch. Zool. Ital.* **21,** 509–539.

Fernando, C. H., and Hanek, G. (1973). Two new species of the genus *Ergasilus* Nordmann, 1932 (Copepoda, Ergasilidae) from Ceylon. *Crustaceana* **25,** 13–20.

Field, L. H. (1969). The biology of *Notophryxus lateralis* (Isopoda: Epicaridea), parasitic on the euphausiid *Nematoscelis difficilis*. *J. Parasitol.* **55,** 1271–1277.

Forest, J., and Saint Laurent, M. de (1967). Campagne de la Calypso au large des Côtes Atlantiques de l'Amerique du sud (1961–1962). 6. Crustaces decapodes: Pagurides. *Ann. Inst. Oceanogr. Paris* **45,** 47–169.

Fricke, von Hans-W., and Hentschel, M. (1971). Die Garnelen-Seeigel-Partnerschaft—eine Untersuchung der optischen Orientierung der Garnele. *Z. Tierpsychol.* **28,** 453–462.

Fryer, G. (1968). The parasitic Crustacea of African freshwater fishes; their biology and distribution. *Proc. Zool. Soc. London* **156,** 45–95.

Gilpin-Brown, J. B. (1969). Host-adoption in the commensal polychaete *Nereis fucata*. *J. Mar. Biol. Assoc. U. K.* **49,** 121–127.

Glynn, P. W. (1968). Ecological studies on the associations of chitons in Puerto Rico, with special reference to sphaermoid isopods. *Bull. Mar. Sci.* **18,** 572–626.

Glynn, P. W. (1976). Some physical and biological determinants of coral community structure in the eastern Pacific. *Ecol. Monogr.* **46,** 431–456.

Gore, R. H., and Shoup, J. B. (1968). A new starfish host and an extension of range for the commensal crab, *Minyocerus angustus* (Dana, 1852) (Crustacea: Porcellanidae). *Bull. Mar. Sci.* **18,** 240–248.

Gosse, P. H. (1860). "A History of the British Sea-Anemones and Corals." Van Voorst, London.

Gotto, R. V. (1969). "Marine Animals. Partnerships and Other Associations." English Univ. Press, London.

Gray, I. E., McCloskey, L. R., and Weihe, S. C. (1968). The commensal crab *Dissodactylus mellitae* and its reaction to sand dollar host-factor. *J. Elisha Mitchell Sci. Soc.* **84,** 472–481.

Haig, J. (1974). Two new species of *Pagurus* from deep water off Peru and Chile (Decapoda, Anomura, Paguridae). *Crustaceana* **27,** 119–130.

Hameed, M. S., and Pillai, N. K. (1973). Description of a new species of *Caligus* (Crustacea: Copepoda) from Kerala. *Zool. Anz., Leipzig* **191,** 114–118.

Hamond, R. (1972). Four new copepods (Crustacea: Harpacticoida, Canuellidae) simultaneously occurring with *Diogenes senex* (Crustacea: Paguridea) near Sydney. *Proc. Linn. Soc. N.S.W.* **97,** 165–201.

Hand, C. (1975). Behaviour of some New Zealand sea anemones and their molluscan and crustacean hosts. *N. Z. J. Mar. Freshwater Res.* **9,** 509–527.

Harbison, G. R., Biggs, D. C., and Madin, L. P. (1977). The associations of Amphipoda Hyperiidea with gelatinous zooplankton—II. Associations with Cnidaria, Ctenophora and Radiolaria. *Deep-Sea Res.* **24,** 465–488.

Hartnoll, R. G. (1967). The effects of sacculinid parasites on two Jamaican crabs. *J. Linn. Soc. London, Zool.* **46,** 275–295.

Hartzell, A. (1967). Insect ectosymbiosis. In "Symbiosis" (S. M. Henry, ed.), Vol. 2, pp. 107–140. Academic Press, New York.

Hastings, R. W. (1972). The barnacle, Conchoderma virgatum (Spengler), in association with the isopod, Nerocila acuminata (Schioedte and Meinert), and the orange filefish, Alutera schoepfi (Walbaum). Crustaceana 22, 274–278.

Henry, S. M., ed. (1966). "Symbiosis," Vols. 1 and 2. Academic Press, New York.

Herberts, C. (1978). Immunochimie—Relation hôte-parasite entre Carcinus mediterraneus et Sacculina carcini; analyse immunochimique et mise en évidence d'une précipitine anti-sacculine. C. R. Hebd. Seances Acad. Sci., Ser. D 286, 725–728.

Hickling, C. F. (1963). On the small deep-sea shark Etmopterus spinax L., and its cirripede parasite Anelasma squalicola (Loven). J. Linn. Soc. London, Zool. 45, 17–24.

Hindsbo, O. (1972). Effects of Polymorphus (Acanthocephala) on colour and behaviour of Gammarus lacustris. Nature (London) 238, 333.

Hipeau-Jacquotte, R. (1973). Étude des crevettes Pontoniinae (Palaemonidae) associées aux mollusques Pinnidae a Tuléar (Madagascar). 3. Morphologie externe et morphologie des pièces buccales. Tethys (Suppl.) 5, 95–116.

Ho, J. (1972a). Copepods of the family Chondracanthidae (Cyclopoida) parasitic on South African marine fishes. Parasitology 65, 147–158.

Ho, J. (1972b). Four new parasitic copepods of the family Chondracanthidae from California inshore fishes. Proc. Biol. Soc. Wash. 85, 523–540.

Ho, J. (1974). A new species of Chondracanthus (Copepoda, Chondracanthidae) parasitic on Peruvian cusk eels. J. Parasitol. 60, 870–873.

Ho, J., and Perkins, P. S. (1977). A new family of cyclopoid copepod (Namakosiramiidae) parasitic on holothurians from southern California. J. Parasitol. 63, 368–371.

Hogue, C. L., and Bright, D. B. (1971). Observations on the biology of land crabs and their burrow associates on the Kenya coast. Contrib. Sci. (Los Angeles Co., Calif., Mus. Nat. Hist.) 210, 1–10.

Holmes, J. C., and Bethel, W. M. (1972). Modification of intermediate host behaviour by parasites. Zool. J. Linn. Soc. Suppl. 1, 51, 123–149.

Holthuis, L. B. (1952). The Decapoda of the Siboga Expedition. Part XI. The Palaemonidae collected by the Siboga and Snellius Expeditions with remarks on other species II. Subfamily Pontoniinae. Siboga Exped. No. 39a, pp. 1–254.

Humes, A. G. (1947). A new harpacticoid copepod from Bornean crabs. J. Wash. Acad. Sci. 37, 170–178.

Humes, A. G. (1958). Antillesia cardisomae, n. gen. and sp. (Copepoda: Harpacticoida) from the gill chambers of land crabs, with observations on the related genus Cancrincola. J. Wash. Acad. Sci. 48, 77–89.

Humes, A. G. (1966). New species of Macrochiron (Copepoda, Cyclopoidea) associated with hydroids in Madagascar. Beaufortia 14, 5–28.

Humes, A. G. (1967). A new species of Scambicornus (Copepoda, Cyclopoida, Lichomolgidae) associated with a holothurian in Madagascar, with notes on several previously described species. Beaufortia 14, 135–155.

Humes, A. G. (1970). Paramacrochiron japonicum n.sp., a cyclopoid copepod associated with a medusa in Japan. Publ. Seto Mar. Biol. Lab. 18, 223–232.

Humes, A. G. (1971a). Sunaristes (Copepoda, Harpacticoida) associated with hermit crabs at Eniwetok Atoll. Pac. Sci. 25, 529–532.

Humes, A. G. (1971b). Cyclopoid copepods (Stellicomitidae) parasitic on sea stars from Madagascar and Eniwetok Atoll. J. Parasitol. 57, 1330–1343.

Humes, A. G. (1972). *Pseudanthessius comanthi* n.sp. (Copepoda, Cyclopoida) associated with a crinoid at Eniwetok Atoll. *Pac. Sci.* **26,** 373–380.

Humes, A. G. (1973a). *Hemicyclops perinsignis,* a new cyclopoid copepod from a sponge in Madagascar. *Proc. Biol. Soc. Wash.* **86,** 315–328.

Humes, A. G. (1973b). *Nanaspis* (Copepoda: Cyclopoida) parasitic on the holothurian *Thelenota ananas* (Jaeger) at Eniwetok Atoll. *J. Parasitol.* **59,** 384–395.

Humes, A. G. (1974a). *Odontomolgus mundulus* n.sp. (Copepoda, Cyclopoida) associated with the scleractinian coral genus *Alveopora* in New Caledonia. *Trans. Am. Microsc. Soc.* **93,** 153–162.

Humes, A. G. (1974b). New cyclopoid copepods associated with an abyssal holothurian in the eastern North Atlantic. *J. Nat. Hist.* **8,** 101–117.

Humes, A. G.(1975). Cyclopoid copepods associated with marine invertebrates in Mauritius. *J. Linn. Soc. London, Zool.* **56,** 171–181.

Humes, A. G., and Hendler, G. (1972). New cyclopoid copepods associated with the ophiuroid genus *Amphioplus* on the eastern coast of the United States. *Trans. Am. Microsc. Soc.* **91,** 539–555.

Humes, A. G., and Ho, J. (1966). New lichomolgid copepods (Cyclopoida) from zoanthid coelenterates in Madagascar. *Cah. ORSTROM, Ser. Oceanogr.* **4,** 3–47.

Humes, A. G., and Stock, J. H. (1973). A revision of the family Lichomolgidae Kossmann, 1977, cyclopoid copepods mainly associated with marine invertebrates. *Smithson. Contrib. Zool.* No. 127, pp. 1–368.

Illg, P. L. (1970). Report on ascidicole Copepoda collected during the Melanesia Expedition of the Ôsaka Museum of Natural History, Ôsaka, Japan. *Publ. Seto Mar. Biol. Lab.* **18,** 169–188.

Illg, P. L., and Humes, A. G. (1971). *Henicoxiphium redactum,* a new cyclopid copepod associated with an ascidian in Florida and North Carolina. *Proc. Biol. Soc. Wash.* **83,** 569–578.

Jennings, D. H., and Lee, D. L., eds. (1975). Symbiosis. *Symp. Soc. Exp. Biol.* **29.**

Jennings, H. S. (1907). Behavior of the starfish *Asterias forreri* De Loriol. *Univ. Calif. Publ. Zool.* **4,** 53–185.

Jennings, J. B. (1974). Symbioses in the Turbellaria and their implications in studies on the evolution of parasitism. *In* "Symbiosis in the Sea" (W. B. Vernberg, ed.), pp. 127–160. Univ. of South Carolina Press, Columbia, South Carolina.

Jennings, J. B., and Gelder, S. R. (1976). Observations on the feeding mechanism, diet and digestive physiology of *Histriobdella homari* Van Beneden 1858: An aberrant polychaete symbiotic with North American and European lobsters. *Biol. Bull. (Woods Hole, Mass.)* **151,** 489–517.

Jensen, K. (1970). The interaction between *Pagurus bernhardus* (L.) and *Hydractinia echinata* (Fleming). *Ophelia* **8,** 135–144.

Johnson, D. S. (1967). On some commensal decapod crustaceans from Singapore (Palaemonidae and Porcellanidae). *Proc. Zool. Soc. London* **153,** 499–526.

Johnson, S. K., and Rogers, W. A. (1972). *Ergasilus clupeidarum* sp. n. (Copepoda: Cyclopoida) from clupeid fishes of the southeastern U.S. with a synopsis of the North American *Ergasilus* species with a two-jointed first endopod. *J. Parasitol.* **58,** 385–392.

Kabata, Z. (1967). *Proclavellodes pillaii* gen. et sp. n. (Copepoda: Lernaeopodidae) from South India. *J. Parasitol.* **53,** 1298–1301.

Kabata, Z. (1969). *Chondracanthus narium* sp.m. (Copepoda: Chondracanthidae), a parasite of nasal cavities of *Ophiodon elongatus* (Pisces: Teleostei) in British Columbia. *J. Fish. Res. Board Can.* **26,** 3043–3047.

Kabata, Z. (1970). "Diseases of Fishes. Book 1: Crustacea as Enemies of Fishes." T. F. H. Publ., Jersey City, New Jersey.

Kabata, Z. (1974). Lepeophtheirus cuneifer sp.nov. (Copepoda: Caligidae), a parasite of fishes from the Pacific Coast of North America. J. Fish. Res. Board Can. 31, 43–47.

Kabata, Z. (1976). Early stages of some copepods (Crustacea) parasitic on marine fishes of British Columbia. J. Fish. Res. Board Can. 33, 2507–2525.

Kabata, Z., and Cousens, B. (1973). The structure of the attachment organ of Lernaeopodidae (Crustacea: Copepoda). J. Fish. Res. Board Can. 29, 1015–1023.

Kannupandi, T. (1976). A study on the cuticle of Stegoalpheon kempi Chopra (Isopoda: Crustacea) in relation to its parasitic mode of life. J. Exp. Mar. Biol. Ecol. 25 87–94.

Kemp, S. (1922). Notes on Crustacea Decapoda in the Indian Museum. XV. Pontoniinae. Rec. Ind. Mus. Calcutta 24, 113–288.

Kensley, B. (1974). Bopyrid Isopoda from southern Africa. Crustaceana 26, 259–266.

Knudsen, J. W. (1967). Trapezia and Tetralia (Decapoda, Brachyura, Xanthidae) as obligate ectoparasites of pocilloporid and acroporid corals. Pac. Sci. 21, 51–57.

Kruczynski, W. L. (1974). Relationship between depth and occurrence of pea crabs, Pinnotheres maculatus, in blue mussels, Mytilus edulis, in the vicinity of Woods Hole, Massachusetts. Chesapeake Sci. 15, 167–169.

Kükenthal, W. (1926–1927). "Part 1. Handbuch der Zoologie." Walter de Gruyter, Berlin and Leipzig.

Kuris, A. M. (1974). Trophic interactions: Similarity of parasitic castrators to parasitoids. Q. Rev. Biol. 49, 129–148.

Lamberts, A. E., and Garth, J. S. (1977). Coral-crab commensalism in xanthids. Pac. Sci. 31, 245–247.

Laubier, L., and Bouchet, P. (1976). Un nouveau copépode parasite de la cavité palléale d'un gastéropode bathyal dans le Golfe de Gascogne, Myzotheridion seguenziae, gen. sp. nov. Arch. Zool. Exp. Gen. 117, 469–484.

Laubier, L., and Lafargue, F. (1974). Le genre Brementia Chatton and Brément, curieux copepods Notodelphyidae ascidicole parasite de Didemnidae. Crustaceana 27, 235–248.

Levine, D. M., and Blanchard, O. J., Jr. (1980). The acclimation of two shrimps belonging to the genue Periclimenes to sea anemones. Bull. Mar. Sci. 30, 460–466.

Limbaugh, C. (1961). Cleaning symbiosis. Sci. Am. 205, 42–49.

Limbaugh, C., Pederson, H., and Chace, F. A., Jr. (1961). Shrimps that clean fishes. Bull. Mar. Sci. 11, 237–257.

Lytwyn, M. W., and McDermott, J. J. (1976). Incidence, reproduction, and feeding of Stylochus zebra, a polyclad turbellarian symbiont of hermit crabs. Mar. Biol. (Berlin) 38, 365–372.

McFarlane, I. D. (1973). Multiple conduction systems and the behaviour of sea anemones. Publ. Seto Mar. Biol. Lab. 20, 513–523.

MacGinitie, G. E. (1930). The natural history of the mud shrimp Upogebia pugettensis (Dana). Ann. Mag. Nat. Hist. 6, 36–44.

MacGinitie, G. E. (1934). The natural history of Callianassa californiensis Dana. Am. Midl. Nat. 15, 166–177.

MacGinitie, G. E., and MacGinitie, M. (1968). "Natural History of Marine Animals." McGraw-Hill, New York.

McLaughlin, P. A. (1974). The hermit crabs (Crustacea Decapoda, Paguridea) of northwestern North America. Zool. Verh. (Leiden) 130, 1–396.

McLean, R. B., and Mariscal, R. N. (1973). Protection of a hermit crab by its symbiotic sea anemone Calliactis tricolor. Experientia 29, 128–130.

McMullen, J. C., and Yoshihara, H. T. (1970). An incidence of parasitism of deepwater king crab, *Lithodes aequispina*, by the barnacle Briarosaccus callosus. *J. Fish. Res. Board Can.* **27**, 818–821.

Madin, L. P., and Harbison, G. R. (1977). The association of Amphipoda Hyperiidea with gelationous zooplankton. I. Associations with Salpidae. *Deep-Sea Res.* **24**, 449–463.

Mainardi, D., and Rossi, A. (1969). Relations between social status and activity towards the sea anemone *Calliactis parasitica* in the hermit crab *Dardanus arrosor. Atti. Acad. Naz. Lincei, Rend. 8th Ser.* **47**, 16–21.

Markham, J. C. (1973). Six new species of bopyrid isopods parasitic on galatheid crabs of the genus *Munida* in the Western Atlantic. *Bull. Mar. Sci.* **23**, 613–648.

Markham, J. C. (1974). A new species of *Pleurocrypta* (Isopoda, Bopyridae), the first known from the Western Atlantic. *Crustaceana* **26**, 267–272.

Markham, J. C. (1975a). Bopyrid isopods infesting porcellanid crabs in the northwestern Atlantic. *Crustaceana* **28**, 257–270.

Markham, J. C. (1975b). Two new species of *Asymmetrione* (Isopoda, Bopyridae) from the Western Atlantic. *Crustaceana* **29**, 255–265.

Markham, J. C. (1975c). Redescription of the parasitic isopod *Stegias clibanarii* Richardson, 1904 (Epicaridea, Bopyridae), with remarks on its systematic position. *Crustaceana* **28**, 225–230.

Monod, T. (1926). Les Gnathidae. Essai monographique. Thèses presentées à la faculté des sciences de l'Université de Paris.

Morton, B. (1972). Some aspects of the functional morphology and biology of *Pseudopythina subsinuata* (Bivalvia: Leptonacea) commensal on stomatopod crustaceans. *Proc. Zool. Soc. London* **166**, 79–96.

Myers, A. A. (1974). A new species of commensal amphipod from East Africa. *Crustaceana* **26**, 33–36.

Newman, W. A. (1960). Five pedunculate barnacles from the Western Pacific, including two new forms. *Crustaceana* **1**, 100–116.

Newman, W. A. (1961a). Notes on certain species of *Octolasmis* (Cirripedia, Thoracica) from deep sea Crustacea. *Crustaceana* **2**, 326–329.

Newman, W. A. (1961b). On certain littoral species of *Octolasmis* (Cirripedia, Thoracica) symbiotic with decapod Crustacea from Australia, Hawaii and Japan. *Veliger* **4**, 99–107.

Orton, J. H. (1921). The mode of feeding and sex phenomena in the pea crab, *Pinnotheres pisum. Nature (London)* **106**, 533–534.

Patton, W. K. (1967). Studies on *Domecia acanthophora*, a commensal crab from Puerto Rico, with particular reference to modifications of the coral host to feeding habits. *Biol. Bull. (Woods Hole, Mass.)* **132**, 56–67.

Patton, W. K. (1968). Feeding habits, behavior, and host specificity of *Caprella grahami*, an amphipod commensal with the starfish *Asterias forbesi. Biol. Bull. (Woods Hole, Mass.)* **134**, 148–153.

Patton, W. K. (1974). Community structure among the animals inhabiting the coral *Pocillopora damicornis* at Heron Island, Australia. *In* "Symbiosis in the Sea" (W. B. Vernberg, ed.), pp. 219–243. Univ. of South Carolina Press, Columbia, South Carolina.

Patton, W. K. (1976). Animal associates of living reef corals. *In* "Biology and Geology of Coral Reefs" (O. A. Jones and R. M. Dean, eds.), Vol. 3, pp. 1–36. Academic Press, New York.

Pearce, J. B. (1966). The biology of the mussel crab, *Fabia subquadrata*, from the waters of the San Juan Archipelago, Washington. *Pac. Sci.* **20**, 3–35.

Peden, A. E., and Corbett, C. A. (1973). Commensalism between a liparid fish, *Careproctus* sp., and the lithodid box crab, *Lopholithodes foraminatus. Can. J. Zool.* **51**, 555–556.

Poinar, G. O., Jr., and Kuris, A. M. (1975). Juvenile *Ascarophis* (Spirurida: Nematoda) parasitizing intertidal decapod Crustacea in California: with notes on prevalence and effects on host growth and survival. *J. Invertebr. Pathol.* **26,** 375–382.

Poinar, G. O., Jr., and Thomas, G. M. (1976). Occurrence of *Ascarophis* (Nematoda: Spiruridea) in *Callianassa californiensis* Dana and other decapod crustaceans. *Proc. Helminthol. Soc. Wash.* **43,** 28–33.

Potts, F. A. (1915a). On the rhizocephalan genus *Thompsonia* and its relation to the evolution of the group. *Carnegie Inst. Washington Pap. Mar. Biol.* **8,** 1–32.

Potts, F. A. (1915b). *Hapalocarcinus*, the gall-forming crab, with some notes on the related genus *Cryptochirus*. *Carnegie Inst. Washington Pap. Mar. Biol.* **8,** 33–69.

Provenzano, A. J. (1971). Rediscovery of *Munidopagurus macrocheles* (A. Milne Edwards 1880) (Crustacea Decapoda; Paguridae) with a description of the first zoeal stage. *Bull. Mar. Sci.* **21,** 256–266.

Raibaut, A. (1968). *Paralaophonte ormieresi* n.sp., copépode harpacticoide trouvé sur les branchies de *Maia squinado* (Herbst) (Crustaceana: Decapoda). *Bull. Soc. Zool. Fr.* **93,** 451–457.

Reddiah, K. (1968). *Pseudomacrochiron stocki* n.g., n.sp., a cyclopoid copepod associated with a medusa. *Crustaceana* **16,** 43–50.

Reese, E. S. (1962). Shell selection of hermit crabs. *Anim. Behav.* **10,** 347–360.

Reinhard, E. C. (1956). Parasitic castration of Crustacea. *Exp. Parasitol.* **5,** 79–107.

Restivo, F. (1971). Su di una nuova specie di *Cabirops* parasita di *Pseudione*. *Pubbl. Stn. Zool. Napoli* **39,** 70–86.

Ricketts, E. F., and Calvin, J. (1968). "Between Pacific Tides," 4th ed. Stanford Univ. Press, Stanford, California. (Revised by J. W. Hedgpeth.)

Ritchie, L. (1975). A new genus and two new species of Choniostomatidae (Copepoda) parasitic on two deep sea isopods. *J. Linn. Soc. London, Zool.* **57,** 155–178.

Ross, A., and Newman, W. A. (1969). A coral-eating barnacle. *Pac. Sci.* **23,** 252–256.

Ross, D. M. (1960). The association between the hermit crab *Eupagurus bernhardus* (L.) and the sea anemone *Calliactis parasitica* (Couch). *Proc. Zool. Soc. London* **134,** 43–57.

Ross, D. M. (1967). Behavioural and ecological relationships between sea anemones and other invertebrates. *Oceanogr. Mar. Biol.* **5,** 291–316.

Ross, D. M. (1971). Protection of hermit crabs (*Dardanus* spp.) from octopus by commensal sea anemones (*Calliactis* spp.). *Nature (London)* **230,** 401–402.

Ross, D. M. (1973). Some reflections on actinian behavior. *Publ. Seto Mar. Biol. Lab.* **20,** 501–512.

Ross, D. M. (1974). Behavior patterns in associations and interactions with other animals. *In* "Coelenterate Biology" (L. Muscatine and H. M. Lenhoff, eds.), pp. 281–312. Academic Press, New York.

Ross, D. M. (1975). The behavior of pagurids in symbiotic associations with actinians in Japan. *Publ. Seto Mar. Biol. Lab.* **22,** 157–170.

Ross, D. M. (1979a). 'Stealing' of the symbiotic anemone, *Calliactis parasitica*, in intraspecific and interspecific encounters of three species of Mediterranean pagurids. *Can. J. Zool.* **57,** 1181–1189.

Ross, D. M. (1979b). A behaviour pattern in *Pagurus bernhardus* L. towards its symbiotic actinian *Calliactis parasitica* (Couch). *J. Mar. Biol. Assoc. U. K.* **59,** 623–630.

Ross, D. M., and von Boletzky, S. (1979). The association between the pagurid *Dardanus arrosor* and the actinian *Calliactis parasitica*. Recovery of activity in "inactive" *D. arrosor* in the presence of cephalopods. *Mar. Behav. Physiol.* **6,** 175–184.

Ross, D. M., and Kikuchi, T. (1976). Some symbiotic associations between anemones and gastropods in Japan. *Publ. Amakusa Mar. Biol. Lab.* **4,** 41–50.

Ross, D. M., and Sutton, L. (1961a). The response of a sea anemone *Calliactis parasitica* to shells of the hermit crab *Pagurus bernhardus*. *Proc. R. Soc. London, Ser. B* **155**, 266–281.

Ross, D. M., and Sutton, L. (1961b). The association between the hermit crab *Dardanus arrosor* (Herbst) and the sea anemone *Calliactis parasitica* (Couch). *Proc. R. Soc. London, Ser. B* **155**, 282–291.

Rotramel, G. L. (1975). Observations on the commensal relations of *Iais californica* (Richardson, 1904) and *Sphaeroma quoyanum* H. Milne Edwards, 1840 (Isopoda). *Crustaceana* **28**, 247–256.

Rumpus, A. E., and Kennedy, C. R. (1974). The effect of the acanthocephalan *Pomphorhynchus laevis* upon the respiration of its intermediate host, *Gammarus pulex*. *Parasitology* **68**, 281–284.

Saint Laurent, M. de (1972). Sur la famille des Parapaguridae Smith, 1882: Description de *Typhlopagurus foresti* gen.nov., sp.nov., et de quinze especes ou sous-especes nouvelles de *Parapagurus* Smith (Crustacea, Decapoda). *Bijd. Dierkd.* **42**, 97–123.

Sakai, T. (1939). "Studies on the Crabs of Japan. IV. Brachygnatha, Brachyrhyncha," pp. 365–741. Yokendo, Tokyo.

Sastry, A. N., and Menzel, R. W. (1962). Influence of hosts on the behavior of the commensal crab *Pinnotheres maculatus* Say. *Biol. Bull. (Woods Hole, Mass.)* **123**, 388–395.

Schöne, H. (1976). Zur Biologie der dekapoder Krebse. *Publ. Wiss. Filmen.* **9**, 461–556.

Sheader, M. (1977). The breeding biology of *Idotea pelagica* (Isopoda: Valvifera) with notes on the occurrence and biology of its parasite *Clypeoniscus hanseni* (Isopoda: Epicaridea). *J. Mar. Biol. Assoc. U. K.* **57**, 659–674.

Sleeter, T. D., and Coull, B. C. (1973). Invertebrates associated with the marine wood boring isopod, *Limnoria tripunctata*. *Oecologia* **13**, 97–102.

Soyer, J. (1973). *Paramphiascopsis paromolae* n.sp., copépode harpacticide recolté sur les lamelles branchiales du crustacé decapode *Paromola cuvieri* (Risso). *Crustaceana* **24**, 90–96.

Steffan, A. W. (1967). Ectosymbiosis in acquatic insects. *In* "Symbiosis" (S. M. Henry, ed.), Vol. 2, pp. 207–289. Academic Press, New York.

Stock, J. H. (1967). Copepoda associated with invertebrates from the Gulf of Awaba. 3. The genus *Pseudanthessius* Claus, 1889 (Cyclopoida, Lichomolgidae). *K. Ned. Akad. Wet., Proc. Ser. C* **70**, 232–248.

Stock, J. H. (1971a). *Micrallecto uncinata* n.gen., n.sp., a parasitic copepod from a remarkable host, the pteropod *Pneumoderma*. *Bull. Zool. Mus., Univ. van Amsterdam* **2**, i–xii.

Stock, J. H. (1971b). *Collocherides astroboae* n.gen., n.sp., a siphonostome cyclopoid copepod living in the stomach of basket stars. *Bijdr. Dierk.* **41**, 19–22.

Stock, J. H., and Humes, A. G. (1970). On four new notodelphyid copepods, associated with an octocoral, *Parerythropodium fulvum* (Forskål), in Madagascar. *Zool. Anz.* **184**, 194–212.

Tattersall, O. S. (1962). Report on a collection of Mysidacea from South African off-shore and coastal waters (1957–59) and from Zanzibar (1961). *Proc. Zool. Soc. London* **139**, 221–247.

Telford, M., and Daxboeck, C. (1978). *Porcellana sayana* Leach (Crustacea: Anomura) symbiotic with *Strombus gigas* (Linnaeus) (Gastropoda: Strombidae) and with three species of hermit crabs (Anomura: Diogenidae) in Barbados. *Bull. Mar. Sci.* **28**, 202–205.

Temnikow, N. K. (1974). Epicaridium larvae in *Mytilus californianus* (Mollusca: Bivalvia). *Veliger* **16**, 413–414.

Tomlinson, J. T. (1955). The morphology of an acrothoracican barnacle, *Trypetesa lateralis*. *J. Morphol.* **96**, 97–114.

Turquier, Y. (1972). Contribution à la connaissance des cirripèdes acrothoraciques. *Arch. Zool. Exp. Gen.* **113**, 499–551.

Uspenskaya, A. V. (1953). Life cycle of nematodes of the genus Ascarophis van Beneden (Nematodes-Spirurata). *Zool. Zh.* **32**, 828–832.

Uchida, T. (1960). Carcinactis ichikawai n.gen., n.sp., an actinarian commensal with the crab *Dorippe granulata*. *Jpn. J. Zool.* **12**, 595–601.

Utinomi, H. (1970). New and rare commensal pedunculate cirripeds from Amakusa Islands, Western Kyusyu, Japan. *Publ. Seto Mar. Biol. Lab.* **18**, 156–157.

Uzmann, J. R. (1967). Juvenile *Ascarophis* (Nematoda: Spiruroidea) in the American lobster, *Homarus americanus*. *J. Parasitol.* **53**, 218.

Vader, W. (1967). Notes on Norwegian marine amphipods 1–3. *Sarsia* **29**, 283–294.

Vader, W. (1970a). The amphipod, *Aristias neglectus* Hansen, found in association with Brachiopoda. *Sarsia* **43**, 13–14.

Vader, W. (1970b). Amphipods associated with the sea anemone, *Bolocera tuediae*, in western Norway. *Sarsia* **43**, 87–98.

Vader, W. (1970c). *Antheacheres duebeni* M. Sars, a copepod parasite in the sea anemone, *Bolocera tuediae* (Johnston). *Sarsia* **43**, 99–106.

Vader, W. (1970d). On the occurrence of a gall-forming copepod in *Actinostola* spp. (Anthozoa). *Sarsia* **43**, 107–110.

Vader, W. (1971). De vlokreeft Podoceropsis nitida, een kostganger van heremietkreeften. *De Levende Natuur* **74**, 134–136.

Vader, W. (1972a). Associations between amphipods and molluscs. A review of published records. *Sarsia* **48**, 13–18.

Vader, W. (1972b). Associations between gammarid and caprellid amphipods and medusae. *Sarsia* **50**, 51–56.

Vader, W. (1978). Associations between amphipods and echinoderms. *Astarte* **11**, 123–134.

Vader, W., and Lönning, S. (1973). Physiological adaptations in associated amphipods. A comparative study of tolerance to sea anemones in four species of Lysianassidae. *Sarsia* **53**, 29–40.

Veillet, A. (1945). Recherches sur le parasitisme des Crabes et des Galathees par les Rhizocephales et les Epicarides. *Ann. Inst. Oceanogr. Paris* **22**, 193–341.

Vernberg, W. B., ed. (1974). "Symbiosis in the Sea." Univ. of South Carolina Press, Columbia, South Carolina.

Verrill, A. E. (1869). On the parasitic habits of Crustacea. *Am. Nat.* **3**, 239–250.

Walker, G. (1974). The occurrence, distribution, and attachment of the pedunculate barnacle Octolasmis mülleri (Coker) on the gills of crabs, particularly the blue crab, *Callinectes sapidus* Rathbun. *Biol. Bull. (Woods Hole, Mass.)* **147**, 678–689.

Warren, P. J. (1974). Some observations on the relationship of the bopyrid parasite Hemiarthrus abdominalis (Krøyer) with *Pandalus borealis* Krøyer. *Crustaceana* **27**, 21–26.

Waterman, T. H., and Chace, F. A., Jr. (1960). General crustacean biology. *In* "The Physiology of the Crustacea" (T. H. Waterman, ed.), Vol. 1, pp. 1–33. Academic Press, New York.

Weinbauer, G., Nussbaumer, V., and Patzner, R. A. (1982). Studies on the relationship between Inachus phalangium Fabricius (Maiidae) and *Anemonia sulcata* Pennant in their natural environment. *Mar. Ecol.* (in press).

Williams, A. B., and Brown, W. S. (1972). Notes on structure and parasitism of Munida uris H. Milne Edwards (Decapoda: Galatheidae) from North Carolina, USA. *Crustaceana* **22**, 303–308.

Wright, H. O. (1973). Effect of commensal hydroids on hermit crab competition in the littoral zone of Texas. *Nature (London)* **241**, 139–140.

Yamaguti, S. (1963). "Parasitic Copepoda and Branchiura of Fishes." Interscience Publ., New York.

Yeatman, H. C. (1970). Copepods from Chesapeake Bay sponges including *Asterocheres jeanyeatmanae* n.sp. *Trans. Am. Microsc. Soc.* **89,** 27–38.

Yonge, C. M. (1957). Symbiosis. *In* "Treatise on Marine Ecology and Paleoecology" (J. W. Hedgpeth, ed.), Vol. 1, pp. 429–442. Memoir 67, Geol. Soc. Am., New York.

Zann, L. P. (1978). Biology of a barnacle (*Platylepas ophiophilus* Lanchester) symbiotic with sea snakes. *In* "The Biology of Sea Snakes" (W. A. Dunson, ed.), pp. 267–286. Univ. Park Press, Baltimore, Maryland.

5

Pelagic Larval Ecology and Development

A. N. SASTRY

THE BIOLOGY OF CRUSTACEA, VOL. 7

I. INTRODUCTION

Crustaceans follow two basic developmental patterns, each having ecological significance: (1) development with a larva hatching at the end of
embryonic development; and (2) complete development within the egg
which hatches to a juvenile stage. Direct development to the juvenile stage
occurs in some Branchiopoda (Cladocera), all Leptostraca, Anaspidacea,
Peracarida, and a few Decapoda. Both types of development occur in crustaceans from both freshwater and marine habitats. Development with a free-
swimming larva is found in smaller size species belonging to Anostraca,
Conchostraca, Ostracoda, and Copepoda living in freshwater ponds, lakes,
and slow-flowing rivers (Kaestner, 1970). In the marine environment, development with a pelagic larva is the predominant type, but Peracarida, which
are mostly marine, develop directly. The developmental pattern to extend
the embryonic period or to eliminate the larva altogether has evolved in
many crustaceans as an adaptation to the environment.

Typically, crustaceans with a pelagic larval stage in the life-cycle hatch
eggs to release a nauplius larva. This occurs in the Ostracoda, Copepoda,
Cirripedia, Euphausiacea, and Decapoda. In some others, embryonic development is extended with the nauplius stage passing within the egg, which
hatches to release a zoea. The larva typically passes through a series of
stages (instars) before reaching the juvenile stage. Thus, crustacean life cycles can be considered to be one or another of the three basic types: nauplius, zoea, and postlarva.[1] In certain cases, the type of larval development
is characteristic of families or genera; however, there are numerous variations on the basic pattern (Gurney, 1942; Waterman and Chace, 1960;
Kaestner, 1970). Many parasitic crustaceans have aberrant life cycles, and
they are not considered in this chapter.

Development from a newly oviposited egg through a pelagic larva to the
juvenile stage is a continuous process. Egg development is sustained on the
yolk within the egg while attached to the female parent in most species or
when freely shed into the water. The pelagic larval phase on the other hand,
is sustained by energy derived from plankton feeding. Because of these
differences in sources of energy utilized and the environmental conditions
under which development occurs, the two phases are examined separately
here.

Studies on pelagic larvae of Crustacea have been conducted from
plankton samples collected in the field (Lebour, 1928; Gurney, 1942) and
from laboratory cultivation under controlled conditions (Costlow and Bookhout, 1960a; Rice and Williamson, 1970; Sastry, 1970, 1975). Field studies

---

[1]The term "postlarva" is used in this chapter to denote the stage that follows completion of
larval development, i.e., the first juvenile stage.

have been concerned with the description of early life cycle stages of species and the descriptions of distribution and abundance of larvae in a geographical area. The development of techniques for cultivation of larvae through all stages to the juvenile has allowed descriptions of the larvae of a species (Costlow and Bookhout, 1959; Rice and Provenzano, 1964; Sastry, 1977) and has also made possible a variety of studies on the biochemistry, physiology, and behavior of the larvae. Both field and laboratory studies have greatly enhanced our understanding of the ecology of pelagic crustacean larvae in recent years.

Pelagic larvae play an important role in the distribution of species, the maintenance of genetic variation within a species, the year-to-year fluctuations in population size, and the structure and dynamics of communities. Quantitative ecological studies (UNESCO, 1968) on larvae that can answer many questions of ecological and evolutionary importance still remain to be done. In this chapter, aspects of egg incubation, larval development, larval responses to the environmental variables, dispersal and recruitment, and the adaptive value of pelagic larva in the life histories are examined to see the directions this area has taken and the advances made in our knowledge of the pelagic larvae of Crustacea.

II. EGG INCUBATION AND HATCHING

The newly oviposited egg of a crustacean is a self-contained system with all the necessary material for synthetic processes associated with embryogenesis and morphogenesis, and all of the compounds required for oxidative metabolism and energy production. The egg contains nutrient reserves in the form of proteinaceous yolk and lipid vesicles scattered throughout the cytoplasm. The protein yolk also contains carotenoid pigments giving a characteristic coloration of eggs (Herring and Morris, 1975). A major portion of the lipid fraction of lipovitellin consists of phospholipids, triglycerols, and glycolipids complexed with the proteinaceous yolk (Holland, 1978). The rate of egg development (incubation period) to a larva is ecologically important to the timing of larval release into the pelagic environment, and it may be influenced by both endogenous factors and interacting environmental factors.

A. Egg Development

1. ENDOGENOUS FACTORS

The developmental events—blastulation, gastrulation, appearance of chromatophores, the first initiation of heart-beat, and the appearance of larval appendages and other morphological features—follow a temporal sequence that does not vary from one species to another (Green, 1965). Egg development can be separated into different stages based on the developmental events or the percentage of yolk volume utilized, but it is a continuous process requiring raw materials and free energy provided by the nutrient reserves laid down in the oocyte during oogenesis. The mechanisms by which reserve materials are utilized at the right time and in the correct amounts, and the molecular and cellular interations involved in regulation of egg development, are of basic interest to development biology, but are not fully understood. Since the sequence of developmental events have a temporal order, the underlying processes that regulate timing may also occur in a sequence. Soll (1979) proposed models to consider the timing of developmental systems. One model considers that there is one common limiting process for the entire sequence of developmental stages, and this timer continues to function throughout the entire sequence (single timer model). A second model considers that there is a rate-limiting process specific for each developmental stage of which (1) the start of each rate-limiting process is regulated by the termination of the previous rate-limiting process (sequential model) or (2) rate-limiting processes for successive stages start and stop independently of one another and therefore function in parallel with one another (parallel model).

The endogenous factors regulating the rate of development of eggs may include genetic factors, composition of nutrient reserves, rate of substrate utilization for maintenance and tissue production, and size of the eggs. During egg development, energy is partitioned to the processes involved in embryogenesis and morphogenesis and metabolic maintenance of the developing system. In barnacles *Semibalanus balanoides, Balanus balanus,* and *Chthamalus stellatus,* the nutrient reserves contained in the eggs in the form of lipids, carbohydrates, and proteins are utilized during the course of development (Barnes, 1965; Dawson and Barnes, 1966; Barnes and Evens, 1967; Barnes and Blackstock, 1975; Achituv and Barnes, 1976). The relative amount of substrates utilized during the course of development results in a change of biochemical composition that can be related to the developmental stage. In addition, changes in DNA content that can be related to cell proliferation and RNA content to protein synthesis also occur during egg development (Achituv and Barnes, 1976). In the barnacle *C. stellatus,* RNA increases throughout development with the greatest increase occurring in the early stages when protein synthesis is accelerated. DNA also increases in the early stages when cell multiplication occurs, but only increases slightly in later stages, suggesting reorganization rather than cell proliferation.

Energetics of developing eggs has been examined for the shrimp, *Crangon crangon* (Pandian, 1967), hermit crab, *Pagurus bernhardus* (Pandian and Schumann, 1967), and lobsters, *Homarus gammarus* and *H. americanus*

(Pandian, 1970a,b). In all four species, nonprotein nitrogen increases due to chitin synthesis, and caloric content decreases with the development of eggs. Protein and lipids account for the greatest amount of energy expended during development. A marked decrease in lipid occurs reflecting a decrease in energy content with development. Other changes occurring during egg development include an increase in salt and ash content and altered membrane permeability leading to an uptake of water and an increase in size of the egg. Holland (1978) calculated the efficiency of conversion of biochemical components of yolk into new tissues of the embryo, assuming that the egg is a closed system of organic nutrients with no major interconversion between major biochemical constituents. These calculations gave an estimate of the amount of biochemical constituents used for growth and metabolism. In a majority of crustacean eggs, the conversion efficiency of the lipids is high. Protein appears to be conserved for the structure of the embryo, while lipid is utilized as the major energy source for growth (Pandian, 1970a,b; Holland, 1978).

The endogenous factors responsible for the interspecific variation in rate of egg development are not well understood. In the boreoarctic barnacles, eggs of *Balanus balanus* develop much faster than eggs of *Semibalanus balanoides* (Barnes, 1965). This difference in rate of development of the species is reflected in the rate at which all the substrates are utilized, even though the proportions of original material utilized are similar (Dawson and Barnes, 1966). In both species there is a marked depletion of triglyceride during development which is the major contribution of lipid substrate utilization. The rate of this depletion is greater in the faster developing eggs of *B. balanus* than in those of *S. balanoides*.

Egg size may also influence the rate of development, but such a relationship does not seem to apply to distantly related species (McLaren, 1965; Wear, 1974). Among closely related species of copepods (McLaren, 1965) and decapods (Wear, 1974), the development rate is inversely related to the egg size. McLaren (1966) found that development time of copepod eggs increases curvilinearly with diameter among smaller eggs and linearly among larger eggs.

2. ENVIRONMENTAL FACTORS

a. Viability. The viability and rate of development of eggs are affected by a variety of environmental factors, such as temperature, salinity, photoperiod, dissolved gases and chemicals in seawater, and the rate changes, intensity, and combinations of these factors. The response of developing eggs to environmental factors may also change during the course of development. However, normal development to a larva takes place within a defined range of environmental conditions characteristic to each species. Generally, tolerance limits for normal development are greater when only a single environmental factor is considered than when multiple environmental factors are considered. When environmental conditions exceed the favorable range, the egg development is adversely affected, resulting in abnormal larva, cessation of development, or death of the embryo. The effects produced by the environmental extremes differ at the upper and lower end of the range. Environmental limits for normal development appear genetically determined and characteristic to a species.

Salinity and temperature limits for viability and normal development have been determined for some barnacles and decapods. In barnacles, salinities below the tolerance range cause cytolysis, whereas higher salinities cause abnormalities in the shape of the larva, the setation, or the caudal region (Crisp and Costlow, 1963). The egg membrane is permeable to salt and water (Pandian, 1970a,b; Barnes and Barnes, 1974), and the abnormalities at higher salinities may be caused by an increase in salt concentration or dissociation of the rate processes of metabolism and differentiation. When larvae hatch from eggs held in salinities outside the tolerance range, they swim feebly in an abnormal fashion, sink to the bottom, or fail to molt to the next stage. Generally, eggs of barnacles inhabiting different environments seem to tolerate and develop in a wide range of salinities (Barnes, 1953; Crisp and Costlow, 1963; Foster, 1970), with the estuarine species tolerating slightly lower salinities than those from other environments (Barnes and Barnes, 1974).

Viability of decapod eggs is affected when the entire development takes place outside the temperature tolerance range (Wear, 1974). At the lower temperature limit, development is slowed down, whereas adverse effects on development are seen at the upper temperature limit. Gastrulation is a sensitive period in development. Generally, the temperature range within which eggs develop to hatch normal larvae is narrower than the range over which eggs can survive and continue development to hatch abnormal larvae.

b. Metabolism.   Metabolism and morphologically distinguishable stages in development appear to be closely linked, and both are affected by temperature. Barnes and Barnes (1959) reported that metabolic rate increases for each successive stage in development of barnacles *Semibalanus balanoides* and *Pollicipes polymerus* at any given temperature. The developing eggs of both species show differential metabolic responses to temperature. The Q_{10} for respiration for all the stages in egg development is higher ($Q_{10} \approx 4.5$) at colder temperatures and lower ($Q_{10} \approx 2.2$) at warmer temperatures. In *S. balanoides,* change in Q_{10} occurs at successively higher temperatures with the advancing stage of development. In contrast, Q_{10} change occurs at the same temperature for all the stages in development of *P. polymerus*. The mechanisms that regulate metabolism (Hazel and Prosser, 1974) and the relationship to the processes of differentiation and rate of development are not known.

   c. *Rate of Development.* The rate of egg development can be affected by temperature (Patel and Crisp, 1960a; McLaren, 1965, 1966; Wear, 1974), salinity (Crisp and Costlow, 1963), and other factors. Many studies determining the effects of temperature have shown that the rate of development increases linearly within the viable temperature range, and as temperature approaches the upper limit, the rate first decreases and then becomes constant before death of the embryo with a further increase of temperature (Patel and Crisp, 1960a,b; Bernard, 1971; Wear, 1974). At the lower temperature boundary, the rate of development slows down greatly or stops.

   The Q_{10} of egg development may also vary over the temperature range. The Q_{10} of development of barnacle eggs is higher at the lower part of the temperature range over which breeding occurs, but the Q_{10} decreases toward the upper part of the viable range, and becomes constant at the higher temperature at which development takes place (Patel and Crisp, 1960a).

   The effect of temperature on developmental rate has been analyzed by using Bĕlehrádek's temperature function (Bĕlehrádek, 1935, 1957) to predict developmental time at temperatures both intra- and interspecifically (McLaren, 1965, 1966; McLaren et al., 1969; Wear, 1974). This function takes the form

$$D = a\,(T - \alpha)^b$$

where a, b, and α are fitted constants, D is duration of development, and T is temperature. A number of other functions can also be used for describing the effects of temperature on the development time (Precht et al., 1973). In decapods, egg development is continuous for some species, while in others a diapause occurs at the gastrulation stage which cannot be shortened by increasing water temperature (Wear, 1974). After diapause, however, the response in egg development to temperature is similar to other species. The presence or absence of diapause may thus cause variation in the length of egg incubation between species. Wear (1974) suggests that diapause evolved in some species to time the larval release with food abundance.

   Temperature acclimation can also influence the rate of egg development, but the available data are mostly limited to copepods. Eggs of *Acartia clausi* resulting from cold acclimated animals develop at a faster rate compared to those from warm acclimated animals when determined at the same temperature (Landry, 1975). Eggs of *A. clausi* from animals collected in the winter (8°–10°C) develop at a faster rate under summer conditions (15°–20°C) compared to those from animals collected in the summer (18°–20°C). Eggs of summer adapted animals have a depressed rate at cold temperatures relative to winter adapted animals. Adaptation of winter acclimated *A. clausi* to summer conditions takes more than one generation at the higher temperatures. In another copepod (*Pseudocalanus* sp.), eggs acclimated to cold in the laboratory develop more rapidly at high and low temperatures than warm acclimated animals (Hart and McLaren, 1978). But eggs from animals collected in the spring develop more slowly than those from animals collected in summer and autumn under the same temperature conditions. Thus, effects of long-term seasonal exposure of this species to low temperature differ from those seen from short-term laboratory acclimation. No seasonal compensation of development rate of the eggs of *Pseudocalanus* to temperature was observed. Differences in developmental rate of the eggs from summer and winter animals may be related to the larger size of females and their eggs in the colder season and size assorted mating and the influence of male size (Hart and McLaren, 1978).

3. GEOGRAPHIC VARIATION

Egg development rate of geographically separated populations may vary as a phenotypic or genotypic response. McLaren (1966) reported that the rate of egg development in response to temperature of the copepods *Pseudocalanus minutus* and *Calanus finmarchicus* varies only slightly for different geographical populations. Crisp (1964) reported that the rate of development of the embryos of *Semibalanus balanoides* determined *in vitro* at 8°C is faster for the Woods Hole, Massachusetts, population than that of Bangor, North Wales. Eggs of the Bangor population transplanted to Newfoundland developed at about the same rate as in their original locality. Barnes and Barnes (1976) found that development time of a North American population of *S. balanoides* remains unchanged when held with a Scottish population under identical laboratory conditions over the entire year. This suggests a strong genetic component influencing rate of development of the North American population. However, there is a general tendency for decreasing development time with decreasing latitude for populations on both the eastern and western Atlantic coasts. The response of the southern population on the European coast resembles that of the extreme southern population on the North American coast. On the east coast of North America, development time of the populations north and south of Cape Cod vary markedly (Barnes and Barnes, 1976), but the metabolism-temperature responses of eggs from populations collected at different locations are similar (Barnes and Barnes, 1959), suggesting that different rate processes respond differently.

4. CLIMATIC ADAPTATION

Egg size varies for different species and for species inhabiting different climatic regions, but it is uncertain how egg size influences the developmental rate. Developmental rate temperature-response patterns of the eggs of northern and southern species of barnacles are not markedly different (Patel and Crisp, 1960a). Developmental rate of the eggs of different species of barnacles is about the same. McLaren *et al.* (1969) reported that the temperature-dependent developmental rate of copepods from the Arctic to Tropics follows a distinct curve for each species when fitted with the Bĕlehádek's temperature function. However, Landry (1975) found that if depression of the rate at the high and low ends of a copepod's temperature range are taken into account, the relative development rate of eggs from arctic-temperate zone copepods within their optimal temperature ranges follows a common temperature-dependent function, rather than individual unique functions. Although factors controlling the rate of egg development are not fully understood, the characteristic rates for different species in relation to their environment are obviously selected for synchronizing the time of larval release with food in the pelagic environment to ensure larval survival.

B. Hatching of Eggs

Timing the hatching of eggs and larval release may involve endogenous factors within the egg and female parent, and interaction between endogenous and exogenous factors. A coordination of the processes within the egg and the parent, and with the environment may involve physiological and behavioral changes for synchronization of larval release. Many aspects of time-synchronization of hatching and larval release are conjectural.

1. HATCHING MECHANISMS

The process of egg hatching in Crustacea may be mechanical, non-mechanical, or both (Davis, 1968; Pandian, 1970a). The mechanisms of hatching may include osmotic uptake of water and bursting of the egg membrane (e.g., copepods). In some species, osmotic hatching may be aided by mechanical factors. The mechanisms in hatching may also include digestion of the whole or limited portions of the egg envelope by enzymes produced by the larvae, stimulation of hatching eggs by metabolites produced by parent (e.g., barnacles), movement of the larva that aids in bursting the egg membrane, and parental assistance in breaking the egg membrane or flushing the eggs.

In barnacles, mechanical processes brought about by the activity of larva within the egg and stimulated by the parent are involved (Crisp, 1956; Crisp and Spencer, 1959). A hatching substance is liberated after the parent has been fed, which appears to be a metabolite formed of lactone or lactum derivative at low pH (Crisp, 1969). This mechanism prevents premature hatching of eggs and liberation of larvae before food is available. In *Balanus balanoides*, the eggs are hatched within the mantle cavity or expelled as eggs from the mantle cavity. The female lobsters *Homarus gammarus* and *Homarus americanus* (Ennis, 1975a; Branford, 1978) and fiddler crabs *Uca* (DeCoursey, 1979) have behaviors associated with the hatching of eggs.

2. HATCHING RHYTHMS

Nocturnal hatching of eggs has been observed for a number of crustaceans. Hatching rhythms have been examined in some detail for the lobsters *Homarus gammarus* (Ennis, 1973a), *Homarus americanus* (Ennis, 1975a; Branford, 1978), and *Nephrops norvegicus* (Moller and Branford, 1979), and the crabs *Sesarma* (Saigusa and Hidaka, 1978) and *Uca* (DeCoursey, 1979). The eggs of the European lobster *H. gammarus* hatch after sunset with the hatching process lasting for about 1-min when ovigerous females are held under natural illumination cycle (Ennis, 1973a). Larvae are released by a female on sucessive nights about the same time each night over a period of 2 weeks. The rhythm persists when lobsters are held under constant darkness, but hatching is delayed for 2 or 3 days when held under constant illumination, and when it occurs it is erratic or arrhythmic.

Endogenous components operating within the egg, mother, or both, and exogenous factors controlling the hatching rhythm with the onset of darkness synchronize the rhythm in *Homarus gammarus* (Ennis, 1973a). Branford (1978) showed that the hatching rhythm in *H. gammarus* is partly controlled by endogenous components in the female. The time when females perform their part of the hatching process in light:dark regime is dependent on temperature, day-length, and previous photoperiodic experience. The time of hatching is influenced by the time of sunset and day-length. The length of hatching interval (the time when lobsters perceive either sunset or lights-off and the time when they perform their part of the hatching process) is controlled by short-term factors (the temperature and the length of day before hatching) and by a long-term factor (the photoperiodic experience). Sunrise and sunset appear to be important events in the control of the hatching interval, but the mechanisms by which lobsters perceive these time intervals are not known. Under artificial conditions, the hatching interval increases when lights-off occurs earlier and decreases when lights-off occurs later than expected. At a given temperature and day-length, lobsters that have experienced long days before hatching perform their part of the hatching process at different times compared to those that have not. This difference becomes more marked as temperature decreases. A model based on two oscillator system (Saunders, 1974) has been proposed by Branford (1978) to explain the hatching rhythm in the lobsters (Fig. 1).

Other decapods also use light for regulating the timing of hatching (Kurata, 1955; Moller and Branford, 1979). The control of hatching rhythm in the American lobster *Homarus americanus* is not precise. The time of larval release at night is erratic and also occurs at different times during the day (Ennis, 1975a). In some species, eggs hatch at night whether held by the female or isolated (e.g., *Pandalus gracilis*: Kurata, 1955). In others, isolated eggs hatch any time (e.g., *Callinectes sapidus*: Lochhead and Newcombe, 1942; *Emerita analoga*: Burton, 1979), suggesting that females may have some control over the release of larvae.

Hatching rhythms in which larval release coincides with the night-time high tide have been reported in the semiterrestrial fiddler crabs *Uca pugila-*

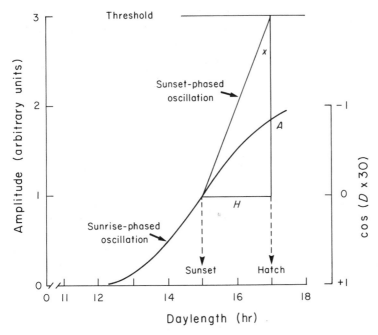

Fig. 1. The model used for predicting the times of hatching by the European lobster *Homarus gammarus*. The example illustrated shows sunset at 15 hr after dawn (*D*, the daylength 15 hr) and a hatch interval (*H*) of 2 hr. $H = \tan x \cdot A; A = 3 - [1 - \cos (Dx30)] = 2 + \cos (Dx30); H = \tan x [2 + \cos (Dx30)]$ hr. Between $D = 12.5$ and 17.5, A increases as D decreases; H also increases as D decreases, but the amount by which H increases depends on the value of x (the temperature and season dependent variable). (From Branford, 1978.)

tor, Uca minax, and *Uca pugnax* (DeCoursey, 1979). This rhythm persists in females held in the laboratory for several days under tide-free conditions before hatching.

Field sampling of larvae has shown that a synchronous mass hatching occurs near the time of nocturnal high tide. In addition to tidal rhythm, lunar factors also appear to be involved in the semilunar hatching rhythm of the littoral crabs *Sesarma haematocheir* and *S. intermedium* (Saigusa and Hidaka, 1978). Both species exhibit temporal synchronization of hatching with the lunar cycle. The peak larval release occurs on the day of and during 4–5 days prior to every syzygy immediately after sunset, and then the peak gradually shifts to later in the evening and reaches a minimum by every half moon. The peak reappears about 2 days after the half moon and again just after sunset.

The adaptive value of hatching rhythms synchronized with the lunar and tidal cycles may be for larval dispersal of littoral crabs (Saigusa and Hidaka, 1978; DeCoursey, 1979). The night-time larval release by the lobsters in small batches over a period of time may be to avoid visual predators and to spread the risk of predation (Branford, 1978). Hatching rhythms have evolved to increase the chances of larval survival, and hence timing of hatching and larval release can vary for different species in relation to the habitat of adults.

III. LARVAL DEVELOPMENT

Larvae hatched from eggs go through a sequence of molts (ecdyses) to reach the postlarval (juvenile) stage by a step-wise transformation (e.g., euphausiids), a transitional metamorphic stage (e.g., Cirripedia, Brachyura), or only a slightly metamorphic stage (e.g., Penaeidae, Nephropidae). Morphological descriptions of larval stages have been reported for a number of crustaceans either from a reconstruction of larval stages collected in plankton samples (Lebour, 1928; Gurney, 1942) or from larvae cultured in the laboratory (Costlow and Bookhout, 1959, 1960b; Rice and Provenzano, 1964; Sastry, 1977).

Larval development involves growth and differentiation and it is underlined by molecular, biochemical, and physiological changes (Sastry, 1978), development of organ systems (Walley, 1969; Factor, 1981), and readily observable morphological features and behaviors. The mechanisms regulating the growth and differentiation of the developing system through an ordered sequence of larval stages characteristic to each species are not fully understood.

A. Developmental Pathways

The basic type of larval development (e.g., nauplius→zoea→postlarva or zoea→postlarva) is fixed for any given species among different groups of crustaceans, but the number of larval instars may be either constant or variable. Generally, brachyuran larvae go through a constant number of larval instars, and this number is unaltered by changes in environmental conditions (Fig. 2). However, variation has been observed in a number of larval instars. For example, extra or abnormal stages have been reported in the development of *Callinectes sapidus* and *Menippe mercenaria,* possibly due to dissociation of molting process from the rate of development (Costlow, 1968).

Larvae of some euphausiids (Mauchline and Fisher, 1969) and palaemonid shrimps (Broad, 1957a,b; Sastry, 1980) go through a variable number of larval instars. In *Euphausia pacifica,* an individual may spend from zero to four instars in any one stage (Ross, 1981). Some euphausiids develop along a fixed pathway (e.g., *Thysanopoda*), while others develop via several pathways (e.g., *Nyctiphanes, Thysanoessa, Meganyctiphanes*). In euphausiids with a variable pattern of development, larvae collected in the same plankton haul may show different combinations of setose and nonsetose pleopods. Different sequences in development are found among larvae of

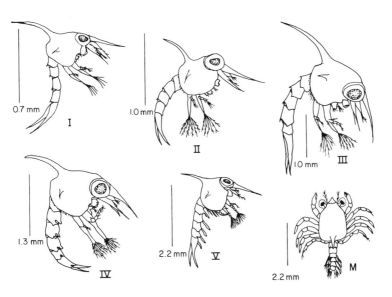

Fig. 2. Larval stages of a brachyuran crab (*Cancer irroratus*). I to V, zoeal stages; M, megalops stage. (From Sastry, 1977.)

the same species collected at different times during the breeding season and in different subsamples within the same geographical area (Mauchline and Fisher, 1969). However, dominant forms (normal pathway) are always found, whereas the relative number of variant forms (minority) is variable in different places and times. A variable development pattern is more often seen among species in coastal and continental shelf areas than in species with more oceanic distributions.

The existence of dominant stages and variants has also been observed in laboratory cultures of euphausiids (LeRoux, 1973, 1974). A considerable morphological variation is seen among individuals that have gone through the same number of instars. Unsuitable algal diet leads to very different pathways of pleopod and setal addition in *Euphausia pacifica* (Ross, 1981). In nature, the difference in development pattern may occur due to diet, geographical location, and temporal variation of the environment (LeRoux, 1973, 1974; Ross, 1981).

Larvae of *Palaemonetes pugio* and *P. vulgaris* that are of the same size but have not molted the same number of times are alike, whereas larvae that have molted the same number of times are not alike (Broad, 1957a,b). The frequency of molting and duration of larval life of *Palaemontes* are influenced by the quality and quantity of available food. Larvae of *P. pugio* reared on a standard diet of *Artemia* nauplii in different temperature and salinity combinations also go through a variable number of instars, but the number of instars is higher at 15°C than at 20°C or 25°C (Sastry, 1980).

Some aspects of the mechanisms controlling and regulating the metamorphosis in brachyuran larvae (Passano, 1960a; Costlow, 1968) and cirripede larvae (Walley, 1969; Crisp, 1974a) have been reviewed. Metamorphosis of brachyuran larvae involves biochemical, physiological, and morphological changes mediated by neuroendocrine factors. Walley (1969) described the processes taking place during metamorphosis of barnacle cyprid larvae. With the exception of naupliar eye, all sense organs and their centers in the nervous system, the main naupliar appendages and their associated musculature, the antennules, the larval mouthparts, and the excretory system are lost. New organ systems develop in advance, but they probably only become functional upon metamorphosis.

B. Growth

The size and/or weight of the developing larvae may increase with molting from one stage to the next. Rice (1968) analyzed the size of larval stages of several decapods and found that "Brooks Law," which states that the sizes of successive larval instars form an exponential series, applies to this group. However, "Przibram's Rule," which states that weight increases between instars by a power of two, does not seem to apply. Larvae of the mud crab *Rhithropanopeus harrisii* show the greatest increase in fresh weight after each larval molt (Frank *et al.*, 1975). Zoeae of the hermit crab *Pagurus criniticornis* show an increase in mean weight after each molt up to the glaucothoe stage, but then weight decreases after metamorphosis to the first crab stage (Vernberg *et al.*, 1981). The weight of larvae increases exponentially for *Cancer irroratus* from hatching to megalopa (Sastry, 1979; Johns and Pechenick, 1980) and for the American lobster *Homarus americanus* with greatest increase between the third and fourth stage (Logan and Epifanio, 1978).

C. Food, Feeding, and Nutritional Requirements

Pelagic larvae of Crustacea are planktotrophic, and the larvae may begin feeding immediately after hatching or at an advanced stage following the early nonfeeding stages, which are sustained by the yolk reserves stored in the body (e.g., copepod *Pseudocalanus minutus*: Corkett and McLaren, 1978; some euphausiids: Mauchline and Fisher, 1969; caridean shrimp: Omori, 1974). The first naupliar and the later cyprid stages of the boreoarctic barnacle *Semibalanus balanoides* are nonfeeding, and the success of metamorphosis is dependent on the energy stored during naupliar development (Crisp, 1974b). Megalopae of the hermit crab *Pagurus bernhardus* do not feed, but successfully metamorphose (Dawirs, 1981), suggesting a dependence on the energy stored during earlier larval development.

The method of feeding and the quality and quantity of food required by the larvae to reach the postlarval stage have not been studied in detail. A general description of the feeding behavior of naupliar larval stages of some copepods and barnacles (Gauld, 1959), calyptopis stage of euphausiids (Mauchline and Fisher, 1969), and phyllosoma stage of the panulirid lobsters (Batham, 1967) has been reported. In nature, larvae of holopelagic copepods, euphausiids, carideans, and penaeids may feed on suspended organic matter in seawater (Mauchline and Fisher, 1969; Marshall and Orr, 1972; Omori, 1974). In the laboratory, unicellular algae or diatoms have been found suitable for larval development of some herbivorous copepods (Corkett and McLaren, 1978). Larvae of some species of barnacles develop to the cyprid stage on unicellular algae or diatoms, whereas others do not develop beyond the naupliar stage (Moyse, 1963).

Brachyuran crab larvae may feed on an algal diet, but a great majority are carnivorous (Lebour, 1922; Broad, 1957a,b; Costlow, 1968; Roberts, 1974). Larvae of some species complete development on an algal diet (Atkins, 1955; Cook and Murphy, 1969), but others fail to reach the postlarval stage (Williams, 1968; Roberts, 1974). Larvae, feeding on an optimal diet, develop to the postlarval stage by molting at a certain frequency, gaining weight and/or increasing in size, and taking a certain length of time. Inadequate diets cause mortality, and if larvae should survive, the development may be partial; if complete development should occur, the frequency of molting and length of larval development is affected (Broad, 1957a,b; Chamberlin, 1962; LeRoux, 1973, 1974; Omori, 1974). Starvation generally prevents molting of the larvae, but larvae of some species molt to the next instar before death (Broad, 1957a,b; Regnault, 1969a,b).

Larvae starved for a certain critical length of time do not recover even when food is subsequently made available possibly due to a physiological weakness from depletion of body reserves (e.g., *Sergia lucens:* Omori, 1974). This critical period is prolonged at the colder temperatures with the decrease in developmental rate to suggest a nutritional stress from energy utilization for metabolism. Larvae of the Alaskan king crab *Paralithodes camtschatica* receiving food within 60 hr after hatching capture the prey, but a delay of feeding for another 24 hr reduces their ability to capture food at temperatures above 4°C (Paul and Paul, 1980). Lack of food of proper size and nutritional value during the period when larvae first begin feeding causes extensive mortalities in the shrimp *Pandalus jordani* (Modin and Cox, 1967) and the crab *Chionoecetes opilio* (Kon, 1979).

Dietary requirements may vary for different larval stages of a species, and for species inhabiting different environments. Larvae of the mud crabs *Rhithropanopeus harrisii* and *Neopanope sayi* complete development on a diet that is inadequate for the larvae of crabs *Callinectes sapidus* and *Libinia emarginata* (Sulkin and Norman, 1976; Bigford, 1978). The larvae of the shallow water crab *Menippe mercenaria* fed on a rotifer diet develop to the third zoeal stage, whereas those of the deep-water crab *Geryon quinquedens* develop to the postlarval stage (Sulkin and Van Heukelem, 1980). In addition, all larval stages of a species may develop on the same diet, although dietary requirements may change for different larval stages of other species.

Nutritional requirements of crustacean larvae have not been studied in detail. Jones *et al.* (1979) studied the fatty acid biosynthesis in the larval stages of *Penaeus japonicus* by feeding larvae microencapsulated diets containing 1–14-carbon palmitic acid and fatty acid diets supplemented with defined fatty acids. Growth to the postlarvae was obtained only when lipid rich in 20:5 ω 3 and 22:6 ω 3 were present in the diet, indicating that they are essential fatty acids for the larvae as well as juveniles. Long chain polyunsaturated ω 3 fatty acids appear to be required, and the zoeae and juveniles appear to have the same metabolic pathways, although there is some difference in relative activity. Dietary requirement of sterols has been demonstrated for the brachyuran crab larvae (Whitney, 1969). Larvae of *Rhithropanopeus harrisii* cannot synthesize sterol precursors, indicating a dietary requirement.

D. Molecular and Biochemical Changes

Little is known of the molecular and metabolic events that control growth and differentiation of the larvae. Differentiation is intimately involved with differential actions of genes affecting control of synthesis of enzymes, enzyme function, and development of specific metabolic pathways. The ontogeny of enzymes provides an indication of the basic changes occurring during differentiation (Masters, 1975). Developmental patterns of multiple forms of enzymes (isozymes) are of interest to developmental biology, and such information could also provide insights into many aspects of the biochemistry of larval development and the mechanisms underlying the adaptations of larvae to the environment.

1. ISOZYMAL VARIATION

Some preliminary results of a study examining the isozymes in larval stages of the mud crab *Rhithropanopeus harrisii* showed a pattern of expression of the different alleles or loci at distinct phases of the life cycle, with some enzyme patterns specific to certain stages (Morgan et al., 1978). The number of isozymes increased with the development of zoeal stages and additional changes occurred during the megalops stage. Changes in isozyme patterns are correlated with the increased differentiation of tissues and organ systems. Gooch (1977) examined the allozyme variation in individuals of the third and fourth zoeal stages and megalops stage of *R. harrisii*, *Sesarma reticulatum*, and *S. cinereum* and found no life-cycle variation in enzyme expression. Barlow (1969) found no pronounced differences in the enzyme patterns during larval development of the American lobster *Homarus americanus*. Lactate dehydrogenase and malate dehydrogenase isozymes were similar in the larvae and juveniles, but quantitative variation was observed in the nonspecific protein.

2. ENZYME ACTIVITY

The activity of a few enzymes has been assayed in the developing larval stages. Specific activity of lactate dehydrogenase, malate dehydrogenase, and glucose-6 phosphate dehydrogenase is higher in the first zoeal stage and again in the fifth zoeal stage and megalops stage of *Cancer irroratus* (Sastry, 1978). In *Rhithropanopeus harrisii*, the activity of glutamic oxaloacetic transaminase involved in the amino acid and nitrogen metabolism increases in the fourth zoeal stage (Frank et al., 1975). The activity of alkaline phosphatase involved in chitin synthesis increases after the molt and then declines prior to the molt in the third and fourth zoeal stages.

3. BIOCHEMICAL CHANGES

Gross changes in the absolute amounts and the relative proportions of DNA, RNA, protein, lipid, and carbohydrates occur during the development of larvae, indicating substrate utilization for energy production and for interconversion and biosynthesis with growth and differentiation (Frank et al., 1975; Sulkin et al., 1975). DNA content of Rhithropanopeus harrisii larvae increases abruptly at the second molt and later increases steadily in the third and fourth zoeal stages, indicating cell proliferation and differentiation (Sulkin et al., 1975). Protein synthetic activity (RNA:DNA) is cyclical in relation to the molt cycle with a peak occurring after each molt. In the first zoeal stage, differentiation is greater, while both growth and differentiation occur in the third and fourth zoeal stages. Early larval stages contain low amounts of lipid, while both lipid and protein content increase in the third zoeal stage. Lipids accumulated in the later stages may be utilized for metabolic energy production as well as for cellular and subcellular activities in preparation for metamorphosis.

Changes in amino acids, which can be involved in protein synthesis, energy production, and osmoregulation also occur during larval development (Costlow and Sastry, 1966; Tucker and Costlow, 1975; Tucker, 1978). In Callinectes sapidus, all the amino acids present in the eggs are also found in the first two zoeal stages, but in the succeeding zoeal and megalops stages the number of amino acids decreases. The crab stage has most of the amino acids present in the early stages, with certain exceptions (Costlow and Sastry, 1966). Proline, taurine, alanine, serine, and glutamic acid increase in the megalops stage of C. sapidus prior to metamorphosis (Tucker and Costlow, 1975). Acceleration of metamorphosis of megalops by eyestalk ablation advances the time of increase in these amino acids compared to that in untreated animals, suggesting that changes are related to the molt cycle.

Quantitative and qualitative changes in fatty acid composition have been observed in the developing larval stages of the crab Cancer irroratus and lobsters Homarus americanus (Sastry, 1980). In the barnacles Semibalanus balanoides, fatty acid composition of the ovary, the cyprid larva, and adult body tissues are quite similar (Dawson and Barnes, 1966). In the copepod Euchaeta japonica, lipid metabolism of the larvae is similar to that of eggs and adults (Lee et al., 1974). Fatty acid composition of the triglycerides is essentially the same during development.

4. SUBSTRATE UTILIZATION

Information available on the energy reserves planktotrophic larvae acquire from plankton feeding, the rate at which they utilize substrates, and the energy requirements for the duration of pelagic life is very limited. Changes in relative composition of the major biochemical constituents and in oxygen and nitrogen ratios provide an indication of the substrates utilized for metabolic energy production. In laboratory cultured larvae of the American lobster *Homarus americanus*, protein is the principal biochemical constituent of the larval stages (Capuzzo and Lancaster, 1979). Carbohydrate content decreases from third larval to postlarval stage, whereas ash and chitin content increases. Lipid content significantly decreases in fourth larval and postlarval stages. Weight-specific oxygen consumption and ammonia excretion rates increase with each larval stage and then decrease in the first postlarval stage. An increase in dependence on protein catabolism by the fourth larval and first postlarval stage is indicated by the decrease in O:N ratios for these stages. Protein catabolism appears to be a principal source of energy, while carbohydrate and lipids are utilized to some extent by the lobster larvae feeding on a diet of *Artemia* nauplii (Capuzzo and Lancaster, 1979). Megalopa larva of the crab *Callinectes sapidus* also utilizes protein rather than lipid as an energy substrate (Holland, 1978). Starvation of the megalops for more than 8 days results in a loss of both protein and carbohydrate.

Other crustaceans may, however, utilize lipid as a source of energy for the larvae (Holland, 1978). Lipid content increases by the late naupliar stage of the barnacle *Semibalanus balanoides* (Lucas *et al.,* 1979). Neutral lipid forms the main energy substrate for the nonfeeding cyprid stage, whereas protein and carbohydrate are utilized to some extent (Holland and Walker, 1975). Energy is partitioned by the cyprid larva for structural components and swimming and exploration (Lucas *et al.,* 1979). The energy utilized for swimming and exploration is mainly derived from lipid reserves, and it is used up within 2.5–4 days after collection from the field. When this energy is used up, the competence for metamorphosis is lost. In cyprids that are prevented from settling, neutral lipids decrease dramatically, and later protein is utilized.

E. Endocrines

Neuroendocrine factors may activate and control the developmental changes through a coordination of the molecular, cellular, metabolic, and morphological processes involved in growth and differentiation of the larvae. There is some evidence of endocrine control of chromatophores, molting, and possibly other aspects of physiology as in adult crustaceans. The chromatophores in decapod larvae are capable of quick color change (Pautsch, 1967) and show a diurnal rhythm and environmental response (Rao, 1968). Chromatophore substances in the extracts of *Sesarma reticulatum* larvae exhibit a cyclical change with a lower activity before and after molting, and higher activity during the intermolt period of all the larval stages (Costlow, 1961). Endocrine control of chromatophores in *Palaemonetes pugio* larvae is similar to that of adults (Borch, 1960).

Periodic molting of larvae as in adult crustaceans indicates endocrine control of growth, differentiation, and molting (Passano, 1960b; Costlow, 1968). Studies conducted during larval development to delineate the function of endocrine systems using eyestalk ablation technique have indicated some differences in larval endocrine systems and their possible mode of action in carideans (Hubschman, 1963) and brachyurans (Costlow, 1963, 1966). Eyestalk ablation has no effect on larval molt cycle or time to reach postlarval stage in the grass shrimp *Palaemonetes pugio* (Hubschman, 1963). Removal of eyestalks either simultaneously or at different times during the larval development of the crab *Sesarma reticulatum* prior to the critical period of day 4 in the megalops stage results in an acceleration of metamorphosis to the crab stage (Costlow, 1966). This acceleration is accompanied by incomplete development or a second megalops stage, which metamorphoses to the crab stage. Eyestalk ablation in the megalops stage of *Callinectes sapidus* within 12 hr after the final zoeal molt also accelerates metamorphosis to the crab stage (Costlow, 1966). The X-organ sinus gland is responsible for the production of molt inhibiting hormone (Passano, 1960b) and appears to function in the megalops stage of these crabs (Costlow, 1968). Bilateral eyestalk ablation in the fourth stage lobster *Homarus americanus* larvae accelerates proecdysis preparation and precocious initiation of ecdysis (Rao et al., 1973). Injection of either ecdysone or 20-hydroxyecdysone also accelerates proecdysal preparation leading to initiation of ecdysis. The number of the hormones involved, their specific mode of action, mechanisms regulating morphological differentiation, and its synchronization with molting frequency to result in the normal pattern of development are still speculative (Costlow, 1966, 1968).

F. Energetics

Energy assimilated by larvae from plankton feeding (planktotrophic) or from storage within the body (nonfeeding) is utilized for formation of new tissues (growth) and maintenance of the developing system. The major energy flows are consumption, egestion, respiration, excretion, loss from molting of exuvia, and storage (growth or production) for the planktotrophic larvae. Partial energy budgets have been constructed for individual larval stages and/or for the whole larval period of some decapod larvae (Reeve, 1969; Mootz and Epifanio, 1974; Logan and Epifanio, 1978; Levine and Sulkin, 1979; Johns and Pechenick, 1980). Growth to the postlarval stage involves storage of assimilated energy in excess of that utilized for maintenance. Larvae may exhibit changes in the rate of food consumption, assimilation efficiency, and metabolic energy expenditure in order to store necessary energy for development and growth to the postlarval stage. The feeding rate of the larvae of the crab *Menippe mercenaria* increases during development to the megalopa stage and then declines prior to onset of metamorphosis (Mootz and Epifanio, 1974). The assimilation efficiency of larvae increases from 41.7% in the first zoeal stage to 84.3% in the megalopa stage. Total energy expenditure for maintenance increases for each succeeding zoeal stage and increase of size, whereas energy loss from exuvia is small. The proportion of assimilated energy utilized for maintenance is relatively high for the first zoeal stage, production and respiration are equal by the third zoeal stage, and energy going into growth exceeds respiration by the fifth zoeal stage. In the megalopa stage, respiration is equal to that of production when energy content of the exuvium is not considered. A similar general pattern of energy flow during larval development is also seen for the mud crab *Rhithropanopeus harrisii* (Levine and Sulkin, 1979).

In contrast to brachyuran crab larvae, the assimilation efficiency of the lobster *Homarus americanus* larvae (81%) does not increase for the successive larval stages and is the same for larvae and juveniles (Logan and Epifanio, 1978). The proportion of assimilated energy lost to routine metabolism and molting remains fairly constant, while that channeled to growth increases during larval life and then declines to a constant level in the postlarval stages. Daily food consumption increases for the successive larval stages. Daily food consumption increases for the successive larval stages, apparently to compensate for growth. In lobsters, which are slightly metamorphic, the changes in energy partitioning and energetic efficiency associated with transition from larval to the postlarval stage are due to changes in rate of growth.

The gross growth efficiency (K_1, the proportion of consumed energy utilized for growth) and net growth efficiency (K_2, the proportion of assimilated energy utilized for growth) may vary for the individual larval stages of a species and for total development of different species. Cumulative gross growth efficiency for the larval development of *Menippe mercenaria* is 30.11% (Mootz and Epifanio, 1974) and *Rhithropanopeus harrisii* is 22.7% (Levine and Sulkin, 1979). The growth efficiency of the larval stages increases during development to the postlarval stage. The gross growth efficiency of *M. mercenaria* and *R. harrisii* increases for the later zoeal stages and then declines for the megalopa stage. The first four zoeal stages of the crab *Cancer irroratus* show similar growth efficiency, and then a dramatic increase occurs in the fifth zoeal stage prior to development into the megalopa stage (Johns and Pechenick, 1980). The growth efficiency of the lobster *Homarus americanus* larvae increases with development and then declines in the postlarval stages (Logan and Epifanio, 1978).

The net growth efficiency of the larvae can be calculated also from the data obtained on the biochemical composition (Holland, 1978). Assuming that assimilation is equal to growth plus respiration and neglecting the energy lost in excretion, the values for growth may be obtained from the increase in protein, lipid, and carbohydrate content of the larvae over a particular period after converting them to caloric equivalents. Calculated this way, the net growth efficiency of the barnacle *Semibalanus balanoides* nauplius stages III to V is between 70 and 79% (Holland, 1978).

G. Duration of Development

The duration of crustacean pelagic larval life varies anywhere from only a few weeks to as long as 6–11 months, e.g., spiny lobsters (Phillips and Sastry, 1980). The duration for individual larval stages and/or total development from hatching to the postlarval stage has been documented for several species from laboratory cultures (Costlow and Bookhout, 1964, 1965; Sastry, 1980). The proportion of total development time is about the same for the first four zoeal stages (11%) of the crab *Cancer irroratus,* but the fifth zoeal stage (16.6%) and the megalopa stage (37.5%) take much longer (Sastry, 1976). The rate at which individual larval stages of a species and larvae of different species develop may vary under given habitat conditions. Rate of larval development is in part genetically determined and it is affected by the rate of energy acquisition and utilization for maintenance as well as environmental factors.

IV. ADAPTATIONS TO THE ENVIRONMENT

Much of our knowledge on the environmental limits for larval development and the physiological and behavioral responses of larvae to environmental changes has been derived from laboratory studies, chiefly with estuarine and coastal species. These studies have shown that limits for complete development vary intra- and interspecifically, and that developing larvae have the potential for altering physiological and behavioral responses to changes in the environment in a manner that could be adaptive for development and survival.

A. Optima and Limits

Development to the postlarval stage occurs within a well defined range of environmental conditions characteristic to a species (Costlow and Bookhout, 1964, 1965; Sastry and Vargo, 1977). Within the favorable range for development, the survival rate of larvae to the postlarval stage is usually maximal under certain combination(s) of the environmental factors. The survival rate of the larvae is reduced when environmental conditions exceed the optimal range until a point is reached beyond which no normal development and/or survival is possible. The optimal range and the limits characteristic for a species, larval development may be genetically determined. For a given species, however, the optimal environmental range for maximum survival and the limits for development may also vary for different larval stages relative to the interacting environmental parameters (Fig. 3). Generally, larval development occurs within a narrow range when multiple environmental factors are considered (Vernberg and Vernberg, 1975). Laboratory determined environmental optima and limits may thus represent only approximations to what may actually affect the survival of larvae in nature.

The effects of environmental factors on the survival of individual larval stages or for complete development to the postlarval stage have been determined for several decapods by employing multifactorial experimental designs with different combinations of temperature and salinity and/or pollutants (Costlow and Bookhout, 1964, 1965; Sastry and Vargo, 1977). These patterns may include (1) minor differences between larval stages (e.g., estuarine mud crab *Rhithropanopeus harrisii,* estuarine grass shrimp *Palaemonetes pugio,* estuarine and intertidal hermit crab *Pagurus longicarpus*), (2) wider tolerance of the earlier stages compared to the later stages (e.g., sublittoral stenohaline crab *Hepatus epheliticus*) or a reverse of this pattern, (3) tolerance of upper end of the range in earlier stages, lower end of the range in later stages, and the entire range in the intermediate stages (e.g., *Sesarma cinereum*), (4) slightly wider tolerance of the earlier and later stages compared to narrower tolerance of the intermediate stages (e.g., *Callinectes sapidus*), (5) tolerance of the lower range in early stages with a shift toward the upper end of the range in later stages (e.g., *Cancer borealis*) or a reverse of this pattern.

Complete development to the postlarval stage with maximum survival may occur for a species within a well-defined optimal combination of temperature and salinity (e.g., *Homarus americanus, Cancer irroratus*) or it may occur in various combinations of temperature and salinity covering a wide range (e.g., *Palaemonetes pugio, Pagurus longicarpus*) (Fig. 4). The limits for complete larval development may be very narrow with a single temperature and salinity combination for some species (e.g., *Cancer borealis, Hepatus epheliticus*), or there may be different combinations of temperature and

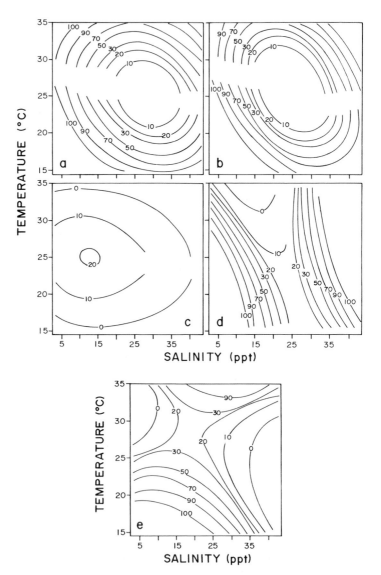

Fig. 3. Estimated percent mortality of larval stages of the brachyuran crab *Sesarma cinereum* based on fitted response surface to observed mortality for larvae reared in different combinations of temperature and salinity. (A–D) First to fourth zoeal stages. (E) Megalops. (Modified from Costlow *et al.*, 1960.)

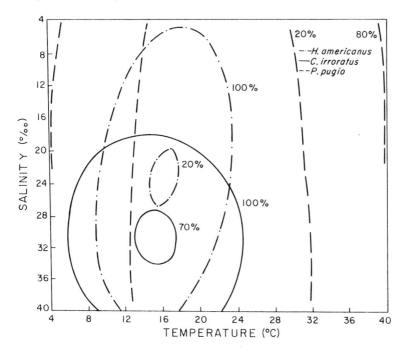

Fig. 4. Percent mortality of the last larval stage of estuarine and sublittoral decapods based on fitted response surfaces to observed mortality for larvae reared in different combinations of temperature and salinity. (From Sastry, 1980.)

salinity at either the upper or lower end of the range for other species (e.g., *C. irroratus, P. longicarpus*).

The optima for survival and the limits for development may vary interspecifically relative to the time and duration of breeding period, the length of larval life, the geographic distributional range, and the habitat and climatic conditions inhabited by a species, but comparative data needed to evaluate the possible ecological relationships is still limited. Larvae of a species breeding in the spring may have a relatively lower temperature limit for complete development compared to those species breeding in the summer (Sastry and Vargo, 1977). Larvae of estuarine and intertidal species may complete development with a higher survival rate over a wider range of temperature and salinity conditions compared to those from sublittoral regions.

The temperature–salinity optima and limits for larval development of a species may be modified by a variety of intrinsic as well as other interacting environmental factors. Thermal history and development stage of the eggs prior to incubation may modify the optima and limits (Sastry and Vargo,

1977). Under identical culture conditions, larvae of *Cancer irroratus* hatched from the eggs obtained from females collected in the spring and incubated at 15°C in 30 ‰ salinity survive better compared to larvae hatched from the eggs of winter or summer animals and incubated at the same temperature and salinity. Larvae from the winter and spring hatches complete development at 10°C in 30 ‰, while those from summer require a temperature minimum of 15°C.

In contrast to the effects of constant environmental conditions, fluctuating temperatures affect the development and survival of larvae differently. The period and amplitude of the fluctuating thermal regime and its nature (sine wave or square wave regime) can also affect the development and survival differently, and the effects are species-specific. Larvae of the estuarine mud crab *Rhithropanopeus harrisii* survive better at the extreme 30°–35°C daily cycle compared to those at constant 30° or 35°C temperatures (Costlow and Bookhout, 1971), whereas larvae of the grass shrimp *Palaemonetes pugio* show no significant differences in survival rate at the daily cyclic and comparable constant temperatures (Sastry, 1980). In contrast, larvae of the sublittoral crab *Cancer irroratus* survive better at 10°–20°C and 15°–25°C daily cycles compared to those at constant 15° or 20°C (Sastry, 1976). Other environmental factors, such as salinity combined with cyclic thermal regimes, can modify effects observed under constant conditions (Christiansen et al., 1977).

Synergistic interactions of multiple environmental factors, the combination of heavy metals or organic pollutants with temperature and/or salinity can affect larval development (Vernberg et al., 1973; McKenney and Costlow, 1981). Toxicity to both cadmium and copper increases inversely with salinity for megalopa larva of *Callinectes sapidus* (Rosenberg and Costlow, 1976). Reduced salinity increases the sensitivity of barnacle nauplii to sublethal concentrations of cadmium (Lang et al.,1981). The survival pattern of *Palaemonetes pugio* throughout larval development is modified by salinity and temperature conditions, but exposure to concentrations of zinc, which are sublethal at optimal temperature and salinity conditions, reduces survival throughout development under stressful levels of temperature and salinity (McKenny and Neff, 1981).

The sensitivity of larvae of different species varies for the same concentration of chemicals when combined with various temperatures and salinities. Larvae of *Callinectes sapidus* are more resistant to malathion compared to the larvae of *Rhithropanopeus harrisii* (Bookhout and Monroe, 1977). Larvae of a species may be also affected differently by different chemicals when combined with the same level of temperature and/or salinity (Bookhout et al., 1976). Synergistic interactions of multiple environmental factors may also vary for larvae cultured under constant and fluctuating thermal regimes. Larvae of *R. harrisii* exposed to a juvenile hormone mimic (ZR-512 Altozar) and cultured throughout larval development at daily cyclic thermal regimes show an increase in mortality rate in the megalops stage (Christiansen et al., 1977). In contrast, the survival of C. sapidus and R. harrisii larvae, both in the presence and absence of cadmium, is enhanced at the cyclic temperature ranges (Rosenberg and Costlow, 1976).

The environmental optima for survival and limits for development may also vary for larvae of geographically separated populations, but precise data are not available. The effects of temperature and salinity on larval development of populations of the mud crab Rhithropanopeus harrisii from Rhode Island (Sastry and Vargo, 1977) and North Carolina (Costlow and Bookhout, 1964, 1965) are different, while no such differences are evident for the populations of Palaemonetes pugio from Rhode Island (Sastry and Vargo, 1977) and Virginia (Roberts, 1971). It is not know whether the differences are phenotypic or truly genetic.

B. Physiological Responses

Within their environmental tolerance limits, developing larvae, like adult poikilotherms, may exhibit alterations in their physiological and behavioral responses which may favor normal rate functions and activities within the mid or normal range (capacity adaptations) and also favor survival under extreme conditions (resistance adaptations) (Prosser, 1975). Both resistance and capacity adaptations of larvae may be influenced by a variety of intrinsic and extrinsic factors such as stage of development, nutritional status, season of larval release, single or multiple environmental factors, amplitude and rate of change of environmental fluctuations, acclimation, and geographical origin of the populations. Larval adaptations are of interest in understanding the ontogeny of mechanisms of adaptation to the environment, and to evaluate the intra- and interspecific variation and the distribution and ecology of the species. Compared to our knowledge of physiological adaptation of adult poikilotherms (F. J. Vernberg and W. B. Vernberg, 1970) and the underlying molecular and biochemical mechanisms (Hazel and Prosser, 1974), the information available for the developing larvae is very limited.

1. RESISTANCE TO THE ENVIRONMENTAL EXTREMES

At either extreme of the tolerance range, there is a point beyond which larvae will not be able to survive (Section IV, A). This lethal zone is dependent on the duration of exposure, intensity of environmental factors under consideration, stage of the larvae, and the levels of other environmental factors (Vernberg and Vernberg, 1975). Acute upper temperature tolerance limits determined at the saturated dissolved oxygen level for the larval stages of *Cancer irroratus* showed little interstage variation (Vargo and Sastry, 1977). However, interstage variation is evident when temperature and low dissolved oxygen stresses are combined, with low oxygen tolerance decreasing as temperature increased. Megalopa stage of *C. irroratus* showed the least temperature dependent low dissolved oxygen tolerance. In other species, both temperature and salinity resistance capacities of the larvae are reduced when exposed to sublethal concentrations of heavy metals at the optimal temperature and salinity for survival (Vernberg *et al.*, 1973; McKenney and Neff, 1981).

Temperature tolerance limits vary for successive larval stages of some species. Acute temperature tolerance limits for the fourth and seventh larval stages of *Palaemonetes pugio* are 2° and 2.3°C higher, respectively, than for other stages (Sastry and Vargo, 1977). The second larval stage of *Homarus americanus* has a much lower tolerance limit than any other stage (Sastry, 1980). Eggs and adults of the barnacle *Semibalanus balanoides* are more tolerant to both high and low temperature extremes than the nauplius and cyprid stages (Crisp and Ritz, 1967). The cyprid is the least tolerant stage in the life-cycle to both high and low temperatures, but after metamorphosis, tolerance of the newly settled juvenile is comparable to that of the adults. The pattern of tolerance to environmental extremes may be characteristic of a species and may have some relationship to their habitat. The upper temperature tolerance limits for the larvae of the sublittoral crab *Cancer irroratus* and the lobster *H. americanus* are lower compared to that of the larvae of estuarine grass shrimp *P. pugio* (Sastry and Vargo, 1977). Between the two sublittoral species, *C. irroratus* larvae have lower temperature tolerance limits than *H. americanus* larvae. For at least some species, it would seem that larvae have capacities for tolerance of greater environmental extremes than they would normally require in their natural environment.

2. OSMOREGULATION

Brachyuran crab larvae exhibit osmoregulatory abilities as an adaptation to salinity changes. Osmoregulation in larval stages, as in adult Crustacea, is closely related to the molting process (Kalber, 1970; Foskett, 1977). Kalber and Costlow (1966, 1968) reported that blood osmotic pressure of *Rhithropanopeus harrisii* increases as each larval molt approaches and decreases 12 hr after molting, and they suggested that this serves as a mechanism to provide an osmotic gradient necessary to ensure water influx at ecdysis. Tucker and Costlow (1975) reported that free amino acids in *Callinectes sapidus* larvae increase with the approach of molting.

In most or all larval stages of brachyuran crabs examined, there is a general tendency for hyperosmoticity in salinities they normally encounter under natural conditions. Foskett (1977) evaluated the data published for different species and found no consistent trend, over the entire range of test salinities, of larval hyperregulation immediately before ecdysis compared to other times in the molt cycle. He concluded that the trend of larvae to hyperregulate over the entire range of salinities encountered in nature may reflect a necessity to maintain an osmotic gradient to ensure water influx, but there is no evidence that an increased hyperosmotic gradient is necessary at the time of molting. Instead, it is suggested that neuroendocrine controls are probably important to ensure water uptake at the molt through regulation of external body surface permeability to salt and water. Larvae lack a heavy exoskeleton, and hyperosmoticity is considered necessary to ensure integrity of the exoskeleton.

The abilities for osmoregulation may vary with the stage of larval development, and the pattern is characteristic of a species (Kalber, 1970). In the littoral crab *Sesarma reticulatum,* the osmoregulatory pattern is established before hatching and does not vary throughout larval development (Foskett, 1977). The adult pattern of hypo- and hyperregulation is not well established even in the megalops stage. In a number of other brachyurans, there is no clear indication of the adult osmoregulatory pattern during larval development. The adult response is generally attained with the appearance of a rigid exoskeleton during the early juvenile stages.

3. METABOLISM

Metabolic rate of the larvae is affected by temperature, salinity, oxygen tension, and chemicals. Metabolic-temperature (M-T) responses have been examined for a single larval stage (W. B. Vernberg and F. J. Vernberg, 1970; Belman and Childress, 1973; Moreira et al., 1981), discontinuous larval stages (Vernberg and Costlow, 1966), or all larval stages (Sastry and McCarthy, 1973; Schatzlein and Costlow, 1978; McKenney and Neff, 1981; Vernberg et al., 1981). The patterns of M-T responses of larval stages are species specific and reflect the metabolically viable tolerance range of a species (Fig. 5). *Cancer irroratus* larvae are metabolically active between 5° and 25°C and show a decline of metabolic rate above 25°C. Larave of this species become progressively stenothermal during zoeal development (Sastry and McCarthy, 1973). In comparison, larvae of the congeneric species *Cancer borealis* are also metabolically active over the entire temperature range of 5°–25°C, but show no depression of the rate at 25°C. The larvae are sensitive ($Q_{10} > 2$) to warmer temperatures in the early zoeal stages and colder temperatures in the later stages. The range of metabolic-temperature independence ($1 < Q_{10} < 2$) gradually shifts from colder temperatures in the early zoeal stages to warmer temperatures in the later stages. The pattern of M-T responses for successive larval stages of the tropical hermit crab *Pagurus criniticornis* shows that the third stage is more sensitive, while

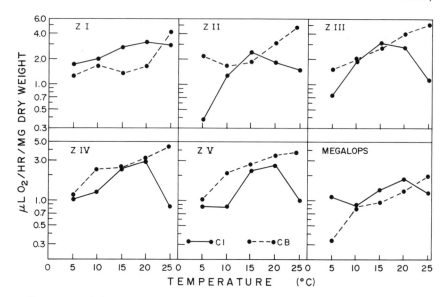

Fig. 5. Metabolism-temperature responses of the larval stages of congeneric species of brachyuran crabs *Cancer irroratus* and *Cancer borealis* cultured in 30 ‰ salinity at 15° and 20°C, the temperature and salinity combinations producing maximal survival rate for the respective species. (Modified from Sastry and McCarthy, 1973.)

glaucothoë stages show the least temperature-dependent variation in metabolic rate (Vernberg et al., 1981). All larval stages show higher Q_{10} values between 10° and 15°C and Q_{10}'s close to unity between 20° and 25°C. The metabolic rate of the larvae declines above 30°C.

Other environmental factors combined with temperature can modify the M-T patterns of the larval stages. Oxygen uptake of a larval stage of the spiny lobster *Panulirus interruptus* and the crab *Cancer productus* varies with oxygen concentration from ambient to 0.5 mg O_2/liter (Belman and Childress, 1973). Both temperature, salinity, and their interactions influence the respiration rate of the larvae of the freshwater shrimp *Macrobrachium holthuisi*, but the effect of temperature is more pronounced (Moreira et al., 1980). The metabolic rate of the larvae is salinity dependent between 7 and 28 ‰, a range larvae encounter in their natural environment, at temperatures between 15° and 30°C. At 20°C, metabolic rate is independent of salinity variation between 7 and 28 ‰, but the rate is elevated at the higher 35 ‰ salinity. M-T response patterns of the developing larval stages of *Palaemonetes pugio* are differentially altered by the salinity in which larvae are reared and by temperature and salinity interactions (McKenney and Neff, 1981). The patterns for *P. pugio* larvae are also disrupted by exposure to zinc with a depression of the rate for certain stages.

Metabolic-temperature responses of the larval stages of a species vary relative to constant or fluctuating culture temperature regimes, and the patterns vary interspecifically (Sastry, 1979, 1980). Larval stages of *Cancer irroratus* cultured at 10°–20°C daily cyclic temperatures show an extension of the M-T independence with a shift toward the higher temperatures relative to larvae cultured at 15°C constant temperature (Fig. 6). The temperature range for depression also shifts from 20°C to 25°C for larvae maintained at constant temperature to 25°–30°C for those subjected to cyclic temperatures. M-T patterns vary for the successive larval stages of *Homarus americanus* and *Palaemonetes pugio* cultured at a cyclic and a comparable constant temperature, but there is no consistent trend as for the larvae of *C. irroratus* (Sastry, 1980).

The mechanisms underlying the metabolic adaptation of developing larvae are not as well understood as they are in adult poikilotherms (Hazel and Prosser, 1974). Specific activities of lactate dehydrogenase, malate dehydrogenase, and glucose-6 phosphate dehydrogenase in the larval stages of *Cancer irroratus* cultured at both 10°–20°C daily cyclic and 15°C constant thermal regimes are higher in the first zoeal stage and again in the later stages (Sastry, 1978; Sastry and Ellington, 1978). However, changes in the specific activities of these enzymes follow different stage-dependent trends relative to the culture regime, indicating that enzyme systems in different metabolic sequences are affected differently in each larval stage by cyclic temperatures (Fig. 7). Differential effects on enzyme activities in different

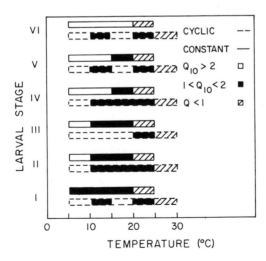

Fig. 6. Metabolism-temperature responses (as Q_{10} values) of *Cancer irroratus* larval stages cultured at 10–20°C daily cyclic and 15°C constant temperatures in 30 ‰ salinity. (From Sastry, 1980.)

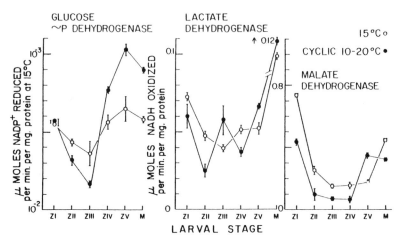

Fig. 7. Specific activities of three enzymes in *Cancer irroratus* larval stages cultured at 10–20°C daily cyclic and 15°C constant temperatures and assayed at 15°C. (From Sastry, 1978.)

metabolic pathways may reorganize metabolism of the larval stages as an adaptation to the culture temperature regimes.

4. ENERGETICS

The influence of environmental factors on the energetics of larvae is not known. However, metabolic energy expenditure is a major cost factor and it varies relative to the stage of development and interacting environmental factors. Larvae may adjust their rates of food consumption and energy expenditure for maintenance relative to the environmental changes and may also conserve energy by metabolic rate compensation to temperature in order to maintain a positive energy balance necessary for development (Sastry, 1979).

5. RATE OF DEVELOPMENT

The rate of larval development may be genetically determined in part, and it may be modified by interacting environmental factors within the tolerance range and as well as by the quality and quantity of available food. The rate of larval development may be different for different species under comparable conditions. Larvae of the congeneric species of brachyuran crabs *Cancer irroratus* and *Cancer borealis* fed on a diet of *Artemia* nauplii take about the same length of time to reach the crab stage when cultured in 35 ‰ at 15°C and 30 ‰ at 20°C temperature and salinity combinations, optimal for survival of the respective species (Sastry and McCarthy, 1973). At the same temperature, zoeae of *C. borealis* develop at a slower rate compared to those of *C. irroratus*.

The duration of larval development has been determined for a number of decapods at different temperatures (Costlow and Bookhout, 1964, 1965). Generally, the rate of larval development increases with temperature in a linear fashion over an intermediate range and decreases at the upper and lower ends of the range (Sastry, 1976). Temperature-dependent developmental time for individual larval stages of the crab *Cancer irroratus* fitted with Bělehrádek's temperature function showed that the curve for each stage is distinct (Fig. 8). The first zoeal stage developed at a slower rate than the later zoeal stages and the megalops at a much slower rate than any of the preceding zoeal stages. The temperature limits for development of zoeal

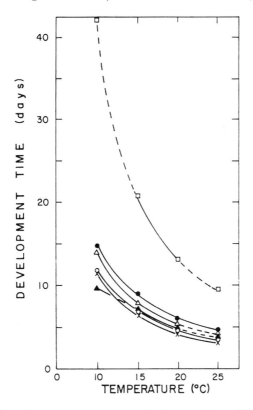

Fig. 8. The effects of constant temperatures on developmental rate of the larval stages of the brachyuran crab *Cancer irroratus* fitted with the Bělehrádek's function. ●——●, First zoeae; X——X, second zoeae; ○——○, third zoeae; ▲——▲, fourth zoeae; △——△, fifth zoeae; □——□, megalops. Dashed lines represent extrapolated values. (From Sastry, 1976.)

stages are broader compared to those for the megalops stage, indicating differential sensitivity of larval stages to temperature.

Rates of development of larval stages are also differentially sensitive to constant and daily cyclic temperatures, and these responses are species specific (Sastry, 1980). The duration of the zoeal and megalops stages of *Cancer irroratus* are shorter and longer, respectively, at a $10°-20°C$ daily cycle than at a constant $15°C$, but the total development is about the same at both culture regimes (Sastry, 1976). The duration of each of the first four zoeal stages as a proportion of the total development to crab is the same at both culture regimes, but the fifth zoeal stage is shortened and the megalops stage lengthened at the cyclic regime. The amplitude and rate of temperature change of the daily cycle also affects the rate of development of the larvae differently.

The effects of salinity on the duration of larval development varies greatly for different species. The duration of larval development of *Palaemonetes pugio*, *Pagurus longicarpus*, and *Cancer irroratus* is unaffected by salinity changes within the tolerance range (Sastry, 1980). Larvae of the stenohaline crab *Hepatus epheliticus* develop at a slower rate in 25 ‰ salinity compared to that in 30 and 35 ‰ salinity (Costlow and Bookhout, 1964, 1965). The duration of complete development of *Callinectes sapidus* is longer at the higher 32 ‰ salinity, but the megalops develops more slowly at lower than higher salinities (Costlow and Bookhout, 1964, 1965). *Homarus americaus* larvae develop slowly in salinities both above and below 25 ‰ (Sastry, 1980).

Other factors combined with temperature and/or salinity may affect the duration of larval development differently from that observed in relation to either temperature or salinity alone. The duration of larval development under optimal temperature and salinity conditions may be modified when heavy metals or organic pollutants are combined. Retardation of development rate on exposure to a pollutant is a generalized response for decapod larvae. The duration of zoeal development of *Rhithropanopeus harrisii* increases progressively with an increase in malathion concentration at any given salinity between 25 and 30 ‰ (Bookhout and Monroe, 1977). Development of *Callinectes sapidus* megalopa is delayed at any salinity on exposure to mercury (McKenney and Costlow, 1981).

6. ACCLIMATION

Larvae may have capacities for acclimation, but detailed studies are rare. The environmental experience of the eggs prior to hatching may alter the capacities for larval adaptation by modifying their survival limits and their rates of development to the postlarval stage (Vernberg and Vernberg, 1975; Sastry and Vargo, 1977). Rosenberg and Costlow (1979) acclimated eggs of the mud crab *Rhithropanopeus harrisii* in the blastula and gastrula stage to 10 or 30 ‰ salinity at 25°–30°C daily cyclic temperatures. After hatching, each zoeal and megalops stage were subjected to an increase in salinity from 10 to 30 ‰ or a decrease in salinity from 30 to 5 ‰ and then reared to postlarval stage. The increase in salinity during zoeal stages had no effect on the survival or development time, but zoeae developed into abnormal megalopae. Salinity increase in the megalops stage or decrease from 30 to 5 ‰ in any stage had no effect on survival of the larval stages. The response to an increase in salinity during zoeal development is manifested later in the megalops stage, probably due to the disturbance caused at intermolt stage at the time of salinity change. These authors have also reported that acclimation of *R. harrisii* eggs to high and low salinity causes death of the resulting larvae when reared in 5 or 35 ‰ at the 30°–35°C daily cycle, but development of larvae through metamorphosis occurs at the 25° to 30°C daily cyclic temperatures.

The rate of oxygen consumption of a larval stage of the crab *Cancer productus* acclimated for 24 hr to 12.5°, 17.4°, and 24.6°C is not affected by temperatures between 12.5° and 17.4°C, but it increases between 17.4° and 24.6°C (Belman and Childress, 1973). Larvae of the spiny lobster *Panulirus interruptus* acclimated under the same conditions as the crab larvae showed no effect of acclimation to colder temperatures and an increase in thermal sensitivity compared to crab larvae. Larvae of *Cancer irroratus* cultured throughout development at 15°C constant or 10°–20°C daily cyclic temperatures show differences in the M-T patterns of each larval stage and in the specific activities of enzymes, suggesting that alterations occur with acclimation to the respective thermal regimes (Sastry, 1978, 1979; Sastry and Ellington, 1978).

7. GEOGRAPHIC VARIATION

Geographically separated adult populations may exhibit biochemical, physiological, and morphological variation (Mayr, 1963; F. J. Vernberg and W. B. Vernberg, 1970; Prosser, 1975). Gene flow between populations (by larval transport) and selection pressures can play an important role in the extent of variation between geographically separated populations. Geographic variation between populations could be phenotypic, ontogenic, or truly genetic, and knowledge of variation through all phases in the life-cycle could enhance our understanding of the basis of variation between populations; but the available information is very limited. Metabolic-temperature responses of the first zoeae of various tropical and temperate zone fiddler crabs (*Uca*) showed no clear intra- and interspecific variation (Vernberg and Costlow, 1966), although at the temperature extremes of 30° and 35°C, M-T curves showed a downward slope for the northern species and an upward slope for the southern species. However, intraspecific differences were observed in metabolism of the megalops and crab stage of latitudinally separated populations of *Uca pugilator* and *U. pugnax*. Gooch (1977) examined the allozyme variation in zoeae, megalopa, and first crab stages of *Rhithropanopeus harrisii* populations from North Carolina, Maryland, and Maine on the east coast of North America and found that 15 allozyme loci are monomorphic with no geographic variability. There appears to be a minimal geographic differentiation between populations. Schneider (1968), studying temperature adaptation of field-collected and laboratory-reared adults of *R. harrisii* from Florida, North Carolina, and Maine, found that population differences were most evident in the field-collected animals, whereas those reared under identical constant laboratory conditions were nearly identical. Nonreversible phenotypic (ontogenic) adaptation appears to have occurred during the course of development of *R. harrisii* populations under field conditions.

C. Behavioral Responses

The three-dimensional pelagic environment of the larvae is characterized by a continuous state of water motion in the horizontal plane and turbulent mixing processes in the vertical plane (Smith, 1976; Warren and Wunsch, 1981). The structure of the pelagic environment is characterized by temperature and salinity properties, light and gravitational fields, and a gradient of hydrostatic pressure. One dominant feature of the pelagic environment is the temporal variation on different time and space scales (Steele, 1981). The environmental variables interacting with the larvae in nature vary locally and regionally, and in estuaries and coastal and oceanic regions.

Although laboratory experiments may not be able to simulate natural environmental conditions, they have allowed analysis of the responses and some understanding of the behavioral repertoire of the larvae. Larvae of cirripedes and brachyurans exhibit taxes (oriented response to stimuli) and kineses (nondirectional responses to stimuli) in response to changes in environmental conditions. Larvae can detect light and gravity, which have vector properties, to orient their position in the water column (Crisp, 1974a). The other parameters, including light intensity, hydrostatic pressure, temperature, salinity, and other chemicals, are scalar quantities and may provide no information unless larvae can detect their gradients in space. However, changes in directional response may be induced by variations, or thresholds, of scalar quantities (Rice, 1967; Crisp, 1974a). Thus, a change in hydrostatic pressure or salinity could trigger a response to light or gravity. Both types of responses, orientation and locomotion, may be combined to permit larvae to reach and maintain themselves in a particular environment and subsequently to reach the adult habitat.

1. SENSORY SYSTEMS

Taxes and kineses exhibited by larvae indicate that they are capable of perceiving their environment for orientation and movement. Some examples follow: an account of the complex eye of *Penaeus duorarum* larvae with an attention to the nervous system has been given (Elofsson, 1969); a functional statocyst is not developed until the end of brachyuran zoeal development (Prentiss, 1901); and larvae respond to changes in hydrostatic pressure (Knight-Jones and Morgan, 1966). In general, however, a detailed knowledge of the receptors and their function and how the various environmental inputs are integrated into sensory and motor activities are lacking.

2. SWIMMING BEHAVIOR

Swimming habits of larvae are closely related to the body form and are influenced by the developmental stage in which larvae are released, the internal physiological state, and the environmental stimuli (Foxon, 1934). Crustacean larvae swim in a variety of attitudes and exhibit orientation and reversal mechanisms. The mechanism of swimming by nauplius larva of different groups is uniform, but their swimming habits are variable (Gauld, 1959). The antennae of nauplii are locomotory in function and the antennules are the main balancing organs. Decapod larvae swim up or down. When not swimming, larvae of the anomuran *Munida rugosa* rest, while brachyuran zoeae sink (Foxon, 1934). Larvae of the European lobster *Homarus gammarus* swim by metachronous beating of the exopodites and by occasionally using short bursts of tail flexes (Neils *et al.*, 1976; MacMillan *et al.*, 1976; Laverack *et al.*, 1976). The first three larval stages swim upward and forward with the abdomen slightly curved or folded under the body. Larvae can also glide. In the fourth stage, the exopodites become nonfunctional and the endopodites become functional for walking.

Swimming speed and sinking rate varies for different larval stages of a species and for different species (Foxon, 1934). The mean sinking rate of zoeal stages increased from early to late stages and then decreased in the megalops stage (Sulkin, 1975). The swimming speed of late phyllosoma larvae is three times greater than that of the early stages (Saisho, 1966). Rimmer and Phillips (1979) calculated the minimal net rates of 13.7 and 13.0 m/hr for ascent and descent, respectively, for the early phyllosoma larva. The rates for mid stage were 16.0 and 16.6 m/hr and for the late stage 19.4 and 20.4 m/hr, respectively.

3. RESPONSES TO ENVIRONMENTAL VARIABLES

In a review, Thorson (1964) discussed the behavioral responses of intertidal and benthic invertebrates (including crustaceans) to light with changes in light intensity, salinity, and temperature. Generally, newly released larvae swim upward to the surface in response to light and gravity. Larvae of some species are indifferent to light and raise to the water surface by geonegative response, while others may be photopositive and continue so throughout their pelagic life. The photobehavior of larvae at the time of settlement correlates with the habitat of adults and ensures the finding and recognizing of the adult habitat in the intertidal and sublittoral regions. The later stage larvae of sublittoral species become photonegative and seek deeper layers when about to settle. The response of the later stage larvae of intertidal species is variable, with some larvae remaining photopositive until they stop swimming and others become indifferent or photonegative.

Behavioral responses of the larvae may change with development, are altered by environmental factors, and are correlated with the adult habitat (Thorson, 1964). Photoresponses of larvae to high light intensities, increased temperatures, and reduced salinity could act together to keep larvae away from the surface layers, especially in stratified estuaries and for larvae of intertidal species. In recent years, a number of workers made detailed laboratory analyses of the behavioral responses of larvae of estuarine crabs and intertidal barnacles to light, gravity, and hydrostatic pressure, and to different combinations of these variables.

a. Light. Larvae exhibit complicated phototactic and photokinetic responses to changes in light conditions. Larvae of a number of barnacles and estuarine intertidal crabs exhibit spectral sensitivity in the 500–600 nm range (Barnes and Klepal, 1972; Forward, 1974; Forward and Costlow, 1974; Forward and Cronin, 1978; Lang et al., 1979), wavelengths best transmitted in the coastal and estuarine waters (Jerlov, 1976). Spectral sensitivity of the barnacle *Balanus improvisus* nauplii changes for the larval stages (Lang et al., 1979), but it does not change for different larval stages of the crab *Rhithropanopeus harrisii* (Forward and Costlow, 1974). However, spectral sensitivity for different species of intertidal crabs varies with a general trend among all species to have either a maximum in the UV-violet and blue-green or uniform sensitivity from the UV-violet to the blue-green region (Forward and Cronin, 1979). Larvae of high intertidal species are sensitive to UV and blue-green (e.g., *Uca pugilator, Uca minax*), while those of low intertidal species (e.g., *Panopeus herbstii*) lack UV sensitivity. Forward and Cronin suggested that larval spectral sensitivity is not adapted to light transmission in the estuarine waters inhabited by the larvae, but probably to the adult intertidal environment.

Larval stages of the mud crab *Rhithropanopeus harrisii* exhibit polarotaxis in the zoeal stages II and IV, but stage I and III zoeae are unresponsive (Via and Forward, 1975). Larval stages appear to perceive and orient to linearly polarized light and may use it as an orientation cue for vertical movements.

The developmental pattern of phototaxis varies depending on species and adult habitat (Thorson, 1964; Forward, 1976a). Phototaxis of larvae may reverse with changes in light intensity, hydrostatic pressure, salinity, and larval feeding. Generally, larvae exhibit negative responses to high light intensity and positive responses to low light intensities. However, all larval stages of the intertidal barnacles (Thorson, 1964) and intertidal brachyuran crabs *Uca pugilator* (Herrnkind, 1968) and *Rhithropanopeus harrisii* (Forward, 1974; Forward and Costlow, 1974) are positively phototaxic to moderate light intensities. Larvae of brachyuran crabs do not reverse from positive to negative at high light intensities and are negatively phototactic at low light intensities. Analysis of phototactic responses of larvae of several other intertidal crabs also showed positive phototaxis to high light intensities and negative phototaxis to low intensities (Forward, 1976b, 1977). Negative phototaxis and directionally oriented movements with a sudden decrease in light intensity have been suggested as a shadow response which may be involved in predator avoidance. The shadow response with negative phototaxis seen in brachyuran zoeae is absent in the barnacle *Balanus improvisus* (Lang et al., 1979). Attempts to relate phototactic behavior of the brachyuran zoeae to diurnal vertical migration have met with limited success (Forward and Cronin, 1978).

Phototactic responses of larvae may be altered by feeding conditions. The fed nauplii of *Semibalanus balanoides* and *Elminius modestus* are more negative than usual, whereas starved nauplii are more positive (Singarajah et al., 1967). First zoeae of the sand crab *Emerita analoga* are strongly phototopositive for the first 24 hr after hatching and later become photonegative if they are fed and photopositive if they are starved (Burton, 1979).

A sudden change in temperature alters the phototactic responses. An increase in temperature makes positive *Balanus* larvae negative, and negative larvae more negative, while a decrease in temperature produces a reverse of this response pattern (Ewald, 1912). Normally negatively phototactic larvae of *Squilla* sp. become positive with a decrease in temperature (Fraenkel, 1931). The sign of phototaxis of *Rhithropanopeus harrisii* larvae is not markedly altered by a sudden change in temperature (Ott and Forward, 1976).

A sudden exposure to low salinity generally induces a negative phototaxis in normally positive larvae (Thorson, 1964). The first and fourth stage zoeae of *Rhithropanopeus harrisii* show an alteration in phototactic and geotactic responses to a sudden change in salinity (Latz and Forward, 1977). Upon sudden exposure to low salinity, the dominant response in the first stage zoeae is negative phototaxis and to higher salinities a positive phototaxis. For stage IV larvae, the dominant response upon exposure to low salinity is positive geotaxis and, at high salinities, an increased level of phototaxis resulting in an upward movement. The stage I and II nauplii of the barnacle *Balanus balanus* exposed to a sudden change in salinity show a temporary reversal to a negative response, but a continuous exposure to different salinities produces a less obvious effect (Lang et al., 1980). Positive phototaxis in the barnacle larvae is regained within 10 min after salinity decrease (Edmondson and Ingram, 1939).

Photokinetic responses vary for the larvae of different species of barnacles, and the responses are modified by nutritional status and salinity (Lang et al., 1980). Using a videocomputer system, Lang et al. (1979) analyzed the phototactic and orthokinetic (velocity) responses of barnacle nauplii. The stage II nauplii of *Balanus improvisus* show an increase in mean linear velocity following exposure to a specific range of light intensities starting at 480 nm, the range being determined in part by the initial state of naupliar light adaptation. Dark adapted larvae show photokinetic responses at reduced light intensities relative to light-adapted nauplii. Mean linear velocity sharply decreases for the stage II nauplii of *B. improvisus*, *B. venustus*, and *B. amphitrite* following the removal of the light stimulus. Stage II nauplii of *Balanus perforatus* have a sinking response following a sudden dark–light transition (Ewald, 1912). Photokinetic responses are absent in stage II nauplii of *Semibalanus balanoides* and *Elminius modestus* (Crisp and Ritz, 1973). Cyprids of these species show reduced activity on exposure to increased light and enhanced activity following a reduction in light intensity.

b. Light and Gravity. Gravitational orientation is always taxic, and the gravitational sense may function in geotactic control of vertical movements in the water column. Light flux and gravity are usually directed vertically, and it is possible to demonstrate orientation to gravity by larvae in vertical columns in darkness. In darkness, brachyuran larvae orient to gravity, even though a functional statocyst is not developed until the last zoeal stage (Prentiss, 1901; Foxon, 1934). All the zoeal stages of the crabs *Panopeus herbstii* and *Cataleptodius floridanus* show negative geotaxis in darkness, whereas the megalops stage shows a positive geotactic response (Sulkin, 1973). Orientation and upward swimming in darkness results in upward movement of the zoeal stages, while active swimming results in downward movement of the megalops stage. Geotactic responses of the zoeal stages of the sublittoral crab *Cancer irroratus* changes abruptly to a positive response in the fifth zoeal stage and megalops stage (Bigford, 1979).

The relative importance of light and gravity as orientating stimuli may change with the progress in larval development to regulate the depth at which larvae maintain themselves in the water column. Light dominates as the orientating stimulus for the first zoeal stages of the crabs *Leptodius floridanus* and *Panopeus herbstii*, while positive geotactic response dominates in the magalops stage (Sulkin, 1973, 1975). In the larvae of *Cancer irroratus*, photopositive movements decrease, while changes in geotactic responses occur for the zoeal stages with a transition to geopositive behavior in the fifth zoeal stage (Bigford, 1979). The megalops stage of some species (*Macropipus* sp. and *Carcinus maenas*) are, however, negatively geotactic (Rice, 1967). The differences in response of the settling stages of these species may be related to the adult habitat.

c. *Water Current.* The influence of water currents on the behavior of larvae has not received much attention. Cyprid larvae of barnacles sink when exposed to current velocities of 10–32 cm/min and redisperse into the water column at current velocities of 35–67 cm/min. Larvae exhibit minimal ability for horizontal swimming (DeWolf, 1973).

d. *Hydrostatic Pressure.* Larvae respond to changes in hydrostatic pressure by increasing activity and swimming upward when pressure increases or by swimming downward or sinking passively when pressure decreases (Knight-Jones and Morgan, 1966; Knight-Jones and Qasim, 1967; Rice, 1967). The significance of the pressure sense varies for different larval stages and also for different species. All the zoeal stages of *Rhithropanopeus harrisii* and stage I and II zoeae of *Cataleptodius floridanus* respond to pressure changes (Wheeler and Epifanio, 1978). Larval stages of these crab species assume a lower mean position in the water column as they develop from stage I zoeae to the megalops stage. The threshold pressure increase required for reaching the mean position in the water column also increases with progress in larval development of *R. harrisii*.

Pressure is a nondirectional stimulus, and responses to changes in pressure have been used to determine the relative importance of light and gravity in orientation (Rice, 1967). Some pressure sensitive larvae orient to light and others to gravity. Those using gravity swim upward in response to an increase in pressure and swim downward or sink in response to a decrease in pressure. When light is a dominant factor, larvae of some species normally positively phototactic swim away from the light when pressure is decreased. These larvae are negatively phototactic at atmospheric pressure. Bently and Sulkin (1977) examined the responses of both positively and negatively phototactic zoeal stages of *Rhithropanopeus harrisii* to increases in hydrostatic pressure by measuring swimming speed. Mean swimming speed increases for the successive zoeal stages at all the pressures tested. Positively phototactic larvae respond to pressure increases with an increase in activity, while negatively phototactic larvae respond with a decrease in activity.

Responses of larvae to gravity and hydrostatic pressure together with passive sinking may allow larval stages of a species to regulate their depth in the water column (Sulkin, 1973). First stage larvae of the sand crab *Emerita analoga* are highly responsive to pressure change and begin active swimming, but the responsiveness decreases by the first zoeae age, resulting in a deeper distribution of the larvae in the water column (Burton, 1979). Responses to pressure seem to provide a depth regulatory mechanism for the larvae, with gravity and light serving as orientating stimuli. Both laboratory and field observations on the larvae of the American lobster *Homarus americanus* indicate that depth regulatory responses may control vertical distribution of the larval stages, even though not within well defined limits (Ennis, 1973b, 1975b).

e. *Chemicals.* Orientation responses of larvae to light may be altered by chemicals (pollutants) introduced into the water. The photopositive response of zoeal stages one to three and megalopa of *Uca pugilator* is suppressed under suboptimal salinities, but larvae become more responsive when exposed to mercury (Vernberg *et al.,* 1973). Phototactic behavior of barnacle larvae is altered on exposure to cadmium and copper (Lang *et al.,* 1981). A marked change in geotactic, phototactic, and pressure responses occur in the larval stages of *Cancer irroratus* exposed to No. 2 fuel oil (Bigford, 1977).

f. *Endogenous Tidal Rhythms.* Larvae of estuarine species may partition their time between the upper low salinity and lower high salinity waters by tidal vertical migrations (Williamson, 1967; DeCoursey, 1976). Cronin and Forward (1979) reported an endogenously controlled migration on a tidal scale for larvae of *Rhithropanopeus harrisii*. Field-caught larvae showed tidal vertical migration under constant conditions in the laboratory, whereas laboratory reared larvae showed no tidal rhythm.

V. DISPERSAL AND RECRUITMENT

Larvae of holopelagic, benthic, intertidal, semiterrestrial, and some terrestrial and freshwater Crustacea develop in marine or brackish water pelagic environments and subsequently are recruited to the populations in the respective adult habitats. Larvae released into the pelagic environment are dispersed from the sites of origin by the prevailing water currents, and hence return of larvae at the completion of development for recruitment is essential for population continuity in an area. Usually, populations of species continue to be represented in the adult habitat from one year to another, suggesting that mechanisms must exist for the return of dispersed larvae for restocking the populations. Recruitment success has important implications to the population stability, community structure and dynamics, and bio-geographical patterns of distribution and evolution, but the mechanisms involved are not well understood for most species.

Dispersal of larvae in the pelagic environment is a function of the length of larval life and the prevailing water currents, but the distance larvae are transported can be influenced by the vertical distribution of larval stages, their behavioral responses to environmental variables in the water column, and the speed and direction of currents at the different depths. Thus, within the geographical area of dispersion, the distribution and abundance of larvae is influenced by the duration of larval stages, the stage-specific behaviors, and the hydrographical mechanisms, resulting in a patchy distribution on different time and space scales. Generally, mortalities of pelagic larvae are considered to be very high (Thorson, 1950). Observations on predation of larvae are mentioned in the literature, but reliable mortality estimates which could provide us with a knowledge of the dynamics of larval populations are unavailable for most species.

The mechanisms of larval return for subsequent recruitment to the populations are also related to the vertical distribution and behavior of larval stages, the water circulation dynamics, and the life history features and adult habitat requirements of the species concerned. The differences in life history features and habitat requirements, in the timing and duration of larval release by populations, and in the regional and geographic hydrographical conditions, can result in an array of patterns of dispersal and recruitment of species. However, the basic mechanisms of larval–hydrographical interactions resulting in dispersal and recruitment appear the same whether a species is holopelagic, benthic, intertidal, semiterrestrial, or terrestrial in the adult stage.

A pelagic dispersal stage in the life cycle raises many questions of ecological importance common to all species, but definitive answers are lacking. Some of these questions are: (1) What is the relationship between length of larval life and the distances larvae are transported by water currents? (2) What is the maximum distance larvae of a species can be transported that still allows restocking of the population? (3) What proportion of larvae originating from a spawning area are permanently lost by transport by the water currents? (4) Do larvae transported elsewhere form a source of recruits to another population? (5) What is the extent of gene exchange by larval transport between contiguous or geographically separated populations? (6) What is the relationship between larval dispersal and genetic variability of populations? (7) What is the relationship between larval dispersal and adult distributional pattern? (8) What are the mechanisms involved in dispersal, retention, and restocking of the populations? (9) What is the relationship between larval mortalities and the influence of hydrographical conditions to the success in recruitment and dynamics of the populations? Comprehensive quantitative studies on the ecology of larvae describing their temporal and spatial distribution and abundance along with sufficiently detailed descriptions of the hydrographical features are generally lacking. In this section, some aspects of larval dispersal and mechanisms of recruitment are considered relative to the adult habitat.

A. Holopelagic Crustacea

Generally, adult holopelagic crustaceans in neritic and oceanic regions are maintained within specific water masses distributed over a certain geographical area (Johnson and Brinton, 1963; Reid et al., 1978). Holopelagic species have well defined ranges of vertical and geographical distribution. Within the species-specific range, the adults perform diurnal vertical migration over a certain depth range (Bainbridge, 1961) and aggregate for breeding (Mauchline and Fisher, 1969), with the timing and duration of larval release varying in different parts of the geographical range.

A complex array of patterns of larval dispersal and recruitment are found for holopelagic Crustacea in relation to (1) their vertical distribution, (2) ontogenic vertical and horizontal migrations, (3) diurnal vertical migration of the larvae, (4) prevailing hydrographical features, and (5) life history and habitats requirements of the adults. Generally, larvae of the holopelagic Crustacea are found nearer the surface compared to the adults. In Suruga Bay, Japan, larvae released by the sergestid shrimp *Sergia lucens* in the inner most part of the bay are dispersed seaward, with the protozoeae and zoeae occurring in the surface layers (Omori, 1974). Larvae are retained within the bay by clock-wise circulation of water, influenced by the Kuroshio current and by the landward transport by tides and eddies caused by jet streams in estuaries. In the Scotia Sea, Antarctica, the pattern of distribution and abundance of larval stages of the euphausiid *Thysanoessa macrura* varies in different areas in relation to hydrographical conditions (Makarov, 1979). Larvae of latitudinally separated populations breeding at different times drift from one area to another by the mixing of water masses which produces different age groups within a plankton sample.

Larvae of some holopelagic Crustacea perform ontogenic vertical migration resulting in a change in the vertical distribution of larval stages (Mauchline and Fisher, 1969; Omori, 1974). Bathymetric distribution of the early larval stages of euphausiids (nauplius, metanauplius, and calyptopis) is influenced by the depth at which females lay eggs and the depth at which eggs hatch to release larvae (Mauchline and Fisher, 1969). When eggs are released in deep waters, early larval stages move up in the water column so that the furcilia stage occurs near the surface. The pattern of ontogenic vertical migration varies for different species depending on the distributional range of the adults (Einarsson, 1945). Coastal and shelf species living close to the bottom lay eggs there, and the larvae migrate to the surface layers and develop. Species with more oceanic distributions live as adults in surface layers and spawn there. Mesopelagic species spawn in that region, but the early stages move upward and develop in the surface layers and later juveniles move into the deeper waters. Bathypelagic species complete their entire life cycle in that region and exhibit no ontogenic vertical migration. Pelagic shrimp also exhibit ontogenic vertical migration, but it varies for different species inhabiting different regions (Omori, 1974). Epipelagic and upper mesopelagic shrimp lay their eggs in the euphotic zone and larvae hatch and develop there. Meso- and bathypelagic shrimp spawn and hatch eggs in the upper most regions of the vertical distributional range of females, but the eggs and larvae move upward so that larval development takes place within the euphotic zone or layers immediately below it. Juveniles and adults move later into deeper waters. Some bathypelagic species complete their entire life cycle in the same region without ever moving to the surface layers.

Larvae of holopelagic Crustacea may also exhibit horizontal ontogenic movements. On the Oregon coast, the pelagic shrimp *Sergestes similis* spawns in inshore areas during the winter, but the population shifts to offshore areas during the summer. This change in pattern of distribution and abundance from nearshore to offshore areas has been termed "horizontal ontogenic migration," due to active swimming with the aid of currents (Omori and Gluck, 1978).

Later larval stages of euphausiids perform diurnal vertical migration, but it varies for the larvae of different species in relation to the vertical range of adults (Mauchline and Fisher, 1969). Early larval stages of euphausiids (nauplius, metanauplius, and calyptopis) move continuously toward the surface without performing diurnal vertical migration, but the second calyptopis and all the later stages perform diurnal vertical migration (as the adults do), except for the limited vertical range. When larvae occur in surface layers and adults live in deeper waters, a gradual change in depth of maximum abundance occurs for the larval stages as they develop to juveniles and adults. Similarly, the eggs of the pelagic shrimp *Sergestes similis* hatch at depth and the resulting nauplii ascend to the surface layers and the protozoeal and zoeal stages remain in waters above 100 m (Omori and Gluck, 1978). Larvae migrate downward by performing diurnal vertical migration beginning in the first protozoeal stage, and it becomes more pronounced for each succeeding stage. Larvae gradually increase their depth distribution to inhabit the deep water as early juveniles and adults. Diurnal vertical migration of the later stages of holopelagic Crustacea, in addition to conferring several other advantages (Longhurst, 1976), also appears to serve as a mechanism for larval retention by the currents and countercurrents within the distributional range of the adults for subsequent restocking of the populations.

B. Benthic Crustacea

Adult benthic Crustacea live in a two-dimensional environment in the estuarine and coastal-shelf regions, and recruitment to these populations is from planktonic larvae in the pelagic environment. Except for this difference, associated with the changes in life-style from a pelagic larval to benthic juvenile and adult habitat, the basic hydrographical mechanisms influencing the dispersal and retention of larvae within the geographical area of the adults and the subsequent recruitment to the populations appear to be the same as for holopelagic Crustacea. However, the features of water circulation in the estuarine and coastal-shelf regions inhabited by the benthic Crustacea are more complex and vary regionally and geographically (Wooster and Reid, 1963; Bumpus, 1973; Wrytki, 1973). Hence, considerable differences in the patterns of larval dispersal, distribution, and abundance, and variations in the basic mechanisms of larval retention and recruitment, are found in different areas and in relation to the life history features of the species. Generally, restocking of benthic crustacean populations occurs from the larvae retained within the geographical range of the species, either by larval return and settlement, or by juvenile immigration to the adult habitat.

1. COASTAL-SHELF REGIONS

The pelagic larval life of many species usually lasts a few weeks, and larvae of many species may not be transported to far off distances. On the Western Kamchatka shelf, larval decapods are distributed in shallower depths, and their distribution is related to the spawning area of adults on the shelf and near shore currents (Makarov, 1969). Larvae increase in numbers as one proceeds from the coast to offshore, but further offshore the numbers decline. At the intermediate depths on the shelf, a zone along the coast contains an increased number of larvae of most decapods as a "larval belt" due to the influence of water currents along the coast and tidal currents. Rice and Williamson (1977) also found that larvae of coastal species are absent in offshore plankton samples collected along the same latitude between the Portuguese and Moroccan coast and Josephine and Meteor Seamounts. Larvae of the American lobster, *Homarus americanus* (Scarratt, 1964, 1973) and the Norwegian lobster *Nephrops norvegicus* in the Irish Sea (Hillis, 1968) are dispersed by the local hydrographical conditions, but they generally remain in hatching areas of adults without a long distance transport. A survey of brachyuran crab larvae between Cape Kennedy and Cape Hatteras on the east coast of North America showed that larvae of some coastal and estuarine species are found in plankton samples collected as far as the axis of the Gulf Stream 96.6 km from the coast, but larval numbers decrease with distance from the coast (Nichols and Keney, 1963). On the Oregon coast, larvae of *Cancer magister* are found 3–10 miles offshore within the range of adult distribution (Lough, 1976).

The pelagic larval life of the panulirid and scyllarid lobsters is particularly long (6–11 months), and larvae have been collected at great distances from the sites of origin, but they are also found in high densities in areas inhabited by the adults (Johnson, 1971, 1974; Phillips and Sastry, 1980). On the West coast of Australia, newly hatched stage I phyllosoma larvae of *Panulirus cygnus* are transported by the surface drift to the offshore portion of the continental shelf, and the relative abundance of stage II and III larvae increases with distance from the coast (Phillips et al., 1978, 1979). Phyllosoma larvae are retained within the continental shelf by hydrographical processes such as eddies and the direction of water movements at different depths. In the same on the west coast of Australia, larvae of *Scyllarus bicuspidatus*, released at the same time of the year as *P. cygnus*, are not transported as far as the larvae of the later species (Phillips et al., 1981). Thus, subtle differences in the length of larval life, vertical distribution, and behavior of larval stages could result in different patterns of dispersal for different species.

The temporal and spatial distributions of larvae can be influenced by the vertical distribution and diurnal vertical migration of the larval stages and the prevailing features of water circulation at different depths. Meroplanktonic larvae of some benthic crustaceans perform only limited diurnal vertical migration (e.g., *Homarus americanus*: Scarratt, 1973; *Panulirus interruptus*: Johnson, 1974), whereas others undergo diurnal vertical migration with light as an important controlling factor influencing the depth distribution (Fig. 9). On the West coast of Australia, the differential vertical distribution and diurnal vertical migration of phyllosoma larvae of *Panulirus cygnus* subjects the early larval stages to the offshore wind-driven transport and the mid- and late-stage larvae to circulation features underlying the immediate surface layers, which returns them to areas near the continental shelf edge (Phillip et al., 1978).

The primary influence of hydrographical processes on dispersal and subsequent return of larvae is indicated for benthic Crustacea inhabiting different geographical regions characterized by different features of water circulation. In hydrographically enclosed areas such as the Gulf of Maine, with a gyral circulation (Redfield, 1941; Iles and Sinclair, 1982), larvae retained

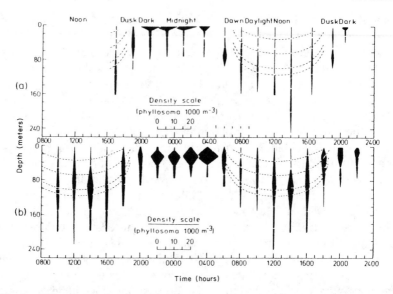

Fig. 9. Diurnal vertical migration of late stage (VI to IX) phyllosoma larvae of *Panulirus cygnus* (a) at new moon; (b) near full moon. Dashed lines represent isolumes for selected quantum values (μEm^{-2} sec^{-1}) during the day; horizontal lines across distribution figures indicate sampling depths. (From Rimmer and Phillips, 1979.)

within the water mass may subsequently settle on the bottom to restock the populations. In some regions, coastal water circulation includes unidirectional currents and countercurrent systems parallel to the shore, as well as offshore surface and onshore bottom drift (Bumpus, 1973). The surface currents may disperse early stage larvae, and subsurface currents may return the later stages for recruitment. On the Oregon coast, early larvae of *Cancer magister* released in February are retained in near shore areas by longshore and onshore currents and prevented from offshore transport by the northward flowing Davidson Current (Lough, 1976). By the transition period in March, when the Davidson Current is replaced by a current flowing south and southwest, larvae develop to late zoeae and megalops and occupy deeper layers in the water column. These later stages are transported toward the coast by bottom flow. The megalops settle and metamorphose by April or May before intense coastal upwelling in June and July, thus reducing the chances of being carried offshore by the resulting Ekman offshore transport.

The pattern of prevailing water currents is also a critical factor in the recruitment to the populations of panulirid lobsters in coastal areas of Southern California (Johnson, 1974). Recruitment of lobsters depends on the larvae being retained within the vicinity of the coast by eddies, current, and countercurrent systems. The settling puerulus stage develops within the vicinity of the coast and is not found in offshore plankton samples. On the west coast of Australia, the phyllosoma of *Panulirus cygnus* develops into the puerulus stage in areas near the continental shelf-edge, and completes larval development while swimming over the shelf with prevailing water currents to reach inshore reef areas for settlement (Phillips *et al.*, 1978).

A different pattern of dispersal and recruitment is seen for the spiny lobster *Panulirus argus* in the Caribbean and Florida regions of the North American coast. Two possible mechanisms have been suggested for the source of recruits for the western North Atlantic populations (Menzies and Kerrigan, 1978). One possibility is the open system of main currents in the Caribbean and the Atlantic Ocean, which could involve larval transport and settlement at any location throughout the distributional range of the species. The other possibility is the very large trans-Atlantic closed system of larval transport and recruitment in which larvae are retained in the region in which they are released by eddies. In inshore areas of South Florida, late phyllosoma are absent, and larvae which developed into the puerulus stage offshore migrate inshore (Sweat, 1968). Sims and Ingle (1967) suggested that recruitment of *P. argus* in South Florida is from other regions in the Caribbean. Richards and Goulet (1976) made computer calculations from surface drift data to determine the possible paths of larval dispersal and recruitment of *P. argus*. These calculations showed that a significant portion of the recruits in the Bahamas comes upstream by the current system, principally from the upper Antilles, while larvae from the stocks in the Bahamas are swept northward by currents, and a significant portion is lost in the North-Atlantic Drift.

2. ESTUARINE REGIONS

Benthic Crustacea may live in estuaries as larvae, adults, or both. Dispersal of larvae of estuarine dependent species is influenced by the water circulation features of the regions where they are released. Larvae released within the estuarine system by freshwater or semiterrestrial species or those completing their whole life cycle in the estuaries are subject to the water circulation features of the estuary. In an estuary with a two-layered circulation, the distances larvae are transported and the chances of larval retention within the estuary are related to the length of larval life, the vertical distribution and behavior of larval stages, the tolerances and requirements of the larvae, and net seaward flow. Larvae of species migrating to the mouths of estuaries for releasing the larvae are subject to the water circulation features characteristic of the adjacent coastal region. Hence, a variety of patterns of larval dispersal and mechanisms of recruitment are found for the estuarine-dependent species (Sandifer, 1973, 1975).

In the Chesapeake–York–Pamunsky estuarine system of the mid-Atlantic coast of North America, larval decapods have been collected in plankton samples in areas where adults are distributed along the estuarine gradient (Sandifer, 1973). Larvae of the fiddler crab *Uca* sp. are found upstream in areas bordering the salt marshes where adults live. Larvae of species spending their whole life cycle in the estuary are the most abundant component of meroplankton and are found throughout the estuary (e.g., several carideans and brachyuran crabs). Larvae of *Callinectes sapidus* (which lives its adult life in the estuary, but migrates to the mouth of the bay to release its larvae) are not found within the estuary. Larvae of *Cancer irroratus*, which breeds offshore, are found only in coastal waters, suggesting that these larvae are not entrained into the estuary.

Within the Chesapeake–York–Pumunsky system, the later larval stages of the species spending their whole life cycle within the estuary are found in bottom layers in the water column (Sandifer, 1973). Similarly, larvae of freshwater shrimp *Macrobrachium carcinus, M. rosenbergii,* and *M. acanthurus,* which require brackish water for development (Choudhury, 1970), also maintain a deeper position in the water column (Hughes and Richard, 1973). The deeper position in the water column subjects them to upstream bottom flow and appears to retain larvae within the estuaries, even though early larval stages occupying a shallower position in the water column may be transported seaward by the surface water flow.

Restocking of populations that spend their whole life cycle within the estuary is accomplished by retaining larvae within the estuarine system (Sandifer, 1975). In contrast, restocking of estuarine populations of species such as *Callinectes sapidus* occurs by immigration of juveniles or adults from the adjacent coastal regions.

The mechanisms of restocking estuarine populations by retention of larvae within the system reduces the possibilities for gene exchange with populations in other estuaries compared to populations recruited by juvenile or adult immigration. The relationships of larval dispersal and retention, and the mechanisms of restocking populations to gene exchange and genetic divergence is not fully understood for any species (Section IV, B). Information relevant to examine this question—larval dispersal (Nichols and Keney, 1963), retention and restocking of populations (Sandifer, 1975), water circulation features (Bumpus, 1973; Scheltema, 1975), allozyme genetics of populations (Gooch, 1977), and physiological variation (Schneider, 1968)—is available for *Rhithropanopeus harrisii,* but no definitive picture has yet emerged to say whether or not there is gene exchange between populations. Phenotypic variation is indicated by the allozyme genetics and physiological variation studies (Section IV, B), while isolation of populations is indicated by the studies on larval distribution and water circulation features.

C. Intertidal Crustacea

Larvae of intertidal Crustacea inhabiting the estuarine and coastal regions are dispersed by the water circulation features of the respective regions, but recruitment is generally from larvae retained within the distributional range of the adults. Larvae of *Emerita analoga* inhabiting the wash zone of sandy beaches on the west coast of North America, have a pelagic life of about 4 months, but they too are retained close to the coast by the surface and subsurface currents running along the coast and by eddies (Efford, 1970). Larvae return to the beach as megalopae and metamorphose into juveniles. At times, anomalies in the features of water circulation along the coast transport larvae to areas beyond the distributional range of the species and establish isolated fringe populations. Success in maintaining such populations is, however, influenced by other factors such as competition, predation, food availability, and reproductive success. Thus, larval dispersal by the prevailing water circulation features determines the distributional range of a species, occasionally expanding the species' geographical range.

Studies examining larval dispersal and retention of sessile barnacles have suggested two possible mechanisms: (1) passive transport and retention by hydrographical processes (DeWolf, 1973); and (2) vertical distribution and behavior in the water column relative to water circulation in the estuaries (Bousfield, 1955). Transport and retention of intertidal barnacle larvae in the Miramichi Estuary, Canada, is determined by the circulation in relation to the depth at which larvae swim. Nauplius larvae are generally found near the surface at high and low water and deeper during the mid-ebb and flood tides. Successive larval stages occur at increasing depths at all stages of the tidal cycle. Larval stages maintain different depths relative to each other by swimming during the tidal cycle. In this estuary, early stages are carried downstream by the surface flow and the later stages, occupying a deeper position, are transported upstream by the bottom flow. In the western Wadden Sea, The Netherlands (a well mixed estuary with strong turbulence associated with the high current velocities in tidal areas), barnacle larvae are retained by the physical processes of water circulation rather than by the swimming behavior induced by tide-coupled factors as in Miramichi Estuary (DeWolf, 1973).

The settlement behavior of barnacle cyprids has been particularly well studied (Crisp, 1974a, 1976a). Settlement involves attachment (testing the substrate), exploration, and cementing to the substratum. Studies have shown that cyprid larvae are at first photopositive and congregate near the sea surface, but later become photonegative and swim down to seek a suitable substrate for settlement. After settling on a surface, the cyprids

become indifferent to light and explore the surface until a suitable substratum is found. Settlement is induced by the presence of already settled
individuals of the same species (gregariousness) and, to a lesser degree, by
closely related species (Knight-Jones, 1953). The gregariousness is due to a
factor (arthropodin) present in the bodies of the individuals of the same or
other species of barnacles (Crisp and Medows, 1962; Crisp, 1965). Gabbott
and Larman (1971), from an electrophoretic examination of partially purified extracts of *Semibalanus balanoides,* found that arthropodin is a
glycoprotein or a mucopolysaccharide complex.

D. Semiterrestrial and Terrestrial Crustacea

Larval dispersal, retention, and restocking of the populations of semiterrestrial and terrestrial decapods is also influenced by the water circulation
features of the adjacent estuarine and coastal regions. Larvae of *Uca* sp. and
Sesarma reticulatum are hatched in low salinity waters and transported
down stream in estuaries, while the later stages concentrate near the bottom
and are returned upstream with the moving water layer as megalops or
young crabs to settle in the marsh habitat of adult populations (Sandifer,
1975). The mangrove crab *Aratus pisonii* living in the deeper mangrove
swamps migrates to adjacent waters at the end of egg incubation period to
release larvae (Warner, 1967). At the end of the pelagic phase, which may
last up to a month, the young crabs settle into the swamps. Females of the
land crab *Gecarcinus lateralis* migrate down to the beach and release larvae
in the surf zone (Bliss, 1979). Larvae dispersed in the coastal waters return at
the end of the pelagic phase to emerge on the beach as megalopae or young
crabs and later migrate back to the land. Two species of grapsid crabs
Sesarma jarvisi and S. cookei exhibit unique adaptations to terrestrial life on
the mountain slopes of Jamaica (Abele and Means, 1977). Their entire life
cycle is completed on land, with the larvae probably hatching and developing in rock rubble during periods of extensive rain fall.

VI. THE ROLE OF PELAGIC LARVAE IN LIFE HISTORIES

Our knowledge of the selection pressures in the evolution of life history
patterns and of the component reproductive traits of Crustacea inhabiting
different environments and climatic regions are speculative (Sastry, 1983).
Crustaceans develop with a pelagic larva at one extreme, direct development at the other, and in between with an early pelagic lecithotrophic larva
which begins feeding at an advanced stage in development. Peracarida as a
group develop directly in all the habitats, environments, and climatic re-

gions in which they are found. Other crustaceans with a planktotrophic pelagic larva also occupy different habitats, environments, and climatic regions, but there are certain general trends among these species relating the pattern of development to the environment and climatic region. The life history features of crustaceans with a pelagic larva—life span, adult body size, age at first maturity, feeding type and food preferences, and semelparity or iteroparity—also vary for the different species. Thus, phylogenetic relationships as well as selection pressures, which may include resource availability, environmental constancy and/or predictability, habitat stability, and competition and predation on different phases of the life cycle, may influence the pattern of development. The combinations of selection pressures may differ between species, environments, and climatic regions.

Planktotrophic larval development is the most predominant pattern of development for a great majority of crustaceans in the tropical and temperate regions. In the polar regions, only a few species of intertidal barnacles in the arctic and benthic decapods (e.g., hippolytid prawn *Chorismus antarcticus,* crangonid shrimp *Notocrangon antarcticus:* Makarov, 1973; Maxwell, 1977) in the antarctic have a pelagic larva, whereas planktotrophic larval stages are common in the life-cycles of holopelagic species.

Gradients are also seen in the pattern of development of holopelagic Crustacea inhabiting the epipelagic to bathypelagic regions (Mauchline and Fisher, 1969; Omori, 1974; Marshall, 1979). The copepods, euphausiids, and penaeid and caridean shrimp in the epipelagic region develop with a pelagic larva. In the lower meso- and bathypelagic regions, the representatives of these groups may develop with lecithotrophic early pelagic stages which switch over to planktotrophic mode at an advanced stage of larval development. Some of the species completing their whole life cycle in these regions release larvae at a large size and advanced stage, which quickly adapt to the adult mode of feeding. Larvae of other species undergo a continuous ontogenic vertical migration to the surface layers and begin feeding, later returning to the depths which adults inhabit (Section V). Generally, species inhabiting the variable environment of the epipelagic zone produce a larger number of small eggs compared to those species inhabiting the relatively food poor and constant environment of the lower meso- and bathypelagic regions.

Variation in the pattern of development is also found among decapods inhabiting freshwater and terrestrial environments (Bliss, 1979). Species returning to the marine or brackish water have a pelagic larva, while freshwater and terrestrial species seeking fresh water develop directly. However, smaller size copepods, ostracodes, and conchostracans inhabiting freshwater ponds, lakes, and slow flowing rivers have a pelagic larva (Kaestner, 1970).

The pelagic larval stage is considered advantageous in the life cycles of benthic species (Thorson, 1950; Mileikovsky, 1971) for (1) gene exchange between populations to maintain genetic variability and adaptability (Crisp, 1974a, 1976b; Scheltema, 1975), (2) colonization of areas remote from the main population, (3) overcoming the possibilities of extinction from unpredictable habitat distrubances within the distributional range of a species (Crisp, 1974b), (4) competition for space (Hines, 1979), and (5) ensuring survival with high juvenile mortalities from predation in the adult habitat (Thorson, 1950). The pelagic larval stage is also disadvantageous since (1) larval mortalities from planktonic predation are generally high, (2) larvae may be transported to unfavorable areas by water currents, (3) food conditions may unpredictably vary in time and space, and (4) juvenile recruitment may be uncertain (Thorson, 1950). On balance, however, pelagic larvae must be considered advantageous to the survival of a species, since populations of species with this pattern of development continue to be represented in an area from one year to another, and overwhelming numbers of benthic marine invertebrates have retained pelagic larvae in their life cycles. Pelagic larval development appears to be a selective compromise between the risks of larval survival and the advantages of dispersal in the pelagic environment as a component in the overall life history strategy of a species.

A few theoretical models have been proposed to explain the patterns of development in benthic marine invertebrates by making simplified assumptions necessary for mathematical analysis. These models have attempted to assess the relative advantage of planktotrophic pelagic larva compared to lecithotrophic (pelagic) larva and/or direct development based on (1) energy considerations (Vance, 1973a,b; Christiansen and Fenchel, 1979), (2) genetic advantage to the spread of sibling larvae (Strathmann, 1974), and (3) advantages of dispersal (Crisp, 1976b). Models incorporating both energy considerations and dispersal advantage, as well as other life history features such as habitat, environment, and climatic conditions, have yet to be developed to evaluate the selection pressures in evolution of a particular pattern of development and its adaptive value to the survival of a species.

Vance (1973a) compared planktotrophic and direct development, and intermediate patterns, based on mortality during the larval phase and specifications of fecundity of adults as a function of egg size and development type, assuming fixed reproductive effort. These models predicted that planktotrophic larval development is more efficient when planktonic predation is high and planktonic food is abundant, and lecithotrophic development is more efficient when planktonic food is rare and predation is low. Direct development is more efficient when planktonic predation is high and/ or development times are long. Mortalities in the larval phase are generally considered high, but no quantitative data on stage-specific mortality rates are available for any species. Christiansen and Fenchel (1979) constructed a model of the evolution of egg size and predicted that size-dependent mortality (as planktonic larvae) leads to an evolutionary stable minimum offspring size for a given fixed amount of reproductive effort.

Comparative data on the reproductive effort of invertebrates with different patterns of development is very limited. Available data are conflicting (Menge, 1975; Grahame, 1977). The only study on crustaceans, by Clarke (1979), compared the reproductive effort (measured as grams wet weight of eggs per gram wet weight of females) of polar species *Notocrangon antarcticus* and *Chorismus antarcticus,* which produce relatively few larger eggs, with that of temperate species *Pandalus montagui* and *Crangon crangon,* which produce large numbers of smaller size eggs. The reproductive effort of polar species is significantly less compared to that of temperate species. Reproductive effort has been suggested as a basis for *r*- and *K*-selection (Pianka, 1970; Stearns, 1976, 1977). Christiansen and Fenchel (1979) indicate that *r*- and *K*-selection may not be a primary determinant of the developmental type and the number of eggs produced per female. Instead, they suggest that for a given reproductive effort, the conflicting forces of the amount of energy devoted to the individual offspring and the chances of survival of offspring are important considerations for making comparisons of reproductive traits, and the points of balance between these selection pressures determine the evolutionarily stable offspring size.

Underwood (1974) pointed out that energy considerations alone may be inadequate to explain the different patterns of development at various latitudes and depths because of the obvious evolutionary importance of dispersal. Crisp (1976b) developed a computer model to assess the advantages of different patterns of development and dissemination and concluded that larval dispersal is advantageous to the survival of a species for colonization of varied environments, even if there is no trophic advantage to the species as a whole in having a pelagic larva.

It is unclear whether reproductive effort and size of eggs are species constants. It is not known whether selection pressures in subdivision of reproductive effort of a species (to produce a large number of small eggs or a small number of large eggs) are influenced by the food availability, the habitat and environment (including competition and predation) of the adults, and/or the survival of the offspring and the dispersal advantages. The optimal life history strategy of a species is a selective compromise of a combination of abiotic and biotic factors in the environment to maximize the probabilities for genetic representation in future generations (Sastry, 1983). Thus, selection pressures on the different phases in the life-cycle may have to be understood to gain insight into the adaptive value of pelagic larvae and their retention in the life cycle of species inhabiting different environments and climatic regions.

The selection of a developmental pattern with a pelagic larva in the life cycle that can utilize the planktonic food sources and disperse in the three-dimensional pelagic environment may have allowed these species to occupy a range of habitats over a wide geographical area and to maintain genetic variability. This life history strategy appears opportunistic for the utilization of planktonic food resources and for colonization of varied environments to ensure species survival. One consequence of the developmental pattern with a planktotrophic larva is the fluctuation in population size of a species caused by unpredictable changes in the pelagic environment. The short-term local fluctuations in population size in ecological time appear to have been compensated by the long-term advantage of species continuity in geologic time.

VII. CONCLUSIONS

A considerable body of information has accumulated over the years on pelagic larvae of Crustacea from studies conducted both in the laboratory and field, but a number of important questions concerning larval development and ecology remain unanswered. Available information on the processes involved in regulating larval development through an ordered sequence of stages under given environmental conditions is fragmentary. Studies on molecular, cellular, and metabolic processes, on the development of organ systems and sensory systems during growth and differentiation, and on the influence of neuroendocrine factors in coordination of the developmental changes can contribute toward a better understanding of the process of larval development. Information of this nature may also allow us to better understand the inter- and intraspecific variation, and the mechanisms by which developing larvae perceive and adapt to a variable pelagic environment by appropriate physiological and behavioral responses.

The technique of artificial cultivation of larvae under controlled conditions from hatching to the postlarval stage has allowed investigators to describe the morphological features of larval stages of a species, to determine the effects of environmental factors and establish the conditions favorable for producing maximal survival rates, to establish the duration and limits for larval development, and to conduct studies on physiological and behavioral responses of larvae to environmental variables. Much of our knowledge on pelagic larvae of Crustacea is limited, however, to species inhabiting the coastal and estuarine environments of the temperature regions. The application of artificial cultivation techniques to larvae of holopelagic, benthic, and intertidal species inhabiting different environments and climatic regions would provide us with comparative data to evaluate the possible ecological relationships and to examine the questions of evolutionary significance.

The development of artificial cultivation techniques has also permitted the use of larval forms as bioassay organisms to assess the effects of pollutants and to produce juvenile stocks of some commercially important species for aquaculture purposes.

The field approach of collection of larvae in plankton samples has provided some descriptive information on the distribution and abundance of larvae of some coastal and estuarine species over limited geographical areas. However, studies describing the temporal and spatial distribution and abundance of larvae, employing quantitative time-series sampling techniques over large enough areas with sufficiently detailed descriptions of hydrographical features and other environmental variables in the pelagic environment, are very few. Quantitative ecological studies on larvae can help answer many questions concerning gene exchange and the maintenance of genetic variability between populations, the mechanisms of larval dispersal, retention, and restocking of populations, the year-to-year fluctuations in population size of species, and the structure and diversity of communities. The selection pressures involved in development of a pelagic larva in the life cycle of species inhabiting different habitats, environments, and climatic regions remain to be fully understood.

ACKNOWLEDGEMENTS

My graduate student, Jeffrey G. Jones, particularly deserves thanks for his assistance in the final preparation of this manuscript.

REFERENCES

Abele, L. G., and Means, D. B. (1977). *Sesarma jarvisi* and *Sesarma cokeii* Montane, Terrestrial Grapsid crabs in Jamaica (Decapoda). *Crustaceana* **32,** 91–93.

Achituv, Y., and Barnes, H. (1976). Studies in the biochemistry of cirripede eggs. V. Changes in the general biochemical composition during development of *Chthamalus stellatus* (Poli). *J. Exp. Mar. Biol. Ecol.* **22,** 263–267.

Atkins, D. (1955). The post-embryonic development of British *Pinnotheres. Proc. Zool. Soc. London.* **124,** 687–715.

Bainbridge, R. (1961). Migrations. *Physiol. Crustacea* **2,** 431–463.

Barlow, J. (1969). Studies on molecular polymorphisms in the American lobster, (*Homarus americanus*). Ph.D Diss., 181 pp. Univ. of Maine, Orono.

Barnes, H. (1953). The effect of lowered salinity on some barnacle nauplii. *J. Anim. Ecol.* **22,** 328–330.

Barnes, H. (1965). Studies in the biochemistry of cirrapede eggs. I. Changes in the general biochemical composition during development of *Balanus balanoides* and *B. balanus*. *J. Mar. Biol. Assoc. U. K.* **45,** 321–339.

Barnes, H., and Barnes, M. (1959). The effect of temperature on the oxygen uptake and rate of development of egg masses of two common cirripedes, *Balanus balanoides* (L.) and *Pollicipes polymerus*. J. B. Sowerby. *Kiel. Meersfrosch.* **15,** 242–251.

Barnes, H., and Barnes, M. (1974). The response during development of the embryos of some common cirrepedes to wide changes in salinity. *J. Exp. Mar. Biol. Ecol.* **15,** 197–202.

Barnes, H., and Barnes, M. (1976). The rate of development of the embryos of *Balanus balanoides* (L.) from a number of European and American populations and the designation of local races. *J. Exp. Mar. Biol. Ecol.* **24,** 251–269.

Barnes, H., and Blackstock, J. (1975). Studies in the biochemistry of cirrepede eggs. IV. The free amino acid pool in the eggs of *Balanus balanoides* (L.) and *B. balanus* (L.) during development. *J. Exp. Mar. Biol. Ecol.* **19,** 57–79.

Barnes, H., and Evans, R. (1967). Studies in the biochemistry of cirripede eggs. III. Changes in the amino acid composition during development of *Balanus balanoides* and *B. balanus*. *J. Mar. Biol. Assoc. U. K.* **47,** 171–180.

Barnes, H., and Klepal, W. (1972). Phototaxis in stage I nauplius larvae of two cirripedes. *J. Exp. Mar. Biol. Ecol.* **10,** 267–273.

Batham, E. J. (1967). The first three larval stages and the feeding behavior of phyllosoma of the New Zealand palinurid crayfish, *Jasus edwardsii* (Hutton, 1875). *Trans. R. Soc. N. Z., Zool.* **9,** 53–64.

Bělehrádek, J. (1935). Temperature and living matter. *Protoplasma Monogr.* **8,** 1–277.

Bělehrádek, J. (1957). Physiological aspects of heat and cold. *Annu. Rev. Physiol.* **19,** 59–82.

Belman, B. W., and Childress, J. J. (1973). Oxygen consumption of the larvae of the lobster *Panulirus interruptus* (Randall) and the crab *Cancer productus* (Randall). *Comp. Biochem. Physiol.* **44A,** 821–828.

Bently, E., and Sulkin, S. D. (1977). The ontogeny of barokinesis during the zoeal development of the xanthid crab *Rhithropanopeus harrisii* (Gould). *Mar. Behav. Physiol.* **4,** 275–282.

Bernard, M. (1971). La forme elliptique de la relation temperature-durée de development embryonnaire chez les copépodes pelagiques et ses proprietes. *Eur. Symp. Mar. Biol., Proc., 4th* pp. 203–209.

Bigford, T. E. (1977). Effects of oil on behavioral responses to light, pressure and gravity in larvae of the rock crab *Cancer irroratus*. *Mar. Biol. (Berlin)* **43,** 137–148.

Bigford, T. E. (1978). Effect of several diets on survival, development time, and growth of laboratory reared spider crab, *Libinia emarginata*, larvae. *Fish. Bull.* **76,** 59–64.

Bigford, T. E. (1979). Ontogeny of light and gravity responses in rock crab larvae (*Cancer irroratus*). *Mar. Biol. (Berlin)* **52,** 69–76.

Bliss, D. E. (1979). From sea to tree: Saga of a land crab. *Am. Zool.* **19,** 385–410.

Bookhout, C. G., and Monroe, R. J. (1977). Effects of malathion on the development of crabs. *In* "Physiological Responses of Marine Biota to Pollutants" (F. J. Vernberg, A. Calabrese, F. P. Thurberg, and W. B. Vernberg, eds.), pp. 3–19. Academic Press, New York.

Bookhout, C. G., Costlow, J. D., and Monroe, R. (1976). Effects of methoxychlor on larval development of mud crab and blue crab. *Water, Air, Soil Pollut.* **5,** 349–365.

Borch, E. S. (1960). Endocrine control of the chromatophores of the zoeae of the prawn, *Palaemonetes vulgaris*. *Biol. Bull. (Woods Hole, Mass.)* **119,** 305.

Bousfield, E. L. (1955). Ecological control of the occurrence of barnacles in the Miramichi estuary. *Bull. Natl. Mus. Can.* **137,** 1–69.

Branford, J. R. (1978). The influence of day-length, temperature, and season on the hatching rhythm of *Homarus gammarus. J. Mar. Biol. Assoc. U. K.* **58,** 639–658.

Broad, A. C. (1957a). Larval development of *Palaemonetes pugio* Holthius. *Biol. Bull. (Woods Hole, Mass.)* **112,** 144–161.

Broad, A. C. (1957b). The relationship between diet and larval development of *Palaemonetes pugio. Biol. Bull. (Woods Hole, Mass.)* **112,** 162–170.

Bumpus, D. F. (1973). A description of the circulation on the continental shelf off east coast of the United States. *Prog. Oceanogr.* **6,** 111–157.

Burton, R. S. (1979). Depth regulatory behavior of the first zoea larvae of the sand crab *Emerita analoga* Stimpson (Decapods: Hippidae) *J. Exp. Mar. Biol. Ecol.* **37,** 255–270.

Capuzzo, J. M., and Lancaster, B. A. (1979). Some physiological and biochemical considerations of larval development in the American lobster, *Homarus americanus* Milne-Edwards. *J. Exp. Mar. Biol. Ecol.* **40,** 53–62.

Chamberlain, N. A. (1962). Ecological studies of the larval development of *Rhithropanopeus harrisii* (Xanthidae, Brachyura). *Chesapeake Bay Inst., Johns Hopkins Univ., Tech. Rep.* **28,** 1–47.

Choudhury, P. C. (1970). Complete larval development of the palaemonid shrimp *Macrobrachium acanthurus* (Weigmann, 1836), reared in the laboratory. *Crustaceana* **18,** 113–132.

Christiansen, F. B., and Fenchel, T. M. (1979). Evolution of marine invertebrate reproductive patterns. *Theor. Pop. Biol.* **16,** 267–282.

Christiansen, M. E., Costlow, J. D., and Monroe, R. (1977). Effects of the juvenile hormone mimic ZR-512 (Altozar) on larval development of the mud-crab *Rhithropanopeus harrisii* in various salinities at cyclic temperatures. *Mar. Biol. (Berlin)* **39,** 281–288.

Clarke, A. (1979). On living in cold water: K-strategies in Antarctic benthos. *Mar. Biol. (Berlin)* **55,** 111–119.

Cook, H. L., and Murphy, M. A. (1969). The culture of larval penaeid shrimp. *Trans. Am. Fish. Soc.* **98,** 751–754.

Corkett, C. J., and McLaren, I. A. (1978). The biology of *Pseudocalanus. Adv. Mar. Biol.* **15,** 1–231.

Costlow, J. D. (1961). Fluctuations in hormone activity in Brachyuran larvae. *Nature (London)* **192,** 183–184.

Costlow, J. D. (1963). The effect of eyestalk extirpation on metamorphosis of megalops of the blue crab, *Callinectes sapidus* Rathbun. *Gen. Comp. Endocrinol.* **3,** 120–130.

Costlow, J. D. (1966). The effect of eyestalk extirpation on larval development of the crab, *Sesarma reticulatum* Say. *In* "Some Contemporary Studies in Marine Science" (H. Barnes, ed.), pp. 209–224. Allen and Unwin, London.

Costlow, J. D. (1968). Metamorphosis in crustaceans. *In* "Metamorphosis—A Problem in Developmental Biology" (W. Etkin and L. I. Gilbert, eds.), pp. 3–41. Appleton, New York.

Costlow, J. D., and Bookhout, C. G. (1959). The larval development of *Callinectes sapidus* Rathbun, reared in the laboratory. *Biol. Bull. (Woods Hole, Mass.)* **116,** 373–396.

Costlow, J. D., and Bookhout, C. G. (1960a). A method for developing brachyuran eggs *in vitro. Limnol. Oceanogr.* **5,** 212–215.

Costlow, J. D., and Bookhout, C. G. (1960b). The complete larval development of *Sesarma cinereum* Bosc reared in the laboratory. *Biol. Bull. (Woods Hole, Mass.)* **118,** 203–214.

Costlow, J. D., and Bookhout, C. G. (1964). An approach to the ecology of marine invertebrate larvae. *Symp. Exp. Mar. Ecol.* No.2, pp. 69–75.

Costlow, J. D., and Bookhout, C. G. (1965). The effect of environmental factors on larval development of crabs. *In* "Biological Problems in Water Pollution" (C. M. Tarzwell, ed.), U.S. Dept. of Health, Educ. and Welfare, Cincinnati, Ohio. (Publ. No. 999-WP-25.)

Costlow, J. D., and Bookhout, C. G. (1971). The effect of cyclic temperatures on larval development in the mud-crab *Rhithropanopeus harrisii*. *Eur. Symp. Mar. Biol., Proc. 4th* pp. 211–220.

Costlow, J. D., and Sastry, A. N. (1966). Free amino acids in developing stages of two crabs, *Callinectes sapidius* Rathbun and *Rhithropanopeus harrisii* (Gould). *Acat. Embryol. Morphol. Exp.* **9,** 44–55.

Costlow, J. D., Bookhout, C. G., and Monroe, R. (1960). The effect of salinity and temperature on larval development of *Sesarma cinereum* (Bosc) reared in the laboratory. *Biol. Bull. (Woods Hole, Mass.)* **118,** 183–202.

Crisp, D. J. (1956). A substance promoting hatching and liberation of young in cirripedes. *Nature (London)* **178,** 263.

Crisp, D. J. (1964). Racial differences between North American and European forms of *Balanus balanoides*. *J. Mar. Biol. Assoc. U. K.* **44,** 33–45.

Crisp, D. J. (1965). Surface chemistry, a factor in the settlement of marine invertebrate larvae, *Botanica Gothoburgensia. Proc. 5th Mar. Biol. Symp.*, Goteborg, pp. 51–65.

Crisp, D. J. (1969). Studies of barnacle hatching substance. *Comp. Biochem. Physiol.* **30,** 1037–1048.

Crisp, D. J. (1974a). Factors influencing the settlement of marine invertebrate larvae. *In* "Chemoreception in Marine Organisms" (P. T. Grant and A. M. Mackie, eds.), pp. 177–265. Academic Press, New York.

Crisp, D. J. (1974b). Energy relations of marine invertebrate larvae. *Thalassia Jugosl.* **10,** 103–120.

Crisp, D. J. (1976a). Settlement responses in marine organisms. *In* "Adaptation to Environment: Essays on the Physiology of Marine Animals" (R. C. Newell, ed.), pp. 83–124. Butterworths, London.

Crisp, D. J. (1976b). The role of the pelagic larva. *In* "Perspectives in Experimental Biology. I. Zoology" (P. S. Davies, ed.), pp. 145–155. Pergamon Press, New York.

Crisp, J. J., and Costlow, J. D. (1963). The tolerance of developing cirrepede embryos to salinity and temperature. *Oikos* **14,** 22–34.

Crisp, D. J., and Medows, P. S. (1962). The chemical basis of gregariousness in cirripedes. *Proc. R. Soc. London, Ser. B* **156,** 500–520.

Crisp, D. J., and Ritz, D. A. (1967). Changes in temperature tolerance of *Balanus balanoides* during its life-cycle. *Helgol. Wiss. Meersunters.* **15,** 98–115.

Crisp, D. J., and Ritz, D. A. (1973). Response of cirripede larvae to light. Experiments with white light. *Mar. Biol. (Berlin)* **23,** 327–335.

Crisp, D. J., and Spencer, C. P. (1959). The control of hatching process in barnacles. *Proc. R. Soc. London, Ser. B* **148,** 278–299.

Cronin, T. W., and Forward, R. B. (1979). Tidal vertical migration: An endogenous rhythmn in estuarine crab larvae. *Science* **205,** 1020–1022.

Davis, C. C. (1968). Mechanisms of hatching in aquatic invertebrate eggs. *Oceanogr. Mar. Biol.* **6,** 325–376.

Dawirs, R. R. (1981). Elemental composition (C, N, H) and energy in the development of *Pagurus bernhardus* (Decapoda: Paguridae) megalopa. *Mar. Biol. (Berlin)* **64,** 117–123.

Dawson, R. M. C., and Barnes, H. (1966). Studies in the biochemistry of cirripede eggs. II. Changes in lipid composition during development of *Balanus balanoides* and *B. balanus*. *J. Mar. Biol. Assoc. U. K.* **46,** 249–261.

DeCoursey, P. J. (1976). Vertical migration of larval *Uca* in a shallow estuary. *Am. Zool.* **16,** 224.

DeCoursey, P. J. (1979). Egg-hatching rhythmns in three species of fiddler crabs. *In* "Cyclic

Phenomenon in Marine Plants and Animals" (E. Nylor and R. G. Hartnoll, eds.), pp. 399–406. Pergamon, Oxford.

DeWolf, P. (1973). Ecological observations on the mechanisms of dispersal of barnacle larvae during planktonic life and settling. *Neth. J. Sea Res.* **6,** 1–129.

Edmondson, C. H., and Ingram, W. H. (1939). Fouling organisms in Hawaii. *Occas. Pap. Bernice Pawahi Bishop Mus.* **14,** 251–300.

Efford, I. E. (1970). Recruitment to sedentary marine populations as exemplified by the sand crab, *Emerita analoga* (Decapoda, Hippidae). *Crustaceana* **18,** 293–308.

Einarsson, H. (1945). Euphausiacea. I. North Atlantic species. *Dana Rep.* **27,** 1–185.

Elofsson, R. (1969). The development of the compound eye of *Penaeus duorarum* (Crustacea Decapods) with remarks on the nervous system. *Z. Zellforsch. Mikrosk. Anat.* **97,** 323–350.

Ennis, G. P. (1973a). Endogenous rhythmicity associated with larval hatching in the lobster *Homarus gammarus. J. Mar. Biol. Assoc. U. K.* **53,** 531–538.

Ennis, G. P. (1973b). Behavioral responses to changes in hydrostatic pressure and light during larval development of the lobster *Homarus gammarus. J. Fish. Res. Board Can.* **30,** 1349–1360.

Ennis, G. P. (1975a). Observations on hatching and larval release in the lobster *Homarus americanus. J. Fish. Res. Board. Can.* **32,** 2210–2213.

Ennis, G. P. (1975b). Behavioral responses to changes in hydrostatic pressure and light during larval development of the lobster *Homarus americanus. J. Fish. Res. Board Can.* **32,** 271–281.

Ewald, E. F. (1912). On artificial modification of light reactions and the influence of electrolytes on phototaxis. *J. Exp. Zool.* **13,** 591–612.

Factor, J. R. (1981). Development and metamorphosis of the digestive system of larval lobsters, *Homarus americanus* (Decapoda: Nephropidae). *J. Morphol.* **169,** 225–242.

Forward, R. B. (1974). Negative phototaxis in crustacean larvae: possible functional significance. *J. Exp. Mar. Biol. Ecol.* **16,** 11–17.

Forward, R. B. (1976a). Light and diurnal vertical migration: photobehavior and photophysiology of plankton. *In* Photochem. Photobiol. Rev. **1,** 157–209. Plenum, New York.

Forward, R. B. (1976b). A shadow response in a larval crustacean. *Biol. Bull.* (*Woods Hole, Mass.*) **151,** 126–140.

Forward, R. B. (1977). Occurrence of shadow response among brachyuran crab larvae. *Mar. Biol.* (*Berlin*) **39,** 331–341.

Forward, R. B., and Costlow, J. D. (1974). The ontogeny of phototaxis by larvae of the crab *Rhithropanopeus harrisii. Mar. Biol.* (*Berlin*) **26,** 27–33.

Forward, R. B., and Cronin, T. W. (1978). Crustacean larval phototaxis: Possible functional significance. *In* "Physiology and Behavior of Marine Organisms" (D. S. McLusky and A. J. Berry eds.), pp. 253–261. Pergamon, Oxford.

Forward, R. B., and Cronin, T. W. (1979). Spectral sensitivity of larvae from intertidal crustaceans. *J. Comp. Physiol.* **133,** 311–315.

Foskett, J. K. (1977). Osmoregulation in the larvae and adults of the grapsid crab *Sesarma reticulatum* Say. *Biol. Bull.* Woods Hole, Mass.) **153,** 505–526.

Foster, B. A. (1970). Responses and acclimation to salinity of some balanomorph barnacles. *Philos. Trans. R. Soc. London, Ser. B.* **256,** 377–400.

Foxon, G. E. H. (1934). Notes on the swimming methods and habits of certain crustacean larvae. *J. Mar. Biol. Assoc. U. K.* **19,** 829–849.

Frank, J. R., Sulkin, S. D., and Morgan, R. P. (1975). Biochemical changes during larval development of Xanthid crab *Rhithropanopeus harrisii.* I. Protein, total lipid, alkaline phosphatase, and glutamic oxaloacetic transaminase. *Mar. Biol.* (*Berlin*) **32,** 105–111.

Fraenkel, G. (1931). Die merchanik der Orientienung der Tiere im Raum. *Biol. Rev.* **6,** 36–87.

274 A. N. Sastry

Gabbott, P. A., and Larman, N. N. (1971). Electrophoretic examination of parially purified extracts of *Balanus balanoides* containing a settlement inducing factor. *Eur. Symp. Mar. Biol., Proc. 4th* pp. 143–153.

Gauld, D. T. (1959). Swimming and feeding in crustacea larvae: The nauplius larva. *Proc. Zool. Soc. London* **132**, 31–50.

Gooch, J. L. (1977). Allozyme genetics of life cycle stages of brachyurans. *Chesapeake Sci.* **18**, 284–289.

Crahame, J. (1977). Reproductive effort and r- and K-selection in two species of *Lacuna* (Gastropoda: Prosobranchia). *Mar. Biol. (Berlin)* **40**, 217–224.

Green, J. (1965). Chemical Embryology of the crustacea. *Biol. Rev.* **40**, 580–600.

Gurney, R. (1942). Larvae of Decapod Crustacea. Ray Society, (London.) Publ. 129, 306 pp.

Hart, R. C., and McLaren, I. A. (1978). Temperature acclimation and other influences on embryonic duration in the copepod *Pseudocalanus* sp. *Mar. Biol. (Berlin)* **45**, 23–30.

Hazel, J. R., and Prosser, C. L. (1974). Molecular mechanisms of temperature compensation in pokilotherms. *Physiol. Rev.* **54**, 620–677.

Herring, P. J., and Morris, R. J. (1975). Embryonic metabolism of carotenoid pigments and lipid in species of *Acanthephyra* (Crustacea: Decapods). *Eur. Symp., Mar. Biol. Proc. 9th* pp. 299–310.

Herrnkind, W. F. (1968). The breeding of *Uca pugilator* and mass rearing of the larvae with comments on the behavior of the larval and early crab stages. *Crustaceana Suppl.* **2**, 214–224.

Hillis, J. P. (1968). Larval distribution of *Nephrops norvegicus* (L.) in the Irish Sea and North Channel. *ICES C. M. K.* **6**, 1–5.

Hines, A. (1979). The comparative reproductive ecology of three species of intertidal barnacles. *Belle W. Baruch Lib. Mar. Sci.* No. 9, pp. 213–234.

Holland, D. L. (1978). Lipid reserves and energy metabolism in the larvae of benthic marine invertebrates. *Biochem. Biophys. Perspect. Mar. Biol.* **4**, 85–123.

Holland, D. L., and Walker, G. (1975). The biochemical composition of the cypris larva of the barnacle *Balanus balanoides* L. *J. Cons., Cons. Int. Explor. Mer.* **36**, 162–165.

Hubschman, J. H. (1963). Development and function of neurosecretory sites in the eyestalks of larval *Palaemonetes* (Decapoda; Natantia). *Biol. Bull. (Woods Hole, Mass.)* **125**, 96–113.

Hughes, D. A., and Richard, J. D. (1973). Some current-directed movements of *Macrobrachium acanthurus* (Weigmann, 1836) (Decapoda, Palaemonidae) under laboratory conditions. *Ecology* **54**, 927–929.

Iles, T. D., and Sinclair, M. (1982). Atlantic herring: Stock discreteness and abundance. *Science* **215**, 627–633.

Jerlov, N. G. (1976). "Marine Optics." Elsevier, Amsterdam.

Johns, D. M., and Pechenick, J. A. (1980). Influence of the water-accomodated fraction of No.2 fuel oil on energetics of *Cancer irroratus* larvae. *Mar. Biol. (Berlin)* **55**, 247–254.

Johnson, M. W. (1971). The palinurid and scyllarid lobster larvae of the tropical eastern Pacific and their distribution as related to the prevailing hydrography. *Bull. Scripps Inst. Oceanogr.* **19**, 1–36.

Johnson, M. W. (1974). On the dispersal of lobster larvae into the east Pacific barrier (Dacapods, Palinuridae). *Fish. Bull.* **72**, 639–647.

Johnson, M. W., and Brinton, E. (1963). Biological species, water masses, and currents. In "The Seas." II. The Composition of Sea Water, Comparative and Descriptive Oceanography" (M. N. Hill, ed.), pp. 381–412. Wiley (Interscience), New York.

Jones, D. A., Kanazawa, A., and Ono, K. (1979). Studies on the nutritional requirements of the larval stages of *Penaeus japonicus* using micro-encapsulated diets. *Mar. Biol. (Berlin)* **54**, 261–267.

Kaestner, A. (1970). Invertebrate Zoology. *Crustacea* **3**, 1–523.

Kalber, F. A. (1970). Osmoregulation in decapod larvae as a consideration in culture techniques. *Helgol. Wiss. Meeresunters.* **20**, 697–706.

Kalber, F. A., and Costlow, J. D. (1966). The ontogeny of osmoregulation and its neurosecretory control in the decapod crustacean, *Rhithropanopeus harrisii* (Gould). *Am. Zool.* **6**, 221–229.

Kalber, F. A., and Costlow, J. D. (1968). Osmoregulation in larvae of the land crab, *Cardisoma guanhumi* Latreille. *Am. Zool.* **8**, 411–416.

Knight-Jones, E. W. (1953). Laboratory experiments on gregariousness during setting in *Balanus balanoides* and other barnacles. *J. Exp. Biol.* **30**, 584–98.

Knight-Jones, E. W., and Morgan, E. (1966). Responses of marine animals to changes in hydrostatic pressure. *Oceanogr. Mar. Biol.* **4**, 267–299.

Knight-Jones, E. W., and Qasim, S. Z. (1967). Responses of crustacea to changes in hydrostatic pressure. *Proc. Symp. Crustacea, Ser. 2, Mar Biol. Assoc. India, Pt. III*, pp. 1132–1146.

Kon, T. (1979). Ecological studies on larvae of the crab belonging to the genus *Chionoecetes opilio*. I. The influence of starvation on the survival and growth of the Zuwai crab. *Bull. Jpn. Soc. Sci. Fish.* **45**, 7–9.

Kurata, H. (1955). The post-embryonic development of the prawn *Pandalus kessleri*. *Bull. Hokkaido Reg. Fish. Res. Lab.* **12**, 1–15.

Landry, M. R. (1975). Seasonal temperature effects and predicting development rates of marine copepod eggs. *Limnol. Oceanogr.* **20**, 434–440.

Lang, W. H., Forward, R. B., and Miller, D. C. (1979). Behavioral responses of *Balanus improvisus* nauplii to light intensity and spectrum. *Biol. Bull. (Woods Hole, Mass.)* **157**, 166–181.

Lang, W. H., Marcy, M., Clem, P. J., Miller, D. C., and Rodelli, M. R. (1980). The comparative photobehavior of laboratory-hatched and plankton-caught *Balanus improvisus* (Darwin) nauplii and the effects of 24-hour starvation. *J. Exp. Mar. Biol. Ecol.* **42**, 201–212.

Lang, W. H., Miller, D. C., Ritacco, P. J., and Marcy, M. (1981). The effect of copper and cadmium on the behavior and development of barnacle larvae. *In* "Biological Monitoring of Marine Pollutants" (F. J. Vernberg, A. Calabrese, F. P. Thurberg, and W. B. Vernberg, eds.), pp. 165–203. Academic Press, New York.

Latz, M. I., and Forward, R. B. (1977). The effect of salinity upon phototaxis and geotaxis in a larval crustacean. *Biol. Bull. (Woods Hole, Mass.)* **153**, 163–179.

Laverack, M. S., MacMillan, D. L., and Neil, D. M. (1976). A comparison of beating parameters in larval and post-larval locomotory systems of the lobster *Homarus gammarus* (L.). *Philos. Trans. R. Soc. London, Ser. B* **274**, 87–99.

Lebour, M. V. (1922). The food of plankton organisms. *J. Mar. Biol. Assoc. U. K.* **12**, 644–677.

Lebour, M. V. (1928). The larval stages of the Plymouth Brachyura. *Proc. Zool. Soc. London* **2**, 478–560.

Lee, R. F., Nevenzel, J. C., and Lewis, A. G. (1974). Lipid changes during the life cycle of the marine copepod, *Euchaeta japonica* Marukawa. *Lipids* **9**, 891–898.

LeRoux, A. (1973). Observations sur le développement larvaire de *Nyctiphanes couchii* (Crustacea: Euphausiacea) au laboratoire. *Mar. Biol. (Berlin)* **22**, 159–166.

LeRoux, A. (1974). Observations sur le developpement larvaire de *Meganyctiphanes norvegica* (Crustacea: Euphausiacea) au laboratoire. *Mar. Biol. (Berlin)* **26**, 45–56.

Levine, D. M., and Sulkin, S. D. (1979). Partitioning and utilization of energy during the larval development of the xanthid crab, *Rhithropanopeus harrisii* (Gould), *J. Exp. Mar. Biol. Ecol.* **40**, 247–257.

Lochhead, M. S., and Newcombe, C. L. (1942). Methods of hatching eggs of the blue crab. *Virginia J. Sci.* **3**, 76–86.

Logan, D. T., and Epifanio, C. E. (1978). A laboratory energy balance of the larvae and juveniles of the American lobster *Homarus americanus. Mar. Biol. (Berlin)* **47,** 381–389.

Longhurst, A. R. (1976). Vertical migration. Chapter 2. In "The Ecology of the Seas" (D. H. Cushing and J. J. Walsh, eds.), pp. 116–137. Saunders, Philadelphia, Pennsylvania.

Lough, R. G. (1976). Larval dynamics of the Dungeness crab, *Cancer magister,* off the Central Oregon Coast, 1970–1971. *Fish. Bull.* **74,** 353–376.

Lucas, M. I., Walker, G., Holland, D. L., and Crisp, D. J. (1979). An energy budget for the free-swimming and metamorphosing larvae of *Balanus balanoides* (Crustacea: Cirripedia). *Mar. Biol. (Berlin)* **55,** 221–229.

McKenney, C. L., and Costlow, J. D. (1981). The effects of salinity and mercury on developing megalopae and early crab stages of the blue crab, *Callinectes sapidus* Rathbun. In "Biological Monitoring of Marine Pollutants" (F. J. Vernberg, A. Calabrese, F. P. Thurberg, and W. B. Vernberg, eds.), pp. 241–262. Academic Press, New York.

McKenney, C. E., and Neff, J. M. (1981). The ontogeny of resistance adaptation and metabolic compensation to salinity and temperature by the caridean shrimp, *Palaemonetes pugio,* and modification by sublethal zinc exposure. In "Biological Monitoring of Marine Pollutants" (F. J. Vernberg, A. Calabrese, F. P. Thurberg, and W. B. Vernberg, eds.), pp. 205–240. Academic Press, New York.

McLaren, I. A. (1965). Some relationships between temperature and egg size, body size, development rate, and fecundity, of the copepod *Pseudocalanus. Limnol. Oceanogr.* **10,** 528–538.

McLaren, I. A. (1966). Predicting development rate of copepod eggs. *Biol. Bull. (Woods Hole, Mass.)* **131,** 457–469.

McLaren, I. A., Corkett, C. J., and Zillioux, E. J. (1969). Temperature adaptation of copepod eggs from the arctic to the tropics. *Biol. Bull. (Woods Hole, Mass.)* **137,** 486–493.

Macmillan, D. L., Neil, D. M., and Laverack, M. S. (1976). A quantitative analysis of exopodite beating in the larvae of the lobster *Homarus gammarus* (I.). *Philos. Trans. R. Soc. London, Ser. B* **274,** 69–85.

Makarov, R. R. (1969). Transport and distribution of decapod larvae in the plankton of the Western Kamchatka shelf. *Oceanology* **9,** 251–261.

Makarov, R. R. (1973). Larval development of *Notocrangon antarcticus* (Decapoda: Crangonidae) (in Russian). *Zool. Zh.* **52,** 1149–1555.

Makarov, R. R. (1979). Larval distribution and reproductive ecology of *Thysanoessa macrura* (Crustacea: Euphausiacea) in the Scotia sea. *Mar. Biol. (Berlin)* **52,** 377–386.

Marshall, N. B. (1979). "Developments in Deep-Sea Biology," 566 pp. Blandford Press Poole, Dorset.

Marshall, S. M., and Orr, A. P. (1972). "The Biology of a Copepod." Springer-Verlag, Berlin and New York.

Masters, C. (1975). Isozyme realization and ontogeny. In "Isozymes. III. Developmental Biology" (C. L. Markert, ed.), pp. 281–196. Academic Press, New York.

Mauchline, J., and Fisher, L. R. (1969). The biology of Euphasiids. *Adv. Mar. Biol.* **7,** 1–454.

Maxwell, J. G. (1977). The breeding biology of *Chorismus antarcticus* (Pfeffer) and *Notocrangon antarcticus* (Pfeffer) (Crustacea, Decapoda) and its bearing on the problems of the impoverished Antarctic decapod fauna. In "Adaptations within Antarctic Ecosystems" (G. A. Llano, ed.), pp. 335–342. Random House (Smithsonian Inst. Press), New York.

Mayr, E. (1963). "Animal Species and Evolution." Harvard Univ. Press, Cambridge, Massachusetts.

Menge, B. A. (1975). Brood or broadcast? The adaptive significance of different reproductive strategies in the two intertidal sea stars *Leptasterias hexactis* and *Pisaster ochraceus*. *Mar. Biol. (Berlin)* **31**, 87–100.

Menzies, R. A., and Kerrigan, J. M. (1978). Implications of spiny lobster recruitment patterns of the Caribbean—A biochemical genetic approach. *Proc. 31st. Annu. Gulf Caribb. Fish. Inst.* pp. 164–178.

Mileikovsky, S. A. (1971). Types of larval development in marine bottom invertebrates: Their distribution and ecological significance, a re-evaluation. *Mar. Biol. (Berlin)* **10**, 193–212.

Modin, J. C., and Cox, K. W. (1967). Post-embryonic development of laboratory reared ocean shrimp, *Pandalus jordani* Rathbun. *Crustaceana* **13**, 197–219.

Moller, T. H., and Branford, J. R. (1979). A circadian hatching rhythm in *Nephrops norvegicus* (Crustacea: Decapoda). *In* "Cyclic Phenomenon in Marine Plants and Animals" (E. Naylor and R. G. Hartnoll, eds.), pp. 391–397. Pergamon, Oxford.

Mootz, C., and Epifanio, C. E. (1974). An energy budget for *Menippe mercenaria* larvae fed *Artemia* nauplii. *Biol. Bull. (Woods Hole, Mass.)* **146**, 44–55.

Moreira, G. S., McNamara, J. C., Hiroki, K., and Moreira, P. S. (1981). The effect of temperature on the respiratory metabolism of selected developmental stages of *Emerita brasiliensis* Schmitt (Anomura, Hippidae). *Comp. Biochem. Physiol.* **70A**, 627–629.

Moreira, G. S., McNamara, J. S., Moreira, P. S., and Weinrich, M. (1980). Temperature and salinity effects on the respiratory metabolism of the first zoeal stage of *Macrobrachium holthuisi* Genofre and Labao (Decapods Palaemonidae). *J. Exp. Mar. Biol. Ecol.* **48**, 223–226.

Morgan, R. P., Kramarsky, E., and Sulkin, S. D. (1978). Biochemical changes during larval development of the Xanthid crab *Rhithropanopeus harrisii* III. Isozyme changes during ontogeny. *Mar. Biol. (Berlin)* **88**, 223–226.

Moyse, J. (1963). A comparison of the value of various flagellates and diatoms as food for the barnacle larvae. *J. Cons., Cons. Perm. Int. Explor. Mer.* **28**, 175–187.

Neils, D. M., McMillan, D. L., Robertson, R. M., and Laverack, M. S. (1976). The structure and function of thoracic expodites in the larvae of the lobster *Homarus gammarus* (L.). *Philos. Trans. R. Soc. London, Ser. B.* **274**, 53–68.

Nichols, P. R., and Keney, P. M. (1963). Crab larvae (*Callinectes*) in plankton collections from cruises of M/V Theodore N. Gill, South Atlantic coast of the United States, 1953–1954. *Fish and Wild. Ser., Spec. Sci. Rept. Fish.* **448**, 1–14.

Omori, M. (1974). The biology of the pelagic shrimp in the ocean. *Adv. Mar. Biol.* **12**, 233–324.

Omori, M., and Gluck, D. (1978). Life history and vertical migration of the pelagic shrimp *Sergestes similis* off the Southern California coast. *Fish. Bull.* **77**, 183–198.

Ott, F. S., and Forward, R. B. (1976). The effect of temperature upon phototaxis and geotaxis by larvae of the crab *Rhithropanopeus harrisii*. *J. Exp. Mar. Biol. Ecol.* **23**, 97–107.

Pandian, T. J. (1967). Changes in chemical composition and caloric content of developing eggs of the shrimp *Crangon crangon*. *Helgol. Wiss. Meeresunters.* **16**, 216–224.

Pandian, T. J. (1970a). Ecophysiological studies on the developing eggs and embryos of the European lobster *Homarus gammarus*. *Mar. Biol. (Berlin)* **5**, 154–167.

Pandian, T. J. (1970b). Yolk utilization and hatching time in the Canadian lobster *Homarus americanus*. *Mar. Biol. (Berlin)* **7**, 249–254.

Pandian, T. J., and Schumann, K. H. (1967). Chemical composition and caloric content of eggs and zoea of the hermit crab *Eupagurus bernhardus*. *Helgol. Wiss. Meeresunters.* **16**, 225–230.

Passano, L. M. (1960a). The regulation of crustacean metamorphosis. *Am. Zool.* **1**, 89–95.

Passano, L. M. (1960b). Molting and its control. *Physiol. Crustacea* **1**, 473–536.

Patel, B. S., and Crisp, D. J. (1960a). Rates of development of the embryos of several species of barnacles. *Physiol. Zool* **33**, 104–119.

Patel, B. S., and Crisp, D. J. (1960b). The influence of temperature on the breeding and the moulting activities of some warm-water species of operculate barnacles. *J. Mar. Biol. Assoc. U. K.* **39**, 667–680.

Paul, A. J., and Paul, J. M. (1980). The effect of early starvation on later feeding success of King crab zoeae. *J. Exp. Mar. Biol. Ecol.* **44**, 247–51.

Pautsch, R. (1967). Pigmentation and color change in decapod larvae. *Proc. Symp. Crustacea, Ser. 2, Mar. Biol. Assoc. India, Pt. III,* pp. 1108–1123.

Phillips, B. F., and Sastry, A. N. (1980). Larval Ecology. In "The Biology and Management of Lobsters" (J. S. Cobb and B. F. Phillips, eds.), Vol. II, pp. 11–57. Academic Press, New York.

Phillips, B. F. D., Rimmer, D. W., and Reid, D. D. (1978). Ecological investigations of the late stage phyllosoma and puerulus larvae of the western rock lobster *Panulirus longipes cygnus. Mar. Biol. (Berlin)* **45**, 347–357.

Phillips, B. F., Brown, P. A., Rimmer, D. W., and Reid, D. D. (1979). Distribution and dispersal of the phyllosoma larvae of the western rock lobster, *Panulirus cygnus* in the Southeastern Indian Ocean. *Aust. J. Mar. Freshwater Res.* **30**, 773–83.

Phillips, B. F., Brown, P. A., Rimmer, D. W., and Braine, S. J. (1981). Distribution and abundance of late larval stages of the Scyllaridae (slipper lobsters) in the Southeastern Indian ocean. *Aust. J. Mar. Freshwater Res.* **32**, 417–437.

Pianka, E. R. (1970). On "r" and "K" selection. *Am. Nat.* **104**, 592–597.

Precht, H., Christophersen, J., Hensel, H., and Larcher, W. (1973). "Temperature and Life." Springer-Verlag, Berlin and New York.

Prentiss, C. W. (1901). The otocyst of decapod crustacea, its structure, development and function. *Bull. Mus. Comp. Zool.* **36**, 167–281.

Prosser, C. L. (1975). Physiological adaptation in animals. In "Physiological Adaptation to the Environment" (F. J. Vernberg, ed.), pp. 1–18. Intext Educational Publ., New York.

Rao, K. R. (1968). Variations in the chromatophorotropins and adaptive color changes during the life history of the crab, *Ocypode macrocera. Zool. Jahrb., Abt. Allg. Zool. Physiol. Tiere* **74**, 274–291.

Rao, K. R., Fingerman, S. W., and Fingerman, M. (1973). Effects of exogenous ecdysones on the molt cycles of fourth and fifth stage American lobsters *Homarus americanus. Comp. Biochem. Physiol.* **44A**, 1105–1120.

Redfield, A. C. (1941). The effect of the circulation of water on the distribution of the Calanoid community in the Gulf of Maine. *Biol. Bull. (Woods Hole, Mass.)* **80**, 86–110.

Reeve, M. R. (1969). Growth, metamorphosis, and energy conversion in the larvae of the prawn, *Palaemon serratus. J. Mar. Biol. Assoc. U. K.* **49**, 77–96.

Regnault, M. (1969a). Influence de la temperature et de l'origin de l'eau de mer sur le developpement larvaire au laboratoire d'*Hippolyte inermis* Leach. *Vie Milieu, Ser. C* **20**, 137–152.

Regnault, M. (1969b). Etude experimentale de la nutrition d'*Hippolyte inermis* Leach (Decapoda, Natantia) au cours de son developpement larvaire, au laboratoire. *Vi Int. Rev. Ges., Hydrobiol.* **54**, 749–764.

Reid, J. L., Briton, E., Fleminger, A., Venrick, E. L., and McGowan, J. A. (1978). Ocean circulation and marine life. *Oceanog., [Collect. Pap. Gen. Symp. Oceanogr. Assem.], 1976* pp. 65–130.

Rice, A. L. (1967). The orientation of the pressure responses of some marine crustacea. *Proc. Symp. Crustacea, Ser. 2, Mar. Biol. Assoc. India, Pt. III,* pp. 1124–1131.

Rice, A. L. (1968). Growth 'rules' and the larvae of decapod crustaceans. *J. Nat. Hist.* **2**, 525–530.

Rice, A. L., and Provenzano, A. J. (1964). The larval stages of *Pagurus marshi* Benedict (Decapoda: Anomura) reared in the laboratory. *Crustaceana* **7**, 217–235.

Rice, A. L., and Williamson, D. I. (1970). Methods for rearing larval decapod Crustacea. *Helgol. Wiss. Meeresunters.* **20**, 417–434.

Rice, A. L., and Williamson, D. I. (1977). Planktonic stages of Crustacea Malacostraca from Atlantic Seamounts. *Meteor Forschergeb., Reihe D* **26**, 28–64.

Richards, W. J., and Goulet, J. R. (1976). An operational surface drift model used for studying larval lobster recruitment and dispersal. *FAO Fish. Rep.* No. 200, pp. 363–374.

Rimmer, D. W., and Phillips, B. F. (1979). Diurnal migration and vertical distribution of Phyllosoma larvae of the western rock lobster *Panulirus cygnus. Mar. Biol. (Berlin)* **54**, 109–124.

Roberts, M. H. (1971). Larval development of *Pagurus longicarpus* Say reared in the laboratory. III. Behavioral responses to salinity discontinuities. *Biol. Bull. (Woods Hole, Mass.)* **140**, 489–501.

Roberts, M. H. (1974). Larval development of *Pagurus longicarpus* Say reared in the laboratory. V. Effect of diet on survival and molting. *Biol. Bull. (Woods Hole, Mass.)* **146**, 67–77.

Rosenberg, R., and Costlow, J. D. (1976). Synergistic effects of cadmium and salinity combined with constant and cycling temperatures on the larval development of two estuarine crab species. *Mar. Biol. (Berlin)* **38**, 291–303.

Rosenberg, R., and Costlow, J. D. (1979). Delayed response to irreversible non-genetic adaptation to salinity in early development of the brachyuran crab *Rhithropanopeus harrisii*, and some notes on adaptation to temperature. *Ophelia* **18**, 97–112.

Ross, R. M. (1981). Laboratory culture and development of *Euphausia pacifica. Limnol. Oceanogr.* **26**, 235–246.

Saigusa, M., and Hidaka, T. (1978). Semilunar rhythm in the zoea-release activity of the land crabs *Sesarma. Oecologica* **37**, 163–176.

Saisho, T. (1966). Studies on the phyllosoma larvae with reference to the oceanographical conditions. *Mem. Fac. Fish. Kagoshima Univ.* **15**, 177–239.

Sandifer, P. A. (1973). Distribution and abundance of decapod crustacean larvae in the York River estuary and adjacent lower Chesapeake Bay, Virginia 1968–1969. *Chesapeake Sci.* **14**, 235–257.

Sandifer, P. A. (1975). The role of pelagic larvae in recruitment to populations of adult decapod crustaceans in the York River estuary and adjacent lower Chesapeake Bay, Virginia. *Estuarine Coastal Mar. Sci.* **3**, 269–279.

Sastry, A. N. (1970). Culture of brachyuran crab larvae using a recirculating sea water system in the laboratory. *Helgol. Wiss. Meeresunters.* **20**, 406–416.

Sastry, A. N. (1975). An experimental culture-research facility for the American lobster, *Homarus americanus. Eur. Symp. Mar. Biol., Proc., 10th, 1975* pp. 419–435.

Sastry, A. N. (1976). Effects of constant and cyclic temperature regimes on the larval development of a brachyuran crab. *ERDA Symp. Ser. (Conf. 750425)* **40**, 81–87.

Sastry, A. N. (1977). Larval development of the rock crab *Cancer irroratus* Say 1817, reared under laboratory conditions (Decapoda, Brachyura). *Crustaceana* **32**, 115–168.

Sastry, A. N. (1978). Physiological adaptation of *Cancer irroratus* larvae to cyclic temperatures. In "Physiology and Behavior of Marine Organisms" (D. S. McLusky and A. J. Berry, eds.), pp. 57–65. Pergamon, Oxford.

Sastry, A. N. (1979). Metabolic adaptation of *Cancer irroratus* developmental stages to cyclic temperatures. *Mar. Biol. (Berlin)* **51**, 243–250.

Sastry, A. N. (1980). Effects of thermal pollution on pelagic larvae of Crustacea. EPA-600/3-80-064, National Tech. Inf. Service, 52 pp. Springfield, Virginia.

Sastry, A. N. (1983). Ecological aspects of reproduction. In "The Biology of Crustacea: Environmental Adaptations," Vol. 8 (F. J. Vernberg and W. B. Vernberg, eds.), Academic Press, New York.

Sastry, A. N., and Ellinton, W. R. (1978). Lactate dehydrogenase during the larval development of Cancer irrortus: Effect of constant and cyclic thermal regimes. Experientia 34, 308–309.

Sastry, A. N., and McCarthy, J. F. (1973). Diversity in metabolic adaptation of pelagic larval stages of two sympatric species of brachyuran crabs. Neth. J. Sea Res. 7, 434–446.

Sastry, A. N., and Vargo, S. L. (1977). Variations in the physiological responses of crustacean larvae to temperature. In "Physiological Responses of Marine Biota to Pollutants" (F. J. Vernberg, A. Calabrese, F. P. Thurberg, and W. B. Vernberg, eds.), pp. 401–423. Academic Press, New York.

Saunders, D. S. (1974). Evidence of 'dawn' and 'dusk' oscillators in the Nasonia photoperiodic clock. J. Insect Physiol. 20, 77–88.

Scarratt, D. J. (1964). Abundance and distribution of labster larvae (Homarus americanus) in Northumberland Strait. J. Fish. Res. Board Can. 21, 666–680.

Scarratt, D. J. (1973). Abundance, survival, and vertical and diurnal distribution of lobster larvae in Northumberland Strait, 1962–1963 and their relationship with commercial stocks. J. Fish. Res. Board Can. 30, 1819–1824.

Schatzlein, F. C., and Costlow, J. D. (1978). Oxygen consumption of the larvae of the decapod crustaceans Emerita talpoida (Say) and Libinia emarginata Leach. Comp. Biochem. Physiol. 61A, 441–450.

Scheltema, R. S. (1975). Relationship of larval dispersal, gene flow, and natural selection to geographic variation of benthic invertebrates in estuarines and along coastal regions. In "Estuarine Research. I. Chemistry, Biology and the Estuarine System" (L. E. Cronin, ed.), pp. 372–391. Academic Press, New York.

Schneider, D. E. (1968). Temperature adaptations in latitudinally separated populations of the crab, Rhithropanopeus harrisii. Am. Zool. 8, 772.

Sims, H. W., and Ingle, R. W. (1967). Caribbean recruitment of Florida's spiny lobster population. Q. J. Fla. Acad. Sci. 29, 207–242.

Singarajah, K. V., Moyse, J., and Knight-Jones, E. W. (1967). The effect of feeding upon the photactic behavior of cirripede nauplii. J. Exp. Mar. Biol. Ecol. 1, 144–153.

Smith, R. L. (1976). Waters of the sea: The ocean's characteristics and circulation, Chapter 2. In "The Ecology of the Seas" (D. H. Cushing and J. J. Walsh, eds.), pp. 23–58. Saunders, Philadelphia, Pennsylvania.

Soll, D. R. (1979). Timers in developing systems. Science 203, 841–849.

Stearns, S. C. (1976). Life history tactics: A review of the ideas. Q. Rev. Biol. 51, 3–47.

Stearns, S. C. (1977). The evolution of life history traits: A critique of the theory and a review of the data. Annu. Rev. Ecol. Syst. 8, 145–171.

Steele, J. H. (1981). Some varieties in biological oceanography. In "Evolution of Physical Oceanograhy" (B. A. Warren and C. Wunsch, eds.), pp. 376–383. MIT Press, Cambridge, Massachusetts.

Strathmann, R. (1974). The spread of sibling larvae of sedentary marine invertebrates. Am. Nat. 108, 29–44.

Sulkin, S. D. (1973). Depth regulation of crab larvae in the absence of light. J. Exp. Mar. Biol. Ecol. 13, 73–82.

Sulkin, S. D. (1975). Influence of light in the depth regulation of crab larvae. Biol. Bull. (Woods Hole, Mass.) 148, 333–343.

Sulkin, S. D., and Norman, K. (1976). A comparison of two diets in the laboratory culture of the zoeal stages of the brachyuran crabs *Rhithropanopeus harrisii* and *Neopanope* sp. *Helgol. Wiss. Meeresunters.* **28**, 183–190.

Sulkin, S. D., and van Heukelem, W. F. (1980). Ecological and evolutionary significance of nutritional flexibility in planktotrophic larvae of the deep sea red crab *Geryon quinquidens* and the stone crab *Menippe mercenaria*. *Mar. Ecol. Prog. Ser.* **2**, 91–95.

Sulkin, S. D., Morgan, R. P., and Minasian, L. L. (1975). Biochemical changes during larval development of the xanthid crab *Rhithropanopeus harrisii*. II. Nucleic acids. *Mar. Biol. (Berlin)* **32**, 113–117.

Sweat, D. (1968). Growth and tagging studies on *Panulirus argus* (Latrelle) in the Florida Keys. *Fla., Board Conserv. Tech. Pub.* No. 57, pp. 1–30.

Thorson, G. (1950). Reproductive and larval ecology of marine bottom invertebrates. *Biol. Rev. (Woods Hole, Mass.)* **25**, 1–45.

Thorson, G. (1964). Light as an ecological factor in the dispersal and settlement of larvae of marine bottom invertebrates. *Ophelia* **1**, 167–208.

Tucker, R. K. (1978). Free amino acids in developing larvae of the stone crab *Menippe mercenaria*. *Comp. Biochem. Physiol.* **60A**, 169–172.

Tucker, R. K., and Costlow, J. D. (1975). Free amino acid changes in normal and eyestalk-less megalopa larvae of the blue crab, *Callinectes sapidus*, during the course of the molt cycle. *Comp. Biochem. Physiol.* **51A**, 75–78.

Underwood, A. J. (1974). On models for reproductive strategies in marine benthic invertebrates. *Am. Nat.* **108**, 874–878.

UNESCO (1968). Zooplankton sampling. *Monogr. Oceanogr. Methodol.* **2**, 1–99.

Vance, R. R. (1973a). On reproductive strategies in marine benthic invertebrates. *Am. Nat.* **107**, 339–352.

Vance, R. R. (1973b). More on reproductive strategies in marine benthic invertebrates. *Am. Nat.* **107**, 353–361.

Vargo, S. L., and Sastry, A. N. (1977). Acute temperature and low dissolved oxygen tolerances of brachyuran crab (*Cancer irroratus*) larvae. *Mar. Biol. (Berlin)* **40**, 165–171.

Vernberg, F. J., and Costlow, J. D. (1966). Studies on physiological variation between tropical and temperate-zone fiddler crabs of the genus *Uca*. IV. Oxygen consumption of larvae and young crabs reared in the laboratory. *Physiol. Zool.* **39**, 36–52.

Vernberg, F. J., and Vernberg, W. B. (1970). "The Animal and The Environment," 398 pp. Holt, New York.

Vernberg, F. J., and Vernberg, W. B. (1975). Adaptations to extreme environments. *In* "Physiological Ecology of Estuarine Organisms." (F. J. Vernberg, ed.), pp. 165–180. Univ. of South Carolina Press, Columbia, South Carolina.

Vernberg, W. B., and Vernberg, F. J. (1970). Metabolic diversity in oceanic animals. *Mar. Biol. (Berlin)* **6**, 33–42.

Vernberg, W. B., DeCoursey, P. J., and Padgett, W. J. (1973). Synergistic effects of environmental variables on larvae of *Uca pugilator*. *Mar. Biol. (Berlin)* **22**, 307–312.

Vernberg, W. B., Moreira, G., and McNamara, J. C. (1981). The effect of temperature on the respiratory metabolism of the developmental stages of *Pagurus criniticornis* (Dana) (Anomura: Paguridea). *Mar. Biol. Lett.* **2**, 1–9.

Via, S. E., and Forward, R. B. (1975). The ontogeny and spectral sensitivity of polarotaxis in larvae of the crab *Rhithropanopeus harrisii* (Gould). *Biol. Bull. (Woods Hole, Mass.)* **149**, 151–166.

Walley, L. J. (1969). Studies on the larval structure and metamorphosis of *Balanus balanoides* (L.). *Philoso. Trans. R. Soc. London, Ser. B* **256**, 237–280.

A. N. Sastry

Warner, G. F. (1967). The life history of the mangrove tree crab, *Aratus pisoni*. *J. Zool.* **153**, 321–335·

Warren, B. A., and Wunsch, C. (1981). "Evolution of Physical Oceanography." MIT Press, Cambridge, Massachusetts.

Waterman, T. H., and Chace, F. A. (1960). General crustacean biology. In *Physiol. Crustacea* **1**, 1–33.

Wear, R. G. (1974). Incubation in British Decapod Crustacea and the effects of temperature on the rate and success of embryonic development. *J. Mar. Biol. Assoc. U. K.* **34**, 745–762.

Wheeler, D. E., and Epifanio, C. E. (1978). Behavioral responses to hydrostatic pressure in larvae of two species of xanthid crabs. *Mar. Biol. (Berlin)* **46**, 167–174.

Whitney, J. O. (1969). Absence of sterol synthesis in larvae of the mud crab *Rhithropanopeus harrisii* and of the spider crab *Libinia emarginata*. *Mar. Biol. (Berlin)* **3**, 124–135.

Williams, B. G. (1968). Laboratory rearing of the larval stages of *Carcinus maenas* (L.) (Crustacea: Decapoda). *J. Nat. Hist.* **2**, 121–126.

Williamson, D. I. (1967). Some recent advances and outstanding problems in the study of larval crustacea. *Proc. Symp. Crustacea, Ser. 2, Mar. Biol. Assoc. India, Pt. II*, pp. 815–823.

Wooster, W. S., and Reid, J. L. (1963). Eastern boundary currents. In "The Seas" (M. N. Hill, ed.), Vol. 2, pp. 253–280. Wiley, New York.

Wrytki, K. (1973). Physical oceanography of the Indian ocean. In "The Biology of the Indian Ocean" (B. Zeitschel and S. A. Gerlach, eds.), pp. 18–36. Springer-Verlag, Berlin and New York.

6

Biotic Assemblages: Populations and Communities*

BRUCE C. COULL AND SUSAN S. BELL

I. INTRODUCTION

It is not our purpose to reiterate in a single chapter all that has been accomplished on populations and communities of the Crustacea. Rather, we shall attempt to delineate population and community phenomenona of

*Contribution No. 459 from the Belle W. Baruch Institute for Marine Biology and Coastal Research, University of South Carolina.

THE BIOLOGY OF CRUSTACEA, VOL. 7
Copyright © 1983 by Academic Press, Inc.
All rights of reproduction in any form reserved.
ISBN 0-12-106407-7

Crustacea by using examples of three habitats where crustaceans are significant members. We do not infer that crustaceans are insignificant members of other assemblages nor portend that crustaceans are the only members of the type communities. In fact, it is imperative that the reader be fully aware that any attempt to talk of crustacean "communities" per se is erroneous. Crustaceans are but a part of biotic assemblages, even those they dominate. Therefore, interactions with other taxa, as well as other Crustacea, are necessarily important in any discussion of community dynamics.

To illustrate crustacean population/community attributes we have chosen three "type" assemblages, i.e., (1) sandy bottom marine communities, (2) hard-bottom marine communities (including rocky intertidal areas and reef assemblages), and (3) plankton assemblages (marine and fresh water). We chose these habitats because in all three: Crustacea often dominate; data are available; and the habitats are world-wide in distribution, and data are available from a variety of geographically disjunct locations.

It will become readily apparent that we lack much information on the population/community dynamics of crustacean assemblages. This is not due to the lack of effort, but rather to inherent difficulties with organisms that traditionally molt and have a dispersed larval stage. Demographic data are particularly difficult to procure *in situ* because it is impossible to estimate age specific mortality with animals whose cohorts often overlap, have up to 10–15 broods from a single insemination, or whose larvae are carried away from the site of spawning. Field estimates of such parameters are imperative if we are to ever understand the role of crustaceans in aquatic ecosystems. In the following we will summarize what we know of the population/community dynamics of Crustacea, but more importantly, we will point out areas in need of future research on these important members of most aquatic ecosystems.

II. MARINE SANDY BOTTOM HABITATS

A. Descriptions of Habitats and Fauna

While muddy bottom habitats are often dominated by polychaete worms and marine free living nematodes, intertidal and subtidal sandy bottoms are regularly dominated by Crustacea. Although various crustacean taxa may occur in these habitats, amphipods regularly comprise the bulk of the macrofauna (herein defined as those benthos retained on a 0.5-mm sieve) and harpacticoid copepods often dominate the meiofauna (those benthic metazoans passing through a 0.5-mm sieve and retained on sieves with mesh widths smaller than 0.1 mm) (Dahl, 1953; Sanders, 1956, 1958; Coull, 1970; Ivester, 1980).

Dahl (1953) in a review of the macrofauna of sandy beaches states that "most of these species are crustaceans," and he proposed zonation patterns of sandy intertidal areas based on the crustacean elements. Table I lists the worldwide zonation patterns of the crustaceans on sandy beaches as Dahl (1953) perceived them. Dahl's classification, although generally correct, is not universally applicable, and the major problem appears to center on the universality of the midlittoral cirolanid isopod zone (Table I). Colman and Segrove (1955) found it did not exist in England, Wood (1968) did not find it in New Zealand, Fincham (1971) found it mixed with amphipods in England, and, except for Dexter's (1969) finding of *Exosphaeroma diminutum* as the fifth most abundant animal on a North Carolina beach, none of the sandy beach studies along the U.S. East Coast (Croker, 1967, 1977; Sameoto, 1969; Holland and Polgar, 1976; Holland and Dean, 1977) reported the existence of the cirolanid zone. Based on these limited data, it certainly appears that the cirolanid zone is not universal in temperate regions. However, recent tropical studies suggest that the cirolanid genus, *Excirolana*, is well represented on sandy beaches, but usually much higher on the beach than suggested by Dahl (Hedgpeth, 1957; Trevallion et al., 1970; Dexter, 1972, 1976; Glynn et al., 1975). The zonation scheme of Dahl then, with the temperate zone midlittoral exceptions, appears to still be quite valid. Taltrid amphipods of the genera *Talorchestia, Orchestia,* and *Talitrus* tend to dominate the supralittoral fringe in temperate areas, and co-dominate with ghost crabs (*Ocypode* spp.) in the warm temperate and tropical regions (Trevallion et al., 1970). The midlittoral zone as proposed by Dahl (1953) seems to vary greatly as mentioned above, but the sublittoral fringe appears as Dahl proposed with hippid crabs (*Emerita spp*) and/or various amphipods co-occurring. The East Coast of North America seems to be anomalous to the world wide pattern. The overwhelming dominance of haustorid amphipods along the entire intertidal gradient appears to be restricted to the U.S. East Coast. Studies from Maine (Croker, 1977) to Georgia (Croker, 1967), spanning 11° of latitude, all suggest that the beach is dominated and specifically zoned by haustorid amphipods (Croker et al., 1975). Of course, other crustaceans and taxa do coexist, but this unique sandy beach amphipod assemblage dominates. Not only is there haustorid dominance on the open intertidal beach, but on almost any intertidal sandy substrate, low in silt and clay (<10%) (e.g., point bars) within this wide geographic range (Kinner et al., 1974; Holland and Dean, 1977). Table II lists the general intertidal zonation of these ubiquitous amphipods along the East Coast of the United States, and clearly the fauna is zoned as well within

TABLE I

Dominant Sandy Beach Macro-Crustacea and Their Zonation in Different Parts of the World's Ocean[a]

Beach zone	Location				
	Arctic	North temperate	Tropics	South temperate	Antarctic
Subterrestrial fringe	—	Talitrid amphipods ↓	Ocypodid crabs	Talitrid amphipods	?
Midbeach	—		Cirolanid isopods		?
Subtidal fringe	Lysianassid amphipods	Oeocerotid and Haustoriid amphipods	Hippid crabs	Oedocerotid and Phoxocephalid amphipods ↑	?

[a] Modified from Dahl, 1953.

TABLE II

Intertidal Sandy Beach Zonation Patterns of North American East Coast Haustoriid Amphipods[a,b]

Beach zone	Species
Subterrestrial fringe	*Haustorius* sp.
	Neohaustorius schmitzi
Midbeach	*Lepidactylus dytiscus*
	Acanthohaustorius millsi
	Amphioporeia virginiana
	Parahaustorius longimerus
	Pseudohaustorius caroliniensis
Subtidal fringe	*Acanthohaustorius intermedius*
	Bathyporeia parkeri

[a] Modified from Rhodes, 1974.
[b] Table illustrates that zonation is distinct, even within this one group.

the haustorids as Dahl (1953) proposed for all the Crustacea. Although best known from the U.S. East Coast, a similar assemblage of haustorids exists in the British Isles, where several species of *Bathyporeia* and *Urothoe* coexist with *Haustorius arenarius* (Colman and Segrove, 1955; Fincham, 1971). However, in Britain rarely do the amphipods so overwhelm the community, and it is generally "composed of a few long-lived lamellibranch species (*Tellina*. . . .) . . . crustaceans and polychaetes . . ." (McIntyre and Eleftheriou, 1968).

Sandy intertidal meiofauna communities are regularly dominated by nematodes, copepods, gastrotrichs, and turbellarians. On high energy beaches, mystacocarids often become a numerically important taxa. Meiobenthic crustaceans rarely dominate intertidal sand except when the grain size of the sediment is extremely coarse (e.g., shell-gravel sediments). In the shell-gravel habitats, harpacticoid copepods will often outnumber the omnipresent free-living nematodes and comprise the majority of intertidal fauna (Renaud-Debyser and Salvat, 1963; McLachlan, 1977).

The macrofaunal assemblages of subtidal sands also are often characterized by their dominant amphipods. Sanders (1958) first recognized that subtidal habitats in Long Island Sound and Buzzard's Bay, Massachusetts, with less than 35–45% silt–clay, were dominated by ampeliscid amphipods; he designated these sand bottom associations the *Ampelisca spp.* community. Not all such sand bottom assemblages are dominated by amphipods (see Maurer *et al.*, 1978), but they will often comprise a significant portion of the macrobenthos in sandy habitats (e.g., Frankenburg and Leiper, 1977). As in the intertidal regions, the meiofauna of coarse sands is often dominated by harpacticoid copepods (Coull, 1970; Ivester, 1980), and in fact Rieger and Ruppert (1978) report mystacocarids common in shell-gravels at 20 m on the southeastern United States continental shelf.

What are the environmental factors that so often allow macro- and meiobenthic crustaceans to dominate sandy substrates? Obviously, degree of exposure and concommitant temperature fluctuations and surf intensity will be most important in governing what animals can adapt to the sandy beach. For those of the high intertidal (supralittoral) fringe, the major problem faced is desiccation, particularly at low tide. The dominant crustaceans of this zone have solved the desiccation problem in one of three ways: (1) Maintaining themselves in or under the rafted-detritus (wrack beds) of the high tide line where evaporation is decreased (e.g., talitrid amphipods). (2) Burrowing into the sediment to levels where moisture remains in the stitial lacunae (e.g., some haustorid amphipods, mystacocarids, ghost crabs). (3) Migrating up and down the beach slope during each tidal cycle to maintain position in moist sediments as the interstitial water table rises and falls (e.g., some haustorid amphipods, hippid crabs). The desiccation problem, of course, becomes less important toward the mid and low littoral zones, but still, adaptations (particularly burrowing into the moist sand) must be made.

The question still remains as to why crustaceans tend to dominate this habitat: certainly other taxa can physiologically adapt to the problems of exposure and pounding by surf. First, deposit feeding taxa are limited and excluded because of the low amount of organic matter in sediments in such a hydrodynamically active environment. Thus, the beach macrofauna is restricted to suspension feeding animals or those that scavenge the beach (primarily at low tide) for carrion. Second, the constant beat of the surf requires some mechanism to avoid being washed away, and in most cases morphological tapering of the body to allow rapid unimpeded burrowing has been extremely effective. Gammarid amphipods are, of course, ideally built for sands with their dorso-ventrally compressed body, small size, and rapidly moving spadelike legs which allow rapid burrowing (Bousfield, 1970). The hippid crabs with bodies tapered anteriorly to posteriorly, rather than the traditional crab shape of a broadened carapace, are also ideally suited for burrowing away from the beating surf. However, morphological "tapering" and rapid burrowing are not unique to the crustacean elements of the "surf" fauna. Mollusks (e.g., surf clams, *Donax*) have extremely streamlined shells, and those polychaetes that are able to live in the surf are extremely rapid burrowers (e.g., *Glycera*). Third, the rather impermeable exoskeleton of Crustacea reduces desiccation loss as well as serving as protection in a churning sand environment. A large soft-bodied animal would soon become abraded to death in such a habitat, and thus the presence of some "hard" protection, whether that be a chitinous exoskeleton as in the crustaceans or a "shell" as in the mollusks, appears to be requisite for existence in this habitat. Increased protection from wave action often leads to community dominance by polychaete worms (Croker *et al.*, 1975; Croker, 1977), but in wave swept sandy beaches, the well-adapted crustacean element, particularly the amphipods, dominate. Sediment size parameters (e.g., median diameter, percent silt–clay, skewness, and kurtosis) are factors well known to influence the composition of any soft bottom benthic association. Whether intertidally or subtidally, benthic crustacean dominance, or even presence, is closely tied to sediment parameters. Overall, benthic Crustacea will have their highest abundances in sandy substrates where current or wave activity provides sufficient food supply for these predominantly suspension feeding organisms. Once current velocity is reduced and sedimentation of flocculent material occurs, deposit feeding taxa will dominate and suspension feeding Crustacea will become less conspicuous members of the biocenose.

xther environmental factors such as temperature, salinity, pH, Eh*, and O_2 content, also play a role in determining Crustacea abundance and distribution. The effects of these parameters are discussed in detail in other chapters in this and other volumes of this work, and there is no need to reiterate them here.

B. Population Studies

Population growth studies of sandy bottom crustaceans have been primarily descriptive, and the majority of the literature available deals with the ubiquitous amphipods. Since it is impossible to measure cohorts or age classes of naturally occurring crustacean populations (Wenner *et al.*, 1974; Van Dolah *et al.*, 1975; Birklund, 1977), size-class analysis is the technique regularly used. Several authors (Hynes, 1959; Dexter, 1969; Sameoto, 1969; Fincham, 1971; Klein *et al.*, 1975; Gable and Croker, 1977; Van Dolah, 1978; Williams, 1978) provide size-class analyses of amphipod populations, but to date there is no available field-generated life table for any soft bottom crustacean. Size-frequency–growth rate analyses are indeed valuable, since such data provide information on the reproductive success of a population and about persistence of that population under particular environmental circumstances.

Most size frequency histograms indicate that temperate zone sandy bottom crustaceans have one or two reproductive peaks per year, and that the number of eggs carried by an incubating female varies with female length. Reproductive activity appears to be controlled by temperature (e.g., Sameoto, 1969; Dexter, 1971; Gable and Croker, 1977; Bynum, 1978) or photoperiod (Williams, 1978). Figure 1 illustrates a 1-year size-frequency

*Oxidation-reduction potential.

distribution of *Neohaustoris schmitzi*, a species with two generations (winter and summer), and Fig. 2 illustrates the 1-year size-frequency distribution of *Ampelisca brevicornis*, a species with but a single reproductive peak per year. In general, it appears that those species in warm temperate regions (e.g., *Neohaustoris schmitzi*) tend to have bimodal reproductive peaks as in Fig. 1 (Croker, 1967; Dexter, 1971), whereas those in colder regions have a unimodal distribution as in Fig. 2 (e.g., Mills, 1967; Sameoto, 1969; Klein *et al.*, 1975; Williams, 1978).

Attempting to provide more sophisticated population data on a group of organisms "which has neither scales, otoliths, vertebrae nor fin rays, which at each moult loses all hard parts . . . and which over the size range normally captured does not fall into recognizable size or age groups (so that) the estimation of growth and age is particularly difficult" (Wilder, 1953), several authors have developed and proposed techniques utilizable for crustaceans. Wenner *et al.* (1974) suggested that size of first reproduction can be estimated by plotting the cumulative distribution data on probability paper and reading off the 50% intersect (for details see Wenner *et al.*, 1974). Van Dolah *et al.* (1975) and Van Dolah (1978) used a modified egg-ratio technique (developed primarily for a sexually reproducing populations) to calcuate field generated intrinisic rates of natural increase for *Gammarus palustris*. By measuring per capita birth and death rates, "*r*" was calculated. Changes in *r* were correlated with season, migration, and size of female.

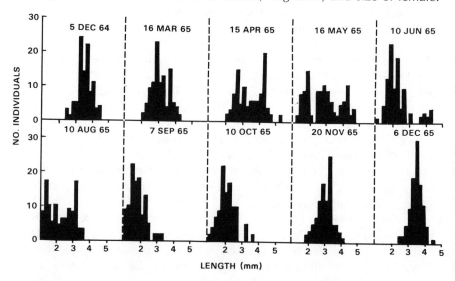

Fig. 1. Size class histograms of *Neohaustorius schmitzi* from North Carolina, USA, for 1 year. Note the presence of two distinct population age classes (an overwintering generation and a May produced generation). (Redrawn from Dexter, 1971.)

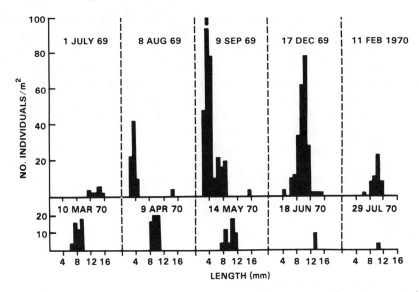

Fig. 2. Size class histograms of *Ampelisca brevicornis* from Germany for a 1-year period. Note but one reproduction per year. (Redrawn from Klein *et al.*, 1975.)

Such analyses can provide significant insight into fundamental population dynamics and certainly need to be continued.

Using a unique system of "cages," Birklund (1977) was able to follow cohort growth of the amphipod *Corophium insidiosum*. He established *in situ* cages which allowed immigration of juvenile amphipods, and then after 10 days, he placed a smaller mesh over the original cage to eliminate immigration/emigration, thus establishing *in situ* growth chambers. The growth cages were collected at 10- to 14-day intervals and growth was measured. Such "true" growth estimates are invaluable in estimating *in situ* production of a particular population and critical to understanding the energetic role of soft bottom crustaceans.

Most production estimates are based primarily on field collected density/biomass estimates, coupled with laboratory observations of life-cycle turnover and number of generations. *In situ* production estimates are in the initial stages. Klein *et al.* (1975) and Birklund (1977) provided secondary production estimates for benthic amphipods using field estimated size classes. Klein *et al.* (1975) found that *Ampelisca brevicornis* production averaged $0.3–0.7 \text{ g} \cdot \text{m}^{-2} \cdot \text{year}^{-1}$ over a 3-year period, and Birklund (1977) reported two *Corophium* species that ranged from 0.2 to 8 $\text{g} \cdot \text{m}^2$ for a 4-month period (we could not calculate an annual rate since Birklund did not provide data for the remainder of the year). Similarily, Feller (1977) and Fleeger and Palmer (1982), using size-class distributions and short-interval field sampling (2–6 days), estimated production of benthic harpacticoid copepods at 1.0 g · C · m^{-2} · year^{-1} ($\bar{x}$ of *Huntemannia jadensis:* Feller, 1977) and 0.87 g · C · /m^{-2} · year^{-1} ($\bar{x}$ of *Microarthridion littorale:* Fleeger and Palmer (1982), with the values varying significantly over the seasons. What is so amazing is that these two studies on phylogenetically and geographically distant animals (Feller—sandy beaches, Puget Sound, Washington; Fleeger and Palmer—salt marshes, South Carolina) came to approximately the same values. Both authors also present the first *in situ* life history data on meiofaunal animals. Fleeger (1979) reported that at stable age distribution and peak reproduction, *Microarthridion* had generation times of 12–15 days and up to 12 generations per year. Feller (1977) however, reported but two generations a year for *Huntemannia* from Washington. Obviously, *in situ* production/life history studies are an important avenue of research to pursue if we are ever to evaluate the energetic role of the benthos.

C. Population Interactions

Crustaceans often play a critical role in marine benthic community dynamics. Tube building amphipods, burrowing callianassids, or fiddler crabs certainly contribute structural heterogeneity to a habitat and biogenic structures (e.g., tubes, burrows, which often serve as refugia—*sensu:* Woodin, 1978) for many other taxa. Conversely, tubes built by other taxa, e.g., polychaetes, often serve as refugia for crustaceans. For example, Young and Rhoads (1971) report that the suspension feeding caprellid amphipod, *Aeginina longicornis,* was able to live and feed in a silt–clay sediment only because polychaete tubes protruded above the sediment–water interface. *Aeginina longicornis* lived on the tubes and avoided having its delicate suspension feeding apparatus clogged by the flocculent sediment surface.

Soft bottom crustaceans have also been the focus of several studies of mechanisms of niche fractionation. As some examples, Croker (1967) demonstrated temporal, spatial, and trophic separation in sympatric haustorid amphipods; MacDonald (1977) illustrated such separation in mud crabs; Ivester and Coull (1975), using SEM, demonstrated mouthpart differentiation in four sympatric amphipods, which they suggested was sufficient to allow co-existence; and similarly, Miller (1961) demonstrated how three congeneric fiddler crabs could co-exist without competing. In fact we can find no empirical evidence in the literature of competition between any soft bottom crustaceans. Certainly, there is competition for shells in hermit crabs (Hazlett, 1974) and competition for food among large malacostracans (e.g., *Homarus, Callinectes*), but evidence for a competitive hierachy among sediment dwellers is lacking.

Predation, on the other hand, appears to be very important in structuring crustacean assemblages (e.g., Van Dolah, 1978). Crustaceans are well known to be preyed upon by a variety of predators from alligators (up to 70% of their diet may be blue crabs: Anonymous, 1978) to juvenile salmon, where harpacticoid copepods comprise large portions of the gut contents (Sibert et al., 1977). Such predation is often responsible for controlling crustacean population densities (Darnell, 1961), as well as structuring the crustacean subassemblage of a habitat.

Crustaceans also appear to be very important as predators/disturbers in structuring other soft bottom assemblages (e.g., Young et al., 1976; Virnstein, 1977; Bell and Coull, 1978). Segerstrale (1978) suggests that the well-known negative correlation between Pontoporeia (Amphipoda) and Macoma (Bivalvia) densities in the Baltic is a direct result of Pontoporeia predation on newly settled Macoma spat. Segerstrale asserted that Pontoporeia is capable of crushing thin-shelled, minute bivalves and is the most likely cause of the inverse correlation between the two animals. Thus, it appears that in a variety of soft bottom habitats, crustaceans play a significant role in structuring the biotic assemblages.

Crustaceans also play an important role as early colonizers in areas subject to perturbation. McCall (1977) found Ampelesca abdita to be one of three species (the other two were polychaetes) that quickly colonized defaunated mud, and he defined the three species as "opportunistic." Because of their small size, rapid development, high recruitment, and high reproductive rates, these species entered a site early, reached peak density quickly, and then died off. Kaplan et al. (1975) also report a crustacean, the crab Neopanope sayi, as an early colonizer in a dredged channel. They assert that it was because N. sayi is such a large, swiftly-moving form. Simon and Dauer (1977), however, report crustaceans as rather slow colonizers in a subtropical area defaunated by red tide. They suggest that if one is interested in "understanding the most rapid response of the fauna to environmental perturbation, the polychaetes would be the group to examine carefully." They do point out, however, that "if one were interested in whether the habitat was restored and at an 'equilibrium' level of species, the molluscs and the amphipods should be considered." Obviously the role of crustaceans in benthic succession is not clear and certainly worthy of continued investigation.

III. HARD SUBSTRATES

Two hard substrate environments, i.e., rocky shores and reefs, are regularly inhabited by crustaceans. Studies in both of these habitats have provided us with much information on population/community dynamics of crustaceans. Barnacles on rocky shores have been the focus of many studies, and in fact, much of the present theoretical ecology literature is based on such barnacle studies (e.g., Connell, 1961a). Reef assemblage studies are a more recent development, but no less important.

A. Rocky Shores

Barnacles, dominant rocky shore inhabitants, are an unique group of crustaceans, since they have planktonic larvae but they spend their adult life as sedentary forms. Due to their sessile characteristics, barnacles possess unusual properties useful for population studies of Crustacea. For example, a population can be mapped and followed over time and, therefore, recruitment of juveniles and population mortality can be determined (Connell, 1961a,b). Furthermore, since most barnacles inhabit the intertidal zone of rocky shores, these organisms comprise a system that is readily accessible and easily manipulated.

Much of the early work on barnacle ecology was done in the United Kingdom (Moore, 1934, 1935, 1938; Moore and Kitching, 1939; Crisp, 1950; Barnes, 1953a,b, 1956; Barnes and Powell, 1953; Southward, 1953; Southward and Crisp, 1954) and in North America (Weiss, 1948; Bousfield, 1954, 1955). Lewis (1964) synthesized much of the available information on rocky shore ecology and found a general tendency for barnacles to dominate in sheltered bay areas where mussels were uncommon. One of the most obvious barnacle traits is that different species tend to proliferate in distinct zones relative to tidal height. Lewis (1964) suggested that such zonation was strongly related to physical controls such as desiccation and temperature.

Connell (1961a,b) investigated the barnacle zonation phenomena to determine whether alternative factors might be responsible for the distribution patterns. In an elegant set of experiments combining monitoring of barnacle populations over time with transplant and cage manipulations, he showed that the upper limits of barnacle distribution were generally set by physical factors, whereas lower limits were generally the result of biological interactions. More specifically, Connell illustrated that for a complex of two barnacle species from Scotland, *Chthamalus stellatus*, which inhabited the high upper intertidal area was able to settle as cyprids in lower zones, but that the lower intertidal dominant, *Semibalanus balanoides*, grew faster and smothered any *Chthamalus stellatus* which settled in these lower regions (see Fig. 3). In a complimentary experiment, Connell (1961b) determined that the upper shore distributions of *S. balanoides* were set by adverse weather conditions, but that intraspecific competition for space and preda-

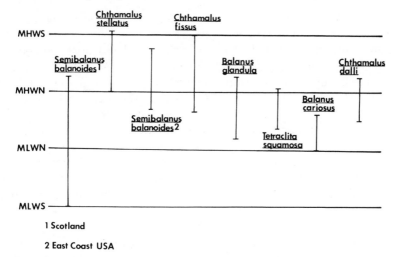

Fig. 3. Approximate tidal distributions of barnacle species discussed in text (MHWS, mean high water tides; MHWN, mean high water neap tides; MLWN, mean low water neap tides; MLWS, mean low water spring tides).

tion by the gastropod *Thais* on large individuals determined the lower limit of distribution. Grant (1977) has recently shown that such density dependent processes (intraspecific and interspecific competition) are also important factors regulating the distribution and abundance of *Semibalanus balanoides* from the East Coast of North America.

In contrast to his findings from Scotland, where intraspecific competition controlled distribution and abundance, Connell (1970) found that predation accounted for most of the mortality of young *Balanus glandula* from Washington, state. Predation, then, importantly contributed to barnacle abundance patterns in this West Coast of North America assemblage. Dayton (1971), also working on the West Coast, showed that adults of *Balanus cariosus*, which inhabited the lower mid-intertidal zone, escaped predation only by growing quickly to large size, while *B. glandula* only had refuge in space (the high intertidal zone) from predation (see Fig. 3). Furthermore, Dayton demonstrated that predation by the gastropod *Thais* on *B. glandula* was critical for successful establishment of a smaller, poorer competitor, *Chthamalus dalli*, in the same high intertidal zone. In addition, limpet grazing selectively affected both species of *Balanus* thereby increasing survival of *Chthamalus dalli*. Thus, in these West Coast experiments, unlike the experiments in Scotland (Connell 1961a,b) and on the East Coast of North America (Grant, 1977), predation and not competition played an important role in regulating barnacle distribution and abundance and, by preventing superior competitors from occupying space, predation also contributed to the coexistence of three barnacle species.

Strathman and Branscomb (1979) conducted additional experiments on settlement of *Balanus glandula* and *Balanus cariosus*. As adults, *B. glandula* is found in the upper mid-intertidal, while *B. cariosus* occupies much lower intertidal areas (Fig. 3). During 1976, when a very large number of larvae settled, Strathman and Branscomb observed that *B. cariosus* larvae did not settle in the higher intertidal area where they were physiologically limited, but larvae of *B. glandula* did settle lower in intertidal areas where predation generally limits distribution. Based on these findings, Strathman and Branscomb suggested that cues for larval settling may well be associated with the physical condition of the environment, but that these cues were probably not well correlated with presence or absence of mobile predators. Other factors regarding successful recruitment of larvae are not well known (e.g., Barnes, 1956). Thus, although we have sophisticated insight into community dynamics of adult barnacles, we need to investigate further the topic of larval ecology.

Hines (1979) provides a comparison of barnacle life history patterns. In his study, Hines compared the population dynamics of *Chthamalus fissus, Balanus glandula,* and *Tetraclita squamosa* from California (see Fig. 3). By following populations of barnacles over time and measuring reproductive parameters (see also Hines, 1978), he was able to construct a life table for each species (see also Connell, 1970) and thus determine individual mortality schedules. A summary of the salient life history features of the three species is presented in Table III. In general the results of Hines' study suggest that factors regulating each population (generally) follow the earlier findings of Connell (1961a, b, 1970) and Dayton (1971) for West Coast barnacles. Such information on life history patterns, as well as information on biological and physical factors regulating communities, makes the barnacle system ecologically one of the best understood crustacean assemblages.

B. Reefs

Decapod crustaceans usually comprise the majority of species and individuals in coral (Abele, 1979) and worm (Gore et al., 1978) reefs. Abele (1979) reported that 80–96% of all individuals and 76–89% of all species associated with *Pocillopora* coral heads were decapod crustaceans. Gore et al. (1978) state that decapod and stomatopod crustaceans on sabellariid worm reefs ". . . account for approximately 90% of the associated macroinvertebrate fauna." Even when crustaceans are not so overwhelmingly dominant, they are usually well-represented members of a reef assemblage (McCloskey, 1970).

TABLE III

Life History Characteristics of Three Barnacle Species from California[a]

Species	Maximum size (mm)	Brooding	Settlement	Generation time (months)	Intrinsic rate of increase
Chthamalus fissus	7	16 Small broods during summer	Erratic year round; survives best at high intertidal levels but overall poorest survivorship of three species	40	1.24
Balanus glandula	20–25	Cold temperatures induce brooding; six broods during winter and spring	Sharp peak of settlement in May–June; survives best at uncrowded high intertidal areas	34	0.818
Tetraclita squamosa	60	Three broods in summer	Settles in fall; only at low tidal levels	64	0.378

[a] Data from Hines, 1979.

Why are decapods/stomatopods so abundant on reefs? Marine benthic decapods use substrates for shelter and/or feeding sites (to graze or capture) and/or as sources of nutrition (Abele, 1974). The complex interwined nature of corals and/or worm tubes increases the number of substrates and thereby increases the number of shelters, feeding sites, and nutritional sources. The presence of several substrates apparently allows more species to coexist through the differential use of each one. One species might use one structure for a shelter, one as a feeding site, and another as a source of nutrition and thus reduce competition for each one (Abele, 1974). As one example, Gore *et al.* (1978) reported that the three dominant crustaceans on sabellariid worm reefs avoided competition by differentially using reef substrates. *Pachycheles monilifer* found shelter in crevices and fed on worm parts suspended in the water by wave action. *Menippe nodifrons* excavated burrows in the worm reef substrate and ranged over the reef to feed on *in situ* worms. *Pachygrapsus transversus* hid in crevices, but ranged freely over the

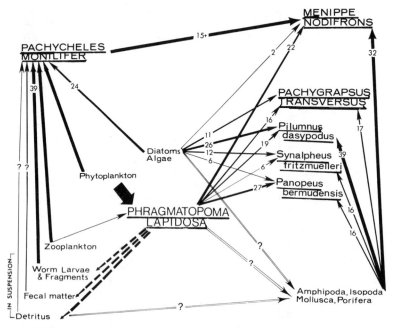

Fig. 4. The trophic subweb postulated for the six most abundant species of decapod crustaceans on central Florida sabellariid worm reefs. The numbers on the lines indicate the percent of observed prey material noted in the gut contents from all combined individuals of a species. Trophic lines without numbers, or with question marks, are interpolated from field observations or from records in the literature. Dashed lines indicate production from the polychaete *Phragmatopoma lapidosa.* (From Gore *et al.,* 1978.)

entire reef and fed as an omnivore on a variety of associated plants and animals. Figure 4 illustrates the postulated food web for the six dominant species on the worm reef. Trophic separation is rather clear for the three dominant species as well as for others. Thus by various niche fractionation mechanisms, 96 species in 52 genera coexisted on the sabellariid reef.

Size of the available habitat is also important in determining the diversity and density of crustaceans in reef habitats. Abele and Patton (1976) found that in most cases small coral heads had fewer species than large heads and smaller population sizes of the individual species, although the density (number per unit area) was slightly higher on the smaller heads. Additionally, fluctuations in the physical environment appear to play a role in determining species diversity, species composition, and density. Abele (1979) has shown that decapods associated with the coral *Pocillopora damicornis* have higher species richness and lower dominance in a physically disturbed (upwelling) area than in a constant environment. Trophic generalists dominate in the physically disturbed region, while trophic specialists, which feed on pocilloporid coral mucus (Patton, 1974), dominate in the more constant environment. Abele (1979) attributed the differing diversities to differential extinction rates in disturbed and constant environments. Since unpredictability would lead to extinction of corals and, thus the associated coral-feeding specialists, this would open up space for colonization by generalists; the case in the disturbed environment. In the constant environment extinction is rare, and therefore the dominants take over, increase their density and frequency of occurrence, and thus, lower diversity.

We know little about the population dynamics of coral-associated crustaceans. Reaka (1979) studied the life history characteristics of 31 populations of coral-inhabiting stomatopods from eight regions of the world between 19°S (Australia) to 30°N (Florida) and provides a most complete account of any coral-inhabiting crustacean population. No recurrent pattern of life history traits emerged and Reaka stated that ''. . . the diversity of life history patterns observed, even within a single lineage and guild, follows directly from the evolutionary flexibility derived from the moulting biology of this group of Crustacea.'' Reaka was able to generalize on the patterns of energy allocation as they varied with body size, latitude, and degree of sympatry. Her table summarizing the patterns of allocation is reproduced here (Table IV) and indicates the diversity of ways energy expenditure is correlated in coral-inhabiting stomatopods. Reproductive patterns are mostly determined by body size and continue to dominate age-specific reproductive events (Table IV). Latitude appeared to have variable effects on both maintenance and reproductive activity, whereas sympatry appeared most importantly to increase interspecific agression e.g., *Gonodactylus zacae* from Costa Rica, which occurred with four congeners, was more aggressive than size

TABLE IV

Patterns of Allocation as They Vary with Body Size, Latitude, and Degree of Sympatry among Populations[a,b]

	Ma	C	Pd	Pa	MR	%G/M	M/l	%G/y	Es	En	%CV	RR	%R/y	E/y	R/l	E/l	L
↑ Body size	↓?	←	↑?	↑?	←	←	—	←	←	←	→	←	—	←	—	←	→
↑ Latitude	→?	—	—	↑?	←	↓?	←	—	—	←	→	↑?	—	—	—	—	—
↑ Sympatry	—	←	—	—	→	—	—	→	—	—	—	—	—	—	—	—	↑?

[a] From Reaka, 1979.

[b] Ma, maintenance cost; C, competition for holes; Pd and Pa, avoidance of predators and parasites; MR and %G/M, molting rates and increments; M/l, molts/life; %G/y, yearly percent volumetric growth; Es, egg size; En, egg number; %CV, percent clutch volume; RR, reproductive rate; %R/y, yearly percent volumetric reproduction; E/y, number of eggs/year; R/l, number of reproductions per life; E/l, number of eggs per life; L, longevity.

matched *G. zacae* from the Gulf of California, which occurred alone (Reaka, 1979). Increased sympatry also increased the molting rate and the yearly growth. Reaka clearly showed the problems and complexities of measuring population attributes, even within a single guild, and thus it should come as no surprise as to why our knowledge and understanding of the population dynamics of coral reef crustaceans, if not all crustaceans, is so severely limited.

A fascinating, predominantly crustacean reef assemblage, the demersal plankton, has been an enigma for us because we did not know whether to include discussion of it in the plankton section (IV) or here. Following Alldredge and King's (1977) arguments that these plankters hide in reef sediments during the day but emerge to swim freely at night, they are best considered resident members of the reef community and thus included in our discussion of reef-associated crustaceans.

The demersal reef plankton includes mysids, gammarid amphipods, ostracods, isopods, harpacticoid copepods, and certain species of calanoid and cyclopoid copepods, decapod larvae, and polychaetes that do not occur in the open water plankton surrounding reefs. Furthermore, the same types and numbers of these "plankton" emerge from the same location night after night and thus maintain their position over localized reef areas. Alldredge and King (1977) suggest several adaptive strategies that allow the demersal plankton to maintain their constant position: (1) Active swimming: Many of the demersal plankton are active, rapidly swimming forms with eyes. Currents within reefs are of low velocity, and many of the demersal taxa thereby maintain themselves by swimming. (2) Hiding in the lee of large object: Several demersal plankters have been observed hiding behind corals and in caves and crevices. (3) Near bottom residence: By staying near the bottom, current velocity is reduced, and two of three studies cited by Alldredge and King (1977) report more demersal plankton near the bottom than in the remainder of the water column; the third study found no difference. (4) Limited emergence patterns: Several demersal plankters limit their exposure to currents and/or predators by swimming only a few hours a night. In fact, some species appear to reenter the substrate when current speeds exceed their swimming speeds. Behavioral adaptations appear to be important in maintaining the demersal reef plankton *in situ*.

Alldredge and King's (1977) findings of high densities and biomass of demersal zooplankton, in contrast to rather low densities of surrounding net zooplankton led them to state, "Resident reef plankton represent an important energy source (and) . . . this reef may be largely independent of oceanic zooplankton." Such a finding has profound ecological significance, and if most reefs are so supported, we certainly must reevaluate our thinking of trophic dynamics in coral reefs.

IV. ZOOPLANKTON

A. Biological Attributes

Crustaceans, particularly copepods and cladocerans, are important members of both freshwater and marine zooplankton communities. Copepods are abundant in freshwater and marine habitats, but most dominant in the latter. Cladocerans assume special importance in lacustrine habitats and are rarely found in marine waters, with the exception of the genera *Evadne, Penilia,* or *Podon,* which are commonly found in brackish waters. Additionally in Antarctic and deeper water depths, euphausiids frequently comprise a high percentage of all zooplankters.

The majority of crustacean zooplankters spend their life cycle in the water column (=holoplanktonic) (resting stages of Cladocera and Copepoda excepted), although many crustaceans that inhabit soft and hard bottom habitats as adults have an obligatory larval stage that is released into the plankton (=meroplanktonic). Planktonic larvae are characteristic of the Decapoda and Cirripedia and the occurrence of larval stages in the plankton is usually seasonally pulsed by temperature. Other crustaceans such as amphipods, cumaceans, mysids, and ostracods are encountered in plankton samples, but are generally not considered planktonic forms.

An open water environment presents special features and problems. Unlike benthic habitats, the fluid medium of water is transparent and there is no visible geometric heterogeneity (Hall et al., 1976). Although there are no physical discontinuities, gradients of oxygen, temperature, and light do exist within the water column. An open water existence demands special physiological and morphological adaptations and, due to the properties and processes which are characteristic of a water column, zooplankton have evolved unique ecological features. For instance, visual feeding is highly developed compared to benthic animals, and mechanisms of niche diversification are correspondingly different.

There are distinct differences between lacustrine and marine openwater habitats (other than salinity) that influence the species composition and biological interactions within each system. Freshwater lakes and ponds are relatively stable with only isolated periods of vertical mixing due to seasonal turnover and/or wind-associated effects. They are essentially "closed" systems, and this has allowed biological interactions within field assemblages to be ascertained with some sophistication. On the other hand, marine systems are greatly influenced by physical factors, and significant water movement exists on both horizontal and vertical axes. On a large scale, water circulation via currents transports great volumes of water (and zooplankton), and water layer stratification is common due to salinity, temperature, and/or density gradients. All of these physical features have a profound effect on zooplankton abundance and distribution and make it almost impossible to identify (let alone experimentally manipulate) discrete zooplankton populations and communities. Patchiness or nonrandom distribution of zooplankton and its food is well known (e.g., Cassie, 1959; Anraku, 1975; Mullin and Brooks 1976; Dagg, 1977), and such patchiness creates difficulties in assessing how representative zooplankton samples are over wide areas and/or time (Wiebe and Holland, 1968; Sameoto, 1975). Turbulence and internal waves have been identified as important generators of zooplankton patchiness (Haury, 1976), but Mullin and Brooks (1976) point out that this same turbulence may be responsible in part for making sufficient food supplies available to zooplankters.

Most studies of zooplankton assemblages have focused on their dominant crustacean members, i.e., cladoceran and copepods, and both of these groups are easily studied under laboratory conditions. Cladocerans are predominantly filter feeders (Edmonson, 1957), and much literature exists on the feeding mechanisms of this group (e.g., Rigler, 1961; Burns and Rigler, 1962; Burns, 1968; Berman and Richman). A few carnivorous cladocerans (*Polyphemus, Leptodora*) tear their prey with their mandibles and then suck in the food (Monakov, 1972). The majority of copepods are also filter feeders or herbivores (see Mullin, 1963, Conover, 1966; Frost, 1972; Richman et al., 1977), but omnivores are also common (Anraku and Omori, 1963; Gauld, 1966; McQueen, 1969; Confer, 1971; Ambler and Frost, 1974; Confer and Cooley, 1977; Landry, 1978a).

Reproductive status of cladocerans and copepods is easily noted, since egg sacs are generally visible within a brood pouch or carried externally, and it is not surprising that there are numerous studies on reproductive ecology of various species. Copepods reproduce sexually by transfer of a male spermatophore to the female, which in turn is capable of producing multiple broods. One mating generally provides a sufficient supply of sperm to last the life of a female and fertilize many broods (Whitehouse and Lewis, 1973). Cladocerans have a specialized reproductive cycle unique among crustaceans in that they may reproduce parthenogenetically and subsequently form diploid female eggs, parthenogenetic diploid male eggs, or haploid resting eggs. During favorable conditions, parthenogenetic reproduction is the rule resulting in a large number of offspring being introduced quickly into a population. However with the onset of adverse conditions, a sexual generation dominates, and resting stage eggs are subsequently produced.

Another feature of zooplankton that has important ecological impact is the phenomenon of vertical migration. This complex behavior thought to be based on cues of light, pressure, or other physical factors allows small crustaceans in both marine and freshwater systems to exploit available food at different levels of the water column and/or avoid predation by visual predators (Lane, 1975). Furthermore, vertical migration may aid to retain larval stages of adult crustaceans in appropriate water depths for dispersal to colonizing sites (Sandifer, 1975; Christy, 1978 or provide an energetic advantage and increased fecundity (McLaren, 1963). For the purposes of our discussion it is sufficient that vertical migration is an important process whereby organisms move to different water layers, thereby complicating not only any interpretation of zooplankton distribution, but also potential biological interrelationships.

B. Ecological Studies

The bulk of ecological study has focused on individual populations. Furthermore, both freshwater and marine zooplankters have been studied extensively under laboratory conditions, and most of our present knowledge about zooplankton's role in open water systems and basic life history strategies have evolved from these studies. Generally, ecological studies on zooplankton have not focused on describing the physical features of a habitat and correlating species abundance patterns and distributions with these features. This may be because freshwater systems are relatively homogeneous, and in such a habitat, it is hard to view marine zooplankton as responding to gradients or evaluate distributions in time and space. But some studies on estuarine zooplankton have looked at communities along a salinity gradient (e.g., Herman et al., 1968). Likewise, there are some lacustrine reports which have investigated near shore species of zooplankton versus open water populations (Gehrs, 1974; Kerfoot, 1977a).

1. POPULATION DYNAMICS AND LIFE HISTORY STUDIES

The cladocerans Daphnia spp. have been the target of many life history and population studies. Laboratory studies have revealed that food and temperature are important influences on growth rate, with food generally limiting growth and brood size, whereas temperature controls frequency of molting, egg development time, and physiological lifespan (Slobodkin, 1954). There have been many studies on many species of Daphnia, and Hebert (1978) provides an up-to-date review of Daphnia population biology. The most pertinent conclusions from his review are:

1. Developmental rates and lifespans of different species are similar.
2. Brood size is variable and depends on food supply.
3. The small, relatively static, mean brood size noted in summer populations of Daphnia suggests that these populations are limited by, and in

equilibrium with, their food supplies. Mortality rates of field populations are considerably higher than in food-limited laboratory populations. Although some mortality has been attributed to predation, highest mortality usually occurs when food levels are low, and therefore
mortality is probably due to starvation.
4. Dietary overlap is high between the filter-feeding *Daphnia* spp.
5. Small species of *Daphnia* predominate in lakes because the larger
species are removed by fish. In ephemeral habitats, large species predominate because of higher birth rates and protection from invertebrate predation.

Other cladocerans have not been as widely studied as *Daphnia*. Kerfoot
(1974) investigated the reproductive ecology and factors regulating
cyclomorphosis in *Bosmina* and found that the lineages of *Bosminia long-
irostris* underwent extreme cyclic changes in egg sizes, a phenomenon
unique among cladocerans. Although females carried small eggs in summer
and large eggs in late fall, Kerfoot did not attribute these changes to nutritional conditions as most other studies would have predicted. Rather, he
argued that the alternation of predation by fish and invertebrates could be
responsible for selection of this reproductive behavior. Furthermore, Kerfoot
(1977a) suggests that predation is extremely important in influencing
cyclomorphosis within the *Bosmina* assemblages. Clearly, these studies emphasize the need for more *in situ* life history studies. There is little information on population dynamics and biology of other cladocerans (but see
Muragan and Sivaramakrishnan, 1973), and this is an area worthy of additional study.

Information on life history phenomena and population dynamics of
copepods in lacustrine environments have come primarily from laboratory-
reared copepods to determine development time, clutch size, and fecundity
(e.g., Smyly, 1970; Whitehouse and Lewis, 1973). Cooley and Mims (1978)
summarized the available literature on freshwater egg development time
and temperature and concluded that no one universal egg development
curve could be applied to either calanoid or cyclopoid copepods, and that
significant variations in development time exist even within a species.

Although there are many studies of seasonal patterns of abundance of
individual copepods, not many studies have dealt with population dynamics
of field assemblages. Vijverberg (1977) investigated the population structure
and life history of copepods in the Netherlands and reported that 91% of the
copepod population went into diapause during the copepod stage, and
cyclopoid diapause eggs are not uncommon (Smyly, 1961). Comita (1972)
attempted to develop population statistics for two copepods, Diaptomus
(*Leptodiaptomus*) *siciloides* and *Mesocyclops edax*, but his methods for

calculation of instar duration were insufficient (Rigler and Cooley, 1974). Rigler and Cooley (1974) subsequently suggested a method for calculation of copepod instar duration from field data, although their method also has an inherent assumption concerning constant mortality rates. Importantly, their findings suggested that there was no universal pattern of mortality rate in calanoid copepods. They emphasized the need for better and more frequent sampling methods so population changes could be more adequately estimated. Gehrs and Robertson (1975) constructed a life table for the copepod *Diaptomus sic i loides* from field samples taken every 2 days combined with laboratory determinations of instar duration. They determined survivorship and mortality. Complimentary laboratory experiments on survivorship indicated that 28% of eggs never reached the N IV stage, while in the field 84% of eggs failed to develop. Once the copepodite stage was reached, survivorship was relatively similar in field and lab populations, suggesting that minimal predation occurred on individual copepodites.

Population studies on marine copepods have generally been limited to laboratory studies of reproductive activity, development times, and fecundity (e.g., McLaren, 1965, 1966, 1978; Mullin and Brooks, 1967; Heinle, 1969; Katona and Moodie, 1969; Geiling and Campbell, 1972 Landry, 1975; Paffenhöffer and Harris, 1976), and demographic data on field populations are rarely available.

A few studies have investigated copepod population dynamics in field assemblages. McLaren (1978) compiled extensive information on marine copepod life-cycles and suggested that researchers should use copepodite stages to follow cohorts in the field. Landry (1978b) combined laboratory and *in situ* experiments with time series sampling of field populations to examine a population of *Acartia clausi* from a shallow lagoon in Washington. He concluded that the seasonal cycle of abundance of *Acartia clausi* was regulated by a consistent pattern of copepodid and adult mortality probably due to the effect of fish predation. However, Landry also added that cannabalism and hatching of eggs in sediments help to regulate population abundance within seasons. D'Apolito and Stancyk (1979) constructed a life table for the pelagic harpacticoid copepod *Euterpina acutifrons* from South Carolina, following a design similar to Landry's (1978b). In contrast to the results of Landry, D'Apolito and Stancyk reported that the highest mortality of *E. acutifrons* occurred between the N VI–CI stages. Thus, there seems to be no universal pattern of copepod mortality, but the use of time series analysis coupled with field and laboratory experiments should prove to be a most valuable approach to further investigation of copepod population dynamics.

Allan (1976) presents an excellent comparative discussion on the life history patterns of crustacean zooplankters, and a summary his findings is presented in Table V. Specifically, he examined the existing information on r_{max} (intrinsic rate of increase) in an attempt to discern relative trade-offs to reproductive potential, predator avoidance, and competitive ability for copepods and cladocerans. Cladocera possessed a greater r_{max} than Copepoda, although both groups relied on a large number of offspring to achieve r_{max}. Allan suggested that Cladocera are susceptible to vertebrate predators, which are strongly pulsed so that the "opportunistic" traits (high-r_{max}) were highly favored evolutionarily. In contrast, he speculated that the lower r_{max} exhibited by the copepods was because copepods, which dominate in marine systems, are probably better able to escape from predators and live in a more climatically "buffered" zone than cladocerans. Since much of the data is based on laboratory results, additional studies of field populations of both freshwater and marine zooplankton should add important information to Allan's comparison.

2. FRESHWATER COMMUNITIES

The study of freshwater zooplankton communities has focused on descriptions of community structure and analysis of patterns of abundance of both copepod and cladoceran assemblages. Commonly, communities are

TABLE V

Summary of Certain Ecological Characteristics of Cladocera and Copepoda[a]

	Cladocera[b]	Copepoda
Biogeography (global)	Freshwater and estuarine	Freshwater, estuarine, and marine
Biogeography (large lake distribution)	Surface and inshore	Deep, open water
Abundance pattern	Vernal peak	Variable
Mode of feeding	Filter feeding via thoracic appendages	Filter and/or raptorial
Food size range (mm)	1–50	5–100
Filtering rate	High	Low
Typical adult body size (mm)	0.3–3.0	0.5–5.0
Largest species (mm)	5.0	17.0
Susceptibility to vertebrate predators	High	Low
Susceptibility to invertebrate predators	Moderate	Moderate to high
r_{max} (1/day)	0.2–0.6	0.1–0.4

[a] Adapted from Allan, 1976.
[b] *Leptodora* not included.

often dominated by either large or small zooplankters. Brooks and Dodson (1965) examined this phenomena and explained the inverse abundance patterns on the basis of predation and competition. It is instructive to examine their hypothesis in detail since subsequent studies have focused on testing these ideas. There argument was structured as follows:

1. Large and small zooplankton are filter feeders which compete for food and large zooplankters are more efficient, hence better competitors, than smaller forms.
2. Predators select larger zooplankton and when predation is of low intensity, the larger zooplankters will dominate a community.
3. If predation is intense, size dependent predation will allow small zooplankter's to become dominant.
4. Moderate predation will allow both size classes to coexist.

The theory of vertebrate (primarily fish) predator size selectivity, based on observations and manipulations is generally accepted and well documented (Macan, 1966; Galbraith, 1967; Brooks, 1968; Dodson, 1970; Anderson, 1972; Werner and Hall, 1974). These predators are known to control herbivores due to their wide prey selection, rapid switching, high feeding rates, and highly plastic feeding behavior (Hall et al., 1976). However, in lake systems, invertebrate predators such as midge larvae *Chaoborus sp.*, the cladoceran *Leptodora,* and predaceous copepods may prey on small herbivorous zooplankters, thus providing an alternative explanation for the observed low abundance of small forms when fish are not present (McQueen, 1969; Allan, 1973; Dodson, 1974; Brandl and Fernando, 1975; Federenko, 1975; Confer and Cooley, 1977; Kerfoot, 1977b). It is difficult to evaluate the hypothesis or its alternative in the laboratory or field. It is believed that invertebrate predators have only a poor control of herbivorous zooplankton, but together with food limitation may drastically limit small zooplankton populations (Hall et al., 1976).

Hall et al. (1976) have provided an extensive review of the size efficiency hypothesis, the numerous related studies, and a good discussion of other possible factors influencing zooplankton community structure. One basic assumption of the hypothesis that may not apply in all instances is that large herbivores are superior competitors to smaller ones. Threlkeld (1976) has cited food limitation and starvation as a possible mechanism to influence size structure of a population, and Lynch (1977) further suggests that food selection of small species does not always overlap greatly with that of larger species, thus questioning one of the basic assumptions of the size efficiency hypothesis. Lynch also points out that the young of the large species may be more likely to compete with the small forms, as other studies have demonstrated (Neill, 1975). These findings point to the need for more critical testing of those factors which influence size of structure of zooplankton assemblages and for continued studies on intrazooplankton competition and resource utilization.

Some investigators have delineated spatiotemporal abundance patterns of major zooplankton groups from freshwater habitats in order to gain insight into zooplankton community structure (Miracle, 1974; Lane, 1975; Makarewicz and Likens, 1975). Lane (1975) examined the abundance and distribution of a guild of cladocerans and copepods from four lakes and reported that in two of the lakes, dominant species segregated by habitat; but in two other lakes, food supplies, and not spatial segregation, probably contributed to the coexistence of these very similar organisms. Makarewicz and Likens (1975), in a similar spatiotemporal study of all zooplankton, suggested that zooplankton fractionated the niche hyperspace as a result of predation and competitive displacement. Also they found, as did Miracle (1974), that potential competitors were usually separated in space and time. Although the various measurements of niche dimensionality and community structure, as well as interpretation of results, remain controversial (Lane, 1978; Makarewicz and Likens, 1978) such studies may give us a better understanding of potential biological interactions which may operate in zooplankton communities.

3. MARINE COMMUNITIES

Information on biological interactions in marine communities is not as abundant as that of freshwater systems, and many studies have focused on characterizing zooplankton distributions in respect to such physical parameters as salinity and/or temperature (Jefferies, 1962, 1966; Lance, 1963; Bayly 1965), or turbulence and water movement (Ketchum, 1954; Barlow, 1955). Furthermore there are many comparative studies which have characterized species composition and abundance of selected locations (e.g., Woodmansee, 1958; Grice, 1960; Jefferies, 1964; Herman et al., 1968; Williams et al., 1968; Bakker and DePauw, 1975; Lonsdale and Coull, 1977; and references within). In many of these studies, predation and/or competition are discussed as potentially important determinants of community structure, although no empirical data is provided.

A few studies have suggested that predation is important in marine zooplankton communities. Hodgkin and Rippingdale (1971) demonstrated in the laboratory that two copepods (*Acartia* and *Sulcanus*) fed on nauplii of a third (*Gladioferens*), and combined with distributional data, Rippingdale and Hodgkin (1974) present some of the first evidence of predator–prey relationships within a marine zooplankton assemblage. Lonsdale et al. (1979) have conducted a similar study on *Acartia tonsa* feeding under laboratory conditions and found that it fed preferentially on nauplii of *Eu-*

rytemora affinis and *Scottolana canadensis* (see also Ambler and Frost, 1974; Landry, 1978a). Clearly, predation by copepods may have an important effect on both size structure and species composition of marine zooplankton communities.

In contrast to the above, McLaren (1978) argues that food may not be a limiting factor in marine systems and that, at least in copepods from Loch Stiven, Scotland, food does not have an effect on the life history strategies of copepods. Furthermore, he suggests that intraspecific and interspecific competition for resources need not be important to marine copepods. Landry (1977) has also argued that physical factors (i.e., upwellings) are important factors influencing trophic pathways and that at any time, the physical environment might consist of various local environments in different successional states. This property, coupled with a complex trophic pathway, suggests that predator–prey relationships may be extremely variable or flexible within the marine system. Therefore, predictions on predator effects on community structure may not be as straightforward as in the freshwater system; it seems that the physical processes in the ocean have a very important effect on development and succession of marine zooplankton communities.

V. PERSPECTIVES

Quantitative data on crustacean populations and communities are severely lacking. While some crustacean assemblages are much better known than others, the role of these animals in aquatic systems is, in general, poorly known. Crustacean workers should not be discouraged. The difficulties inherent in studying animals that molt and whose larvae may "float away" are not easily overcome. Demographic data are particularly hard to obtain, and continued efforts to provide such information are imperative. Field-generated life history data would certainly provide a more natural assessment of what really happens, but the lack of suitable techniques to accurately collect such badly needed data still hampers our estimation of crustacean dynamics.

Determining crustacean community organization is a no less frustrating task. In many cases, crustaceans are but one of many taxa interacting in a habitat. To single out this one taxon, albeit the one crustacean biologists believe the most important, is, in many cases, a naive approach to understanding communities. Certainly, crustaceans are overwhelmingly dominant in some habitats (reefs, water column, sandy beaches) and accordingly, such single taxon analyses are warranted. However, in most other habitats one must include the associated taxa in any evaluation of community structure and function. Efforts must now be directed at discerning interrelationships between Crustacea and associated taxa. As new approaches to the study of crustacean life histories, demography, and species interactions are developed, we should begin to obtain a more refined understanding of the role of crustaceans in ecosystems.

We have reviewed representative information on population dynamics and biological interactions of crustacean assemblages from distinctly different habitats and geographic locations, and throughout this survey a number of important conclusions repeatedly emerge:

1. In general, most of the ecological information on biotic assemblages is descriptive, and a great deal of literature exists on distribution and abundance of Crustacea from many systems.
2. The biological characteristics of crustaceans as well as the physical limitations of their habitats make any field study of population dynamics extremely difficult. Therefore, much of the life history and demographic data have been generated in the laboratory.
3. For those systems that are better studied (i.e., barnacles, freshwater zooplankton) there are great differences in life history features and community organizing processes. This is especially true within similar systems or species guilds from different geographic locations.

REFERENCES

Abele, L. G. (1974). Species diversity of decapod crustaceans in marine habitats. *Ecology* **55,** 156–161.
Abele, L. G. (1979). The community structure of coral associated decapod crustaceans in variable environments. In "Ecological Processes in Coastal and Marine Systems" (R. J. Livingston, ed.), pp. 265–290. Plenum, New York.
Abele, L. G., and Patton, W. K. (1976). The size of coral heads and the community biology of associated decapod crustaceans. *J. Biogeogr.* **3,** 35–57.
Allan, J. D. (1973). Competition and the relative abundances of two cladocerans. *Ecology* **54,** 484–498.
Allan, J. D. (1976). Life history patterns in zooplankton. *Am. Nat.* **110,** 165–180.
Alldredge, A. L., and King, J. M. (1977). Distribution, abundance, and substrate preferences of demersal reef zooplankton at Lizard Island Lagoon, Great Barrier Reef. *Mar. Biol. (Berlin)* **41,** 317–333.
Ambler, J. W., and Frost, B. W. (1974). The feeding behavior of a predatory planktonic copepod, *Tortanus discaudatus. Limnol. Oceanogr.* **19,** 446–451.
Anderson, R. S. (1972). Zooplankton composition and change in an alpine lake. *Verh. Int. Ver. Limnol.* **18,** 264–268.
Anonymous (1978). Endangered Species. *Aquanotes, (Louisiana State Univ., Sea Grant Publ.)* **7,** 1–5.
Anraku, M. (1975). Microdistribution of marine copepods in a small inlet. *Mar. Biol. (Berlin)* **30,** 79–87.

Anraku, M., and Omori, M. (1963). Preliminary survey of the relationship between the feeding habit and the structure of the mouth-parts of marine copepods. *Limnol. Oceanogr.* **8,** 116–126.

Bakker, C., and DePauw, N. (1975). Comparison of plankton assemblages of identical salinity ranges in estuarine tidal and stagnant environments. II. Zooplankton. *Neth. J. Sea Res.* **9,** 145–165.

Barlow, J. P. (1955). Physical and biological processes; determining the distribution of zooplankton in a tidal estuary. *Biol. Bull. (Woods Hole, Mass.)* **109,** 211–225.

Barnes, H. (1953a). Size variations in the cyprids of some common barnacles. *J. Mar. Biol. Assoc. U.K.* **32,** 297–304.

Barnes, H. (1953b). Orientation and aggregation in *Balanus balanus* (L.) Da Costa. *J. Anim. Ecol.* **22,** 141–148.

Barnes, H. (1956). *Balanus balanoides* L. in the Firth of Clyde: The development and annual variation in the larval population and the causative factors. *J. Anim. Ecol.* **25,** 72–84.

Barnes, H., and Powell, H. T. (1950). The development, general morphology, and subsequent elimination of barnacle populations, *Balanus crenatus* and *B. balanoides,* after a heavy initial settlement. *J. Anim. Ecol.* **19,** 175–179.

Bayly, I. A. E. (1965). Ecological studies on the planktonic Copepoda of the Brisbane River estuary. *Aust. J. Mar. Freshwater Res.* **16,** 315–350.

Bell. S. S., and Coull, B. C. (1978). Field evidence that shrimp predation regulates meiofauna. *Oecologia* **35,** 141–148.

Berman, M. S., and Richman, S. (1974). The feeding behavior of *Daphnia pulex* from Lake Winnebago, Wisconsin. *Limnol. Oceanogr.* **19,** 105–109.

Birklund, J. (1977). Biomass, growth and production of the amphipod *Corophium insidiosum* Crawford, and preliminary notes on *Corophium volutator* (Pallas). *Ophelia* **16,** 187–203.

Bousfield, E. L. (1954). The distribution and spawning seasons of barnacles on the Atlantic Coast of Canada. *Bull. Natl. Mus. Can.* **132,** 112–154.

Bousfield, E. L. (1955). Ecological control of the occurrence of barnacles in the Miramichi estuary. *Bull. Natl. Mus. Can.* **137,** 1–69.

Bousfield, E. L. (1970). Adaptive radiation in sand-burrowing amphipod crustaceans. *Chesapeake Sci.* **11,** 143–154.

Brandl, Z., and Fernando, C. H. (1975). Investigations on the feedings of carnivorous cyclopoids. *Verh. Int. Ver. Limnol.* **19,** 2959–2965.

Brooks, J. L. (1968). The effects of prey size selection by lake planktivores. *Syst. Zool.* **17,** 272–291.

Brooks, J. L., and Dodson, S. I. (1965). Predation, body size and composition of plankton. *Science* **150,** 28–35.

Burns, C. W. (1968). Direct observations of mechanisms regulation feeding behavior of *Daphnia,* in lakewater. *Int. Rev. Gesamten Hydrobiol.* **53,** 83–100.

Burns, C. W., and Rigler, F. H. (1967). Comparison of filtering rates of *Daphnia rosea* in lake water and in suspensions of yeast. *Limnol. Oceanogr.* **12,** 492–502.

Bynum, K. H. (1978). Reproductive biology of *Caprella penantis* leach 1814 (Amphipoda:Caprellidae) in North Carolina U.S.A. *Est. Cstl. Mar. Sci.* **7,** 473–485.

Cassie, R. M. (1959). Micro-distribution of plankton. *N. Z. J. Sci.* **2,** 398–409.

Christy, J. H. (1978). Adaptive significance of reproductive cycles in the fiddler crab *Uca pugilator* : A hypothesis. *Science* **199,** 453–455.

Colman, J. S., and Segrove, F. (1955). The fauna in Stoupe Beck sands, Robin Hood's Bay (Yorkshire, North Reading). *J. Anim. Ecol.* **24,** 426–444.

Comita, G. W. (1972). The seasonal zooplankton cycles, production, and transformations of energy in Severson Lake, Minnesota. *Arch. Hydrobiol.* **70,** 14–66.

Confer, J. L. (1971). Intrazooplankton predation by Mesocyclops edax at natural prey densities. Limnol. Oceanogr. **16,** 663–667.

Confer, J. L., and Cooley, J. M. (1977). Copepod instar survival and predation by zooplankton. J. Fish. Res. Board Can. **34,** 703–706.

Connell, J. B. (1961a). The influence of interspecific competition and other factors on the distribution of the barnacle Chthamalus stellatus. Ecology **42,** 710–723.

Connell, J. B. (1961b). Effects of competition, predation by Thais lapillus, and other factors on natural populations of the barnacle Balanus balanoides. Ecol. Monogr. **31,** 61–104.

Connell, J. B. (1970). A predator-prey system in the marine intertidal region. I. Balanus glandula and several predatory species of Thais. Ecol. Monogr. **40,** 49–78.

Conover, R. J. (1966). Factors affecting the assimilation of organic matter by zooplankton and the question of superfluous feeding. Limnol. Oceanogr. **11,** 339–345.

Cooley, J. M., and Mims, C. K. (1978). Prediction of egg development times of freshwater copepods. J. Fish. Res. Board Can. **35,** 1322–1329.

Coull, B. C. (1970). Shallow water meiobenthos of the Bermuda platform. Oecologia **4,** 325–357.

Crisp, D. J. (1950). Breeding and distribution of Chthamalus stellatus. Nature (London) **166,** 311–312.

Croker, R. A. (1967). Niche diversity in five sympatric species of intertidal amphipods. (Crustacea: Haustoriidae). Ecol. Monogr. **37,** 173–200.

Croker, R. A. (1977). Macro-infauna of northern New England marine sand: Long-term intertidal community structure. In "Ecology of Marine Benthos" (B. C. Coull, ed.), pp. 439–450. Univ. of South Carolina Press, Columbia, South Carolina.

Croker, R. A., Hager, R. P., and Scott, K. J. (1975). Macroinfauna of Northern New England marine sand. II. Amphipod-dominated intertidal communities. Can. J. Zool. **53,** 42–51.

Dagg, M. (1977). Some effects of patchy food environments on copepods. Limnol. Oceanogr. **22,** 99–107.

Dahl, E. (1953). Some aspects of the ecology and zonation of the fauna on sandy beaches. Oikos **4,** 1–27.

D'Apolito, L. M., and Stancyk, S. E. (1979). Population dynamics of Euterpina acutifrons (Copepoda, Harpacticoida) from North Inlet, South Carolina with reference to dimorphic males. Mar. Biol. (Berlin) **54,** 251–260.

Darnell, R. M. (1961). Trophic structure of an estuarine community based on studies of Lake Pontchartrain Louisiana. Ecology **42,** 553–568.

Dayton, P. K. (1971). Competition, disturbance, and community organization: The provision and subsequent utilization of space in a rocky intertidal community. Ecol. Monogr. **41,** 351–389.

Dexter, D. M. (1969). Structure of an intertidal sand beach community in North Carolina. Chesapeake Sci. **10,** 93–98.

Dexter, D. M. (1971). Life history of the sandy-beach amphipod Neohaustoris schmitzi (Crustacea: Haustoriidae). Mar. Biol. (Berlin) **8,** 232–237.

Dexter, D. M. (1972). Comparison of the community structures in a Pacific and an Atlantic Panamanian sandy beach. Bull. Mar. Sci., **22,** 449–462.

Dexter, D. M. (1976). The sandy-beach fauna of Mexico. Southwest Nat. **20,** 479–485.

Dodson, S. I. (1970). Complimentary feeding niches sustained by size-selective predation. Limnol. Oceanogr. **15,** 131–137.

Dodson, S. I. (1974). Zooplankton competition and predation: An experimental test of the size-efficiency hypothesis. Ecology **55,** 605–613.

Edmondson, W. T. (1957). Trophic relations of the zooplankton. Trans. Am. Microsc. Soc. **76,** 225–245.

Federenko, A. Y. (1975). Feeding characteristics and predation impact of *Chaoborus* (Diptera: Chaoboridae) larvae in a small lake. *Limnol. Oceanogr.* **20**, 250–258.

Feller, R. J. (1977). Life history and production of meiobenthic harpacticoid copepods in Puget Sound. Ph.D. Thesis, 249 pp. Univ. of Washington, *Seattle*.

Finchman, A. A. (1971). Ecology and population studies on some intertidal and sublittoral sand-dwelling amphipods. *J. Mar. Biol. Assoc. U. K.* **51**, 471–488.

Fleeger, J. W. (1979). Population dynamics of three estuarine meiobenthic harpacticoids (Copepoda) in South Carolina. *Mar. Biol. (Berlin))* **52**, 147–156.

Fleeger, J. W., and Palmer, M. A. (1982). Secondary production of the meiobenthic copepod, *Microarthridion littorale*. *Mar. Ecol. Prog. Ser.* **7**, 157–162.

Frankenberg, D., and Leiper, A. S. (1977). Seasonal cycles in benthic communities of the Georgia continental shelf. *In* "Ecology of Marine Benthos" (B. C. Coull, ed.), pp. 383–398. Univ. of South Carolina Press, Columbia, South Carolina.

Frost, B. W. (1972). Effects of size and concentration of food particles on the feeding behavior of the marine planktonic copepod *Calanus pacificus*. *Limnol. Oceanogr.* **17**, 805–815.

Gable, M. F., and Croker, R. A.(1977). The salt marsh amphipod, *Gammarus palustris* Bousfield, 1969 at the northern limit of its distribution I. Ecology and life cycle. *Est. Cstl. Mar. Sci.* **5**, 123–134.

Galbraith, M. G. (1967). Size-selective predation on *Daphnia* by rainbow trout and yellow perch. *Trans. Am. Fish. Soc.* **96**, 1–10.

Gauld, D. T. (1966). The swimming and feeding of planktonic copepods. *In* "Some Contemporary Studies in Marine Science" (H. Barnes, ed.), pp. 313–334. Allen and Unwin, London.

Gehrs, C. W. (1974). Horizontal distribution and abundance of *Dipatomus clavipes* Schacht in relation to *Potamogeton foliosus* in a pond under experimental conditions. *Limnol. Oceanogr.* **19**, 100–104.

Gehrs, C. W., and Robertson, A. (1975). Use of life tables in analyzing the dynamics of copepod populations. *Ecology* **56**, 665–672.

Geiling, W. T., and Campbell, R. S. (1972). The effect of temperature on the development rate of the major life stages of *Diaptomus pallidus* Herrick. *Limnol. Oceanogr.* **17**, 304–307.

Glynn, P. G., Dexter, D. M., and Bowman, T. E.(1975). *Excirolana braziliensis*, a Pan-American sand beach isopod: Taxonomic status, zonation and distribution. *J. Zool.* **175**, 509–521.

Gore, R. H., Scotto, L. E., and Becker, L. J. (1978). Community composition, stability, and trophic partitioning in decapod crustaceans inhabiting some subtropical sabellarid warm reefs. *Bull. Mar. Sci.* **28**, 221–248.

Grant, W. S. (1977). High intertidal community organization on a rocky headland in Maine, USA. *Mar. Biol. (Berlin)* **44**, 15–25.

Grice, G. D. (1960). Calanoid and cyclopoid copepods collected from the Florida Gulf coast and Florida keys in 1954 and 1955. *Bull. Mar. Sci. Gulf Carib.* **10**, 217–226.

Hall, D. J., Threlkeld, S. T., Burns, C. W., and Crowley, P. H. (1976). The size-efficiency hypothesis and the size structure of zooplankton communities. *Annu. Rev. Ecol. Syst.* **7**, 177–208.

Haury, L. R. (1976). Small-scale pattern of a California current zooplankton assemblage. *Mar. Biol. (Berlin)* **37**, 137–157.

Hazlett, B. A. (1974). Field observation on interspecific agonistic behavior in hermit crabs. *Crustaceana* **26**, 133–138.

Hebert, P. D. N. (1978). The population biology of *Daphnia* (Crustacea, Daphnidae). *Biol. Rev.* **53**, 387–426.

Hedgpeth, J. W. (1957). Sandy beaches. *Mem. Geol. Soc. Am.* **67**, 587–608.

Heinle, D. R. (1969). Temperature and zooplankton. *Chesapeake Sci.* **10**, 186–209.

Herman, S. S., Mihursky, J. A., and McErlean, A. J. (1968). Zooplankton and environmental characteristics of the Patuxent River Estuary. *Chesapeake Sci.* **9**, 67–82.

Hines, A. H. (1978). Reproduction in three species of intertidal barnacles from central California. *Biol. Bull. (Woods Hole, Mass.)* **154**, 262–281.

Hines, A. H. (1979). The comparative reproductive ecology of three species of intertidal barnacles. *Reprod. Ecol. Mar. Invertebr., 1979* pp. 213–234.

Hodgkin, E. P., and Rippingdale, R. J. (1971). Interspecies conflict in estuarine copepods. *Limnol. Oceanogr.* **16**, 573–576.

Holland, A. F., and Dean, J. M. (1977). The community biology of intertidal macrofauna inhabiting sandbars in the North Inlet area of South Carolina, USA. *In* "Ecology of Marine Benthos" (B. C. Coull, ed.), pp. 423–438. Univ. of South Carolina Press, Columbia, South Carolina.

Holland, A. F., and Polgar, T. T. (1976). Seasonal changes in the structure of an intertidal community. *Mar. Biol. (Berlin)* **37**, 341–348.

Hynes, H. N. B. (1959). The reproductive cycle of some British freshwater Gammaridae. *J. Anim. Ecol.* **24**, 353–387.

Ivester, M. S. (1980). Distribution of meiobenthic copepods along a sediment gradient: [Factor and niche analyses.] *Bull. Mar. Sci.* **30**, 634–645.

Ivester, M. S., and Coull, B. C. (1975). Comparative study of ultrastructural morphology of some mouth parts of four haustoriid amphipods. *Can. J. Zool.* **53**, 408–417.

Jefferies, H. P. (1962). Salinity-space distribution of the estuarine copepod genus *Eurytemora*. *Int. Rev. Gesamten Hydrobiol.* **47**, 291–300.

Jefferies, H. P. (1964). Comparative studies on estuarine zooplankton. *Limnol. Oceanogr.* **9**, 348–358.

Jefferies, H. P. (1966). Partitioning of the estuarine environment by two species of *Cancer*. *Ecology* **47**, 477–481.

Kaplan, E. H., Welker, J. R., Krauss, M. G., and McCourt, S. (1975). Some factors affecting the colonization of a dredged channel. *Mar. Biol. (Berlin)* **32**, 193–204.

Katona, S. K., and Moodie, C. F. (1969). Breeding of *Pseudocalanus elongatus* in the laboratory. *J. Mar. Biol. Assoc. U. K.* **49**, 743–747.

Kerfoot, W. C. (1974). Egg-size cycle of a cladoceran. *Ecology* **55**, 1259–1270.

Kerfoot, W. C. (1977a). Competition in cladoceran communities: The cost of evolving defenses against copepod predation. *Ecology* **58**, 303–313.

Kerfoot, W. C. (1977b). Implications of copepod predation. *Limnol. Oceanogr.* **22**, 316–325.

Ketchum, B. H. (1954). Relation between circulation and plankton populations in estuaries. *Ecology* **35**, 191–200.

Kinner, P., Mauer, D., and Leathem, W. (1974). Benthic invertebrates in Delaware Bay: Animal-sediment associations of the dominant species. *Int. Rev. Gesamten Hydrobiol.* **59**, 685–701.

Klein, G., Rachor, E., and Gerlach, S. A. (1975). Dynamics and productivity of two populations of the Benthic tube-dwelling Amphipod. *Ampelisca brevicornis* (Costa). In Helgoland Bight. *Ophelia* **14**, 139–159.

Lance, J. (1963). The salinity tolerance of some estuarine plankton crustaceans. *Limnol. Oceanogr.* **8**, 440–449.

Landry, M. R. (1975). Seasonal temperature effects and predicting development rates of marine copepod eggs. *Limnol. Oceanogr.* **20**, 434–440.

Landry, M. R. (1977). A review of important concepts in the trophic organization of pelagic ecosystems. *Helgol. Wiss. Meeresunters.* **30**, 8–17.

Landry, M. R. (1978a). Predatory feeding behavior of a marine copepod, *Labidocera trispinosa*. *Limnol. Oceanogr.* **23**, 1103–1113.

Landry, M. R. (1978b). Population dynamics and production of a plankton marine copepod, *Acartia clausii*, in a small temperate lagoon on San Juan Island, Washington. *Int. Rev. Gesamten Hydrobiol.* **63**, 77–119.

Lane, P. A. (1975). The dynamics of aquatic systems: A comparative study of the structure of four zooplankton communities. *Ecol. Monogr.* **45**, 307–336.

Lane, P. A. (1978). Zooplankton niches and the community structure controversy. *Science* **200**, 485–461.

Lewis, J. (1964). "The Ecology of Rocky Shores." English Univ. Press, London.

Lonsdale, D. J., and Coull, B. C. (1977). Composition and seasonality of zooplankton of North Inlet, South Carolina. *Chesapeake Sci.* **18**, 272–283.

Lonsdale, D. J., Heinle, D. R., and Siefried, C. (1979). Carnivorous feeding behavior of the adult calanoid copepod *Acartia tonsa* Dana. *J. Exp. Mar. Biol. Ecol.* **36**, 235–248.

Lynch, M. (1977). Fitness and optimal body size in zooplankton populations. *Ecology* **58**, 763–774.

McCall, P. L. (1977). Community patterns and adaptive strategies of the infaunal benthos of Long Island Sound. *J. Mar. Res.* **35**, 221–266.

Macan, T. T. (1966). Predation by *Salmo trutta* in a moorland fishpond. *Verh. Int. Ver. Limnol.* **16**, 1081–1087.

McCloskey, L. R. (1970). The dynamics of the community associated with a marine Scleractinian coral. *Int. Rev. Gesamten Hydrobiol.* **55**, 13–82.

McIntyre, A. D., and Eleftheriou, A. (1968). The bottom fauna of a flatfish nursery ground. *J. Mar. Biol. Assoc. U. K.* **48**, 113–142.

Makarewicz, J. C., and Likens, G. E. (1975). Niche analysis of a zooplankton community. *Science* **190**, 1000–1003.

Makarewicz, J. C., and Likens, G. E. (1978). Zooplankton niches and the community structure controversy. *Science* **200**, 461–463.

McDonald, H. J. (1977). The comparative intertidal ecology and niche relations of *Panopeus herbstii* (Milne-Edwards) and *Eurypanopeus depressus*. Ph. D. Dissertation, Univ. of South Carolina.

McLachlan, A. (1977). Studies on the psammolittoral meiofauna of Algoa Bay, South Africa II. The distribution composition and biomass of the meiofauna and macrofa-na. *Zool. Afr.* **12**, 33–60.

McLaren, I. A. (1963). Effects of temperature on growth of zooplankton, and the adaptive value of vertical migration. *J. Fish. Res. Board Can.* **20**, 685–727.

McLaren, I. A. (1965). Some relationships between temperate and egg size, body size, development rate, and fecundity of the copepod *Pseudocalanus*. *Limnol. Oceanogr.* **10**, 528–538.

McLaren, I. A. (1966). Predicting development rate of copepod eggs. *Biol. Bull. (Woods Hole, Mass.)* **131**, 457–469.

McLaren, I. A. (1978). Generation lengths of some temperate marine copepods: Estimation, prediction, and implications. *J. Fish. Res. Board Can.* **35**, 1330–1342.

McQueen, D. J. (1969). Reduction of zooplankton standing stocks by predaceous *Cyclops biscuspidatus thomasi* in Marion Lake, British Columbia. *J. Fish Res. Board Can.* **26**, 1305–1318.

Maurer D., Watling, L., Leetham, W., and Wethe, C. (1978). Benthic invertebrate assemblages of Delaware Bay. *Mar. Biol. (Berlin)* **45**, 65–78.

Miller, D. C. (1961). The feeding mechanism of fiddler crabs, with ecological considerations of feeding adaptations. *Zoologica* **46**, 89–100.

Mills, E. L. (1967). The biology of an ampeliscid amphipod crustacean sibling species pair. *J. Fish. Res. Board Can.* **24**, 305–355.

Miracle, M. R. (1974). Niche structure in freshwater zooplankton: A principal components approach. *Ecology* **55**, 1306–1316.

Monakov, A. V. (1972). Review of studies on feeding of aquatic invertebrates conducted at the Institute of Biology of Inland waters, Academy of Sciences, USSR. *J. Fish. Res. Board Can.* **29**, 363–383.

Moore, H. B. (1934). The biology of *Balanus balanoides*. I. Growth rate and its relation to size, season and tidal level. *J. Mar. Biol. Assoc. U. K.* **19**, 851–868.

Moore, H. B. (1935). The biology of *Balanus balanoides*. IV. Relation to environmental factors. *J. Mar. Biol. Assoc. U. K.* **20**, 279–308.

Moore, H. B. (1938). The biology of *Purpura lapillus*. Part III. Life history and relation to environmental factors. *J. Mar. Biol. Assoc. U. K.* **23**, 67–74.

Moore, H. B., and Kitching, J. A. (1939). The biology of *Chthamalus stellatus* (Poli). *J. Mar. Biol. Assoc. U. K.* **23**, 521–541.

Mullin, M. M. (1963). Some factors affecting the feeding of marine copepods of the genus *Calanus*. *Limnol. Oceanogr.* **8**, 239–250.

Mullin, M. M., and Brooks, E. R. (1967). Laboratory culture, growth rate, and feeding behavior of a planktonic marine copepod. *Limnol. Oceanogr.* **12**, 657–666.

Mullin, M. M., and Brooks, E. R. (1976). Some consequences of distributional heterogenity of phytoplankton and zooplankton. *Limnol. Oceanogr.* **21**, 784–796.

Muragan, N., and Sivaramakrishnan, K. G. (1973). The biology of *Simocephalus acutitostratus* King (Cladocera: Daphnidae) - Laboratory studies of life span, instar duration, egg production, growth and stages in embryonic development. *Freshwater Biol.* **3**, 77–83.

Neill, W. E. (1975). Experimental studies of microcrustacean competition, community composition, and efficiency of resource utilization. *Ecology* **56**, 809–826.

Paffenhöffer, G. A., and Harris, R. P. (1976). Feeding, growth, and reproduction of marine plankton copepod *Pseudocalanus elongatus* Boeck. *J. Mar. Biol. Assoc. U. K.* **56**, 327–344.

Patton, W. K. (1974). Community structure among the animals inhabiting the coral *Pocillopora damicornis* at Heron Island, Australia. *In* "Symbiosis in the Sea" (W. B. Vernberg, ed.), pp. 219–243. Univ. of South Carolina Press, Columbia, South Carolina.

Reaka, M. L. (1979). The evolutionary ecology of life history patterns in stomapod crustaceans. *Reprod. Ecol. Mar. Invertebr., 1979* pp. 232–260.

Renaud-Debyser, J., and Salvat, B. (1963). Éléments de prospérité des biotopes des sédiments meubles intertidaux et écologie de leurs populations en microfaune et macrofaune. *Vie Milieu, Ser. C* **14**, 463–550.

Rhodes, E. B. (1974). Distribution of *Neohaustoris schmitizi* Bousfield (Amphipoda: Haustoridae) as influenced by physiological tolerances, physical preferences, and environmental conditions. M.S. Thesis, 82 pp. Univ. of South Carolina, Columbia, South Carolina.

Richman, S., Heinle, D. R., and Huff, R. (1977). Grazing by adult estuarine calanoid copepods of the Chesapeake Bay. *Mar. Biol. (Berlin)* **42**, 69–84.

Rieger, R. M., and Ruppert, E. (1978). Resin embedments of quantitative meiofauna samples of ecological and structural studies - description and application. *Mar. Biol. (Berlin)* **46**, 223–235.

Rigler, F. H. (1961). The uptake and release of inorganic phosphorus by *Daphnia magna* Straus. *Limnol. Oceanogr.* **6**, 165–74.

Rigler, F. G., and Cooley, J. M. (1974). The use of field data to derive population statistics of multivoltine copepods. *Limnol. Oceanogr.* **19**, 636–655.

Rippingdale, R. J., and Hodgkin, E. P. (1974). Predation effects on the distribution of a copepod. *Aust. J. Mar. Freshwater Res.* **25,** 81–91.

Sameoto, D. D. (1969). Comparative ecology, life histories, and behavior of intertidal sand-burrowing amphipods (Crustacea:Haustoriidae) at Cape Cod. *J. Fish. Res. Board Can.* **26,** 361–388.

Sameoto, D. D. (1975). Tidal and diurnal effects on zooplankton sample variability in a near shore marine environment. *J. Fish. Res. Board Can.* **32** *(3).*

Sanders, H. L. (1956). The biology of marine benthic communities. *Bull. Bingham Oceanogr. Collect.* **15,** 345–406.

Sanders, H. L. (1958). Benthic Studies in Buzzards Bay I. Animal-sediment relationships. *Limnol. Oceanogr.* **3***(3),* 245–258.

Sandifer, P. A. (1975). The role of pelagic larvae in recruitment to populations of adult decapod crustaceans in the York River estuary and adjacent lower Chesapeake Bay, Virginia. *Est. Cstl. Mar. Sci.* **3,** 269–279.

Segerstale, S. G. (1978). The negative correlation between the abundances of the amphipod *Pontoporeia* and the bivalve *Macoma* in Baltic waters and the factors involved. *Ann. Zool. Fenn.* **15,** 143–145.

Sibert, J., Brown, T. J., Healy, M. C., Kask, B. A., and Naiman, R. J. (1977). Detritus-based food webs: Exploitation by juvenile chum salmon (*Oncorhynchus keta*). *Science* **196,** 649–650.

Simon, J. L., and Dauer, D. M. (1977). Reestablishment of a benthic community following natural defaunation. *In* "Ecology of Marine Benthos" (B. C. Coull, ed.) pp. 139–154. Univ. of South Carolina Press, Columbia, South Carolina.

Slobodkin, L. B. (1954). Population dynamics in *Daphnia obtusa* Kurz. *Ecol. Monogr.* **24,** 69–88.

Smyly, W. J. P. (1961). The life cycle of the freshwater copepod *Cyclops leuckarti* Claus in estuarine water. *J. Anim. Ecol.* **30,** 153–159.

Smyly, W. J. P. (1970). Observations on rate of development, longevity and fecundity of *Acanthocyclops virdis* (Jurine) (Copepoda:Cyclopoida) in relation to type of prey. *Crustaceana* **18,** 21–36.

Southward, A. J. (1953). The ecology of some rocky shores in the south of the Isle of Man. *Proc. Liverpool Biol. Soc.* **59,** 1–50.

Southward, A. J., and Crisp, D. J. (1954). Recent changes in the distribution of the intertidal barnacles *Chthamalus stellatus* Poli and *Balanus balanoides* L. in the British Isles. *J. Anim. Ecol.* **23,** 163–177.

Strathman, R. R., and Branscomb, E. S. (1979). Adequacy of cues of favorable sites used by settling larvae: Relation to cause of death and dispersal in two intertidal barnacles. *Reprod. Ecol. Mar. Invertebr., 1979* pp. 77–90.

Threlkeld, S. T. (1976). Starvation and the size structure of zooplankton communities. *Freshwater. Biol.* **6,** 489–496.

Trevallion, A., Ansell, A. D., Sivadas, P., and Narayanan, B. (1970). A preliminary account of two sandy beaches in South West India. *Mar. Biol. (Berlin)* **6,** 268–279.

Van Dolah, R. F. (1978). Factors regulating the distribution and population dynamics of the amphipod *Gammarus palustris* in an intertidal salt marsh community. *Ecol. Monogr.* **48,** 191–217.

Van Dolah, R. F., Shapiro, L. E., and Rees, C. P. (1975). Analysis of an intertidal population of the amphipod *Gammarus palustris* using a modified version of the egg-ratio method. *Mar. Biol. (Berlin)* **33,** 323–330.

Vijverberg, J. (1977). Population structure, life histories, and abundance of copepods in Tjeukemeer, the Netherlands. *Freshwater Biol.* **7,** 579–597.

Virstein, R. W. (1977). The importance of predation by crabs and fishes on benthic infauna in Chesapeake Bay. *Ecology* **58,** 1199–1217.

Weiss, C. M. (1948). Seasonal and annual variations in the attachment and survival of barnacle cyprids. *Biol. Bull. (Berlin)* **94,** 236–243.

Wenner, A. M., Fusaro, C., and Oaten, A. (1974). Size at onset of sexual maturity and growth rate in crustacean populations. *Can. J. Zool.* **52,** 1095–1106.

Werner, E. E., and Hall, D. J. (1974). Optimal foraging and the size selection of prey by the bluegill sunfish (*Lepomis macrochirus*). *Ecology* **55,** 1042–1052.

Whitehouse, J. W., and Lewis, B. G. (1973). The effect of diet and density on development size and egg production in *Cyclops abyssorum* Sars, 1863 (Copepoda, Cyclopoida.) *Crustaceana* **25,** 225–236.

Wiebe, P. H., and Holland, W. R. (1968). Plankton patchiness: Effects on repeated net tows. *Limnol. Oceanogr.* **13,** 315–321.

Wilder, D. G. (1953). The growth rate of the American lobster (*Homarus-americanus*). *J. Fish. Res. Board Can.* **10,** 371–512.

Williams, J. A. (1978). The annual pattern of reproduction of *Talitrus saltator* (Crustacea:Amphipoda:Taltridae). *J. Zool.* **184,** 231–244.

Williams, R. B., Murdoch, M. B., and Thomas, L. K. (1968). Standing crop and importance of zooplankton in a system of shallow estuaries. *Chesapeake Sci.* **9,** 42–51.

Wood, D. H. (1968). An ecological study of a sandy beach near Auckland, New Zealand. *Trans. R. Soc. N. Z.* **10,** 89–115.

Woodin, S. A. (1978). Refuges, disturbance and community structure: A marine soft-bottom example. *Ecology* **59,** 274–284.

Woodmansee, R. A. (1958). The seasonal distribution of the zooplankton of Chicken Key in Biscayne Bay, Florida. *Ecology* **39,** 247–262.

Young, D. K., Buzas, M. A., and Young, M. W. (1976). Species densities of macrobenthos associated with seagrass: A field experimental study of predation. *J. Mar. Res.* **34,** 577–592.

Young, D. K., and Rhoads, D. C. (1971) Animal-sediment relationships in Cape Code Bay, Massachusetts I. A transect study. *Mar. Biol. (Berlin)* **11,** 242–254.

Systematic Index—Crustacea*

*Note: Names that have been superseded appear in brackets. Parentheses around name of author of scientific name indicate that currently assigned genus is not the original one.

321

327

---

†Included by Bowman and Abele in the Corophiidae (see Volume 1).

Systematic Index
Non-Crustacea

Subject Index

335